# PHOENIX OVER THE NILE

# PHOENIX

# OVER THE NILE

## A HISTORY OF EGYPTIAN AIR POWER 1932–1994

LON O. NORDEEN
AND
DAVID NICOLLE

SMITHSONIAN INSTITUTION PRESS
WASHINGTON AND LONDON

Editor and typesetter: Princeton Editorial Associates
Production editor: Jenelle Walthour
Designer: Linda McKnight

Library of Congress Cataloging-in-Publication Data

Nordeen, Lon O., 1953–
Phoenix over the Nile : a history of Egyptian Air Power, 1932–1994 / Lon O. Nordeen and David Nicolle.
p. cm.
Includes bibliographical references and index.
ISBN 1-56098-626-3 (alk. paper)
1. Egypt. Qūwāt al-Jawwīyah—History. 2. Aircraft industry—Egypt—History. 3. Aeronautics, Military—Egypt—History. I. Nicolle, David. II. Title.
UG635.E42N67 1996
358.4'00962—dc20 95-30047

British Library Cataloguing-in-Publication Data is available.

Manufactured in the United States of America

03 02 01 00 99 98 97 96 5 4 3 2 1

∞ The paper used in this publication meets the minimum requirements of the American National Standard for Information Sciences—Permanence of Paper for Printed Library Materials ANSI Z39.48-1984.

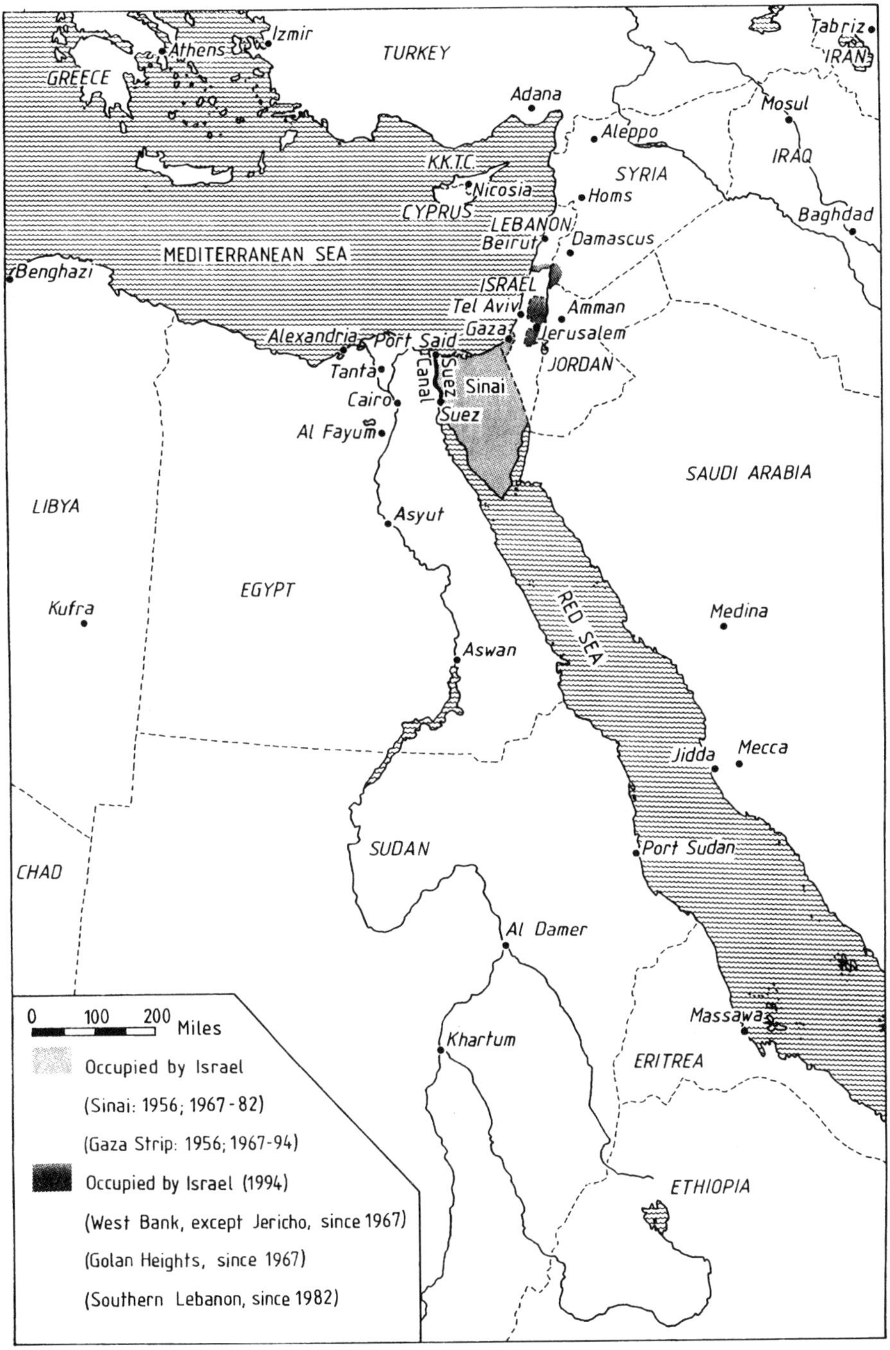


Egypt and neighboring countries.

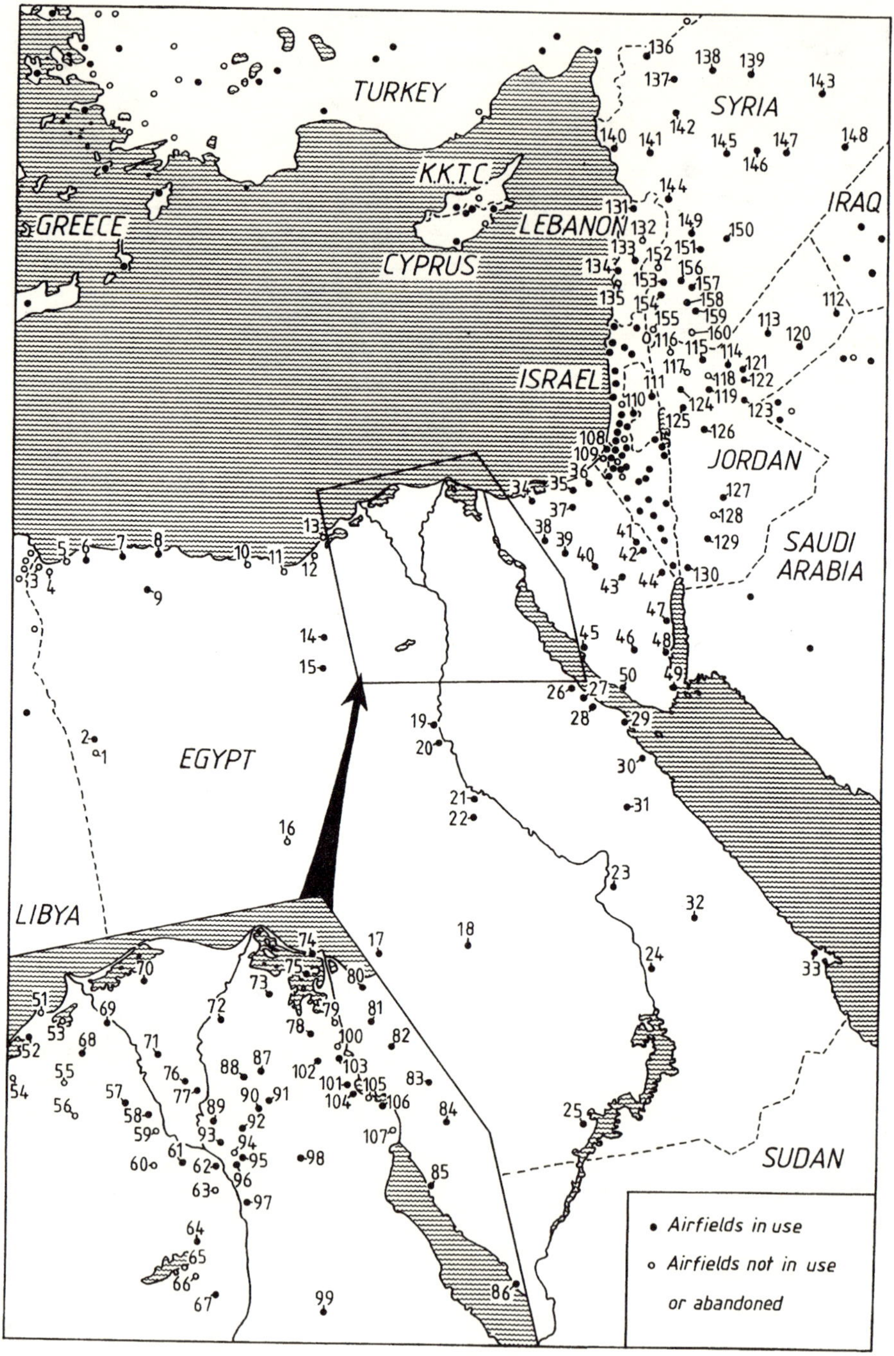
TURKEY
SYRIA
K.K.T.C.
GREECE
LEBANON
IRAQ
CYPRUS
ISRAEL
JORDAN
SAUDI ARABIA
EGYPT
LIBYA
SUDAN
• Airfields in use
∘ Airfields not in use or abandoned

Airfields in Egypt and neighboring regions. The numerous unnamed temporary airstrips used by the British, German, and Italian air force during the Second World War are not included in this map.

*Airfields in Egypt*

1. Siwa Oasis
2. Siwa Oasis North
3. al-Ramlah
4. Sallūm
5. Buqbuq
6. Sīdī Barrānī
7. Inshāṣ (Western Desert)
8. Marsá Maṭrūḥ
9. Abār al-Kanā'is
10. Ḍab'ah
11. al-'Alamayn
12. Burj al-'Arab/al-Ḥammām
13. Dākhilah
14. Ghurd Abū Sannān
15. Ghurd Mubārak
16. Baḥrīyah Oasis
17. Dākhila Oasis
18. New Valley
19. Shūshah
20. al-Minyā
21. Asyūṭ Northeast
22. Asyūṭ
23. Luxor
24. Darāw/ Kawm Umbū
25. Abu Simbel
26. Ra's Ghārib
27. Ra's Shukhayr New
28. Ra's Shukhayr
29. Ra's Jimsah
30. Hurghada (al-Ghardaqah)
31. Wādī Abū Shīḥāt/Wādī Qinā
32. Bi'r Abū Riḥāl
33. Ra's Banās
34. Miṣfaq
35. al-'Arīsh
36. Jurah/Etam
37. Jabal Libnī
38. Bi'r Jifjāfah/ Refidim
39. Bi'r Ḥasanah
40. Nakhl
41. Kuntillah New
42. Kuntillah South
43. al-Thamīd
44. 'Ayn Naqb/Etzion
45. Abū Rudays New
46. St. Catherine's
47. Nuwaybi'
48. Dhahab
49. Ra's Naṣrānī
50. al-Ṭur
51. Abū Qīr
52. Alexandria-Maryūṭ
53. Idkū
54. 'Āmirīyah
55. Wādī Naṭrūn North
56. Wādī Naṭrūn South
57. Jabal al-Bāsūr-Tayariyah
58. Kafr Dā'ūd
59. al-Khaṭāṭibah
60. Jabal Ḥamzī
61. Cairo West
62. Imbābah
63. al-Fayyūm Road Strip
64. Kawm Awshīm
65. Qārūn
66. al-Fayyūm
67. Banī Suwayf
68. Jiyānklīs
69. al-Raḥmānīyah
70. Sayyāḥ al-Sharīf
71. Ṭanṭá-Birma
72. al-Manṣūrah
73. al-Manzilah
74. Jamīl-Port Said
75. (name unknown)
76. Quwaysinā Highway Strip
77. Minshāt Ṣabrī Highway Strip
78. Ṣāliḥīyah
79. al-Ballāḥ
80. Balūza
81. Ḥawḍ al-Bibā
82. Ismā'īlīyah East
83. Umm Khushayb
84. al-Sharqī
85. Ra's Sudr
86. Abū Rudays
87. al-Zaqāzīq
88. Subk Basta
89. Ṭūkh Highway Strip
90. Bilbays
91. Bilbays East
92. Inshāṣ (Delta)
93. Qahā Highway Strip
94. Heliopolis
95. Cairo International
96. Almāẓah
97. Ḥulwān
98. Wādī al-Jandalī
99. Wādī Abū Rīshah
100. al-Firdān
101. Fā'id North
102. Abū Ṣuwayr
103. Ismā'īlīyah
104. Fā'id
105. Kasfarit
106. Kibrīt
107. Shallūfah

*Airfields in non-Israeli Palestine (Gaza Strip and West Bank)*

108. Gaza
109. al-Burayj
110. Qulūnīyah
111. al-Jiftlik

*Airfields in Jordan*

112. Ruwayshid
113. H4 New
114. Prince Ḥasan
115. King Hussein
116. Irbid
117. Khaww
118. Dawson's Field
119. Azraq Highway Strip
120. Abū Ḥusayn
121. Azraq North Highway Strip
122. Azraq
123. al-Ghadaf Highway Strip
124. Amman
125. Queen Alia
126. al-Qaṭrānah Highway Strip
127. King Faisal
128. Ma'ān
129. al-Quwayrah Highway Strip
130. al-'Aqabah

*Airfields in Lebanon*

131. Qulay'āt
132. Ī'āt-Ba'labakk
133. Rīyāq
134. Beirut
135. Sidon

*Airfields in Syria*

136. Mīnakh
137. Aleppo/Nayrab
138. Rasm 'Abbūd
139. 'Ajjūrah
140. Ḥumaymīm
141. Hama
142. Abū al-Ḍuhūr
143. Dayr al-Zawr
144. al-Quṣayr
145. Tiyās (T4)
146. Tadmur/ Palmyra
147. T3
148. T2
149. al-Nāṣirīyah
150. Ṣayqal
151. Ḍumayr
152. Sahl al-Ṣaḥrā'
153. Damascus Highway Strip
154. Mazzah
155. al-'Āl
156. Qabr al-Sitt Highway Strip
157. Damascus International
158. Marj Ruḥayyil
159. Khalkhalah
160. Dar'ā

# Contents


Foreword xiii

Acknowledgments xv

Preface xvii

A Note on Transliteration xix

Ranks xxi

**Introduction** 1

**1 Background (1910–1927)** 9

**2 A Start (1928–1932)** 15

**3 Fledgling Years (1932–1934)** 20

**4 A New Treaty and a World Crisis (1935–1939)** 30

**5 Neutrals at War (1939–1942)** 38

**6 Forgotten Allies (1942–1945)** 51

**7 Threatening Horizons (1945–1948)** 65

**8 The First Offensive (1948)** 74

9 Fighting the Hydra (1948) 91

10 A Losing Battle (1948–1949) 110

11 Reactionaries and Revolutionaries (1949–1953) 123

12 "Czech" Arms (1953–1956) 138

13 The Other Side of Suez (1956) 153

14 A Doubtful Anniversary (1956–1961) 170

15 Wider Horizons (1962–1967) 182

16 The Surprise Assault (1967) 202

17 Egyptian Phoenix (1967) 219

18 Fighting Back (1967–1969) 229

19 Attrition War (1969–1970) 242

20 New Directions (1970–1973) 256

21 Ramadan War (1973) 274

22 Battles for the Bridgeheads (1973) 287

23 Ramadan War Impact (1973) 299

24 New Directions Again (1974–1981) 309

25 Moving Forward (1981–1994) 321

# Foreword

This book presents a history of military aviation in Egypt. While the history of the Egyptian Air Force (EAF) is the prime focus, the book also discusses the development of the Egyptian aviation industry and describes how the air arm, Air Defense force, and foreign services worked together to defend Egyptian air space.

The work includes an overview of military aviation in Egypt and explains the political, cultural, and military factors that shaped and directed the Air Force. Interviews with dozens of current and retired Egyptian Air Force officers and Air Defense personnel complement the historical narrative. Many never before published historic and combat action photographs are also included in the work.

# Acknowledgments

The authors wish to thank the following individuals for their support and assistance:

Fakry El Ashmawi, Col. Ahmad Atef, Brig. Gen. Farouke El Gazawy (ret.), Air Commod. I. H. Gazerine (ret.), Brig. Gen. Kadry Abd El Hamid (ret.), Maj. Gen. Farid F. Harfoush, Air Commod. Mustafa Hafaz (ret.), Maj. Gen. Mustafa al Hinnawi (ret.), Maj. Fawzy M. Hussein, Shirly King, Maj. Gen. Mohadded Nabil El Messiry (ret.), Col. Samir Aziz Michael (ret.), Air Vice Marshal A. M. Mikaati (ret.), the helpful personnel of the UK Public Records Office, London, Alaa E. Shaker, Air Marshal Saad El Din Sherif (ret.), Maj. Gen. Nobil Shoaky (ret.), Air Vice Marshal V. H. Tait (ret.), Maj. Gen. Abdel Moneim El Tawil (ret.).

David Isby deserves special mention for all his editing and the critical comments that vastly improved the manuscript.

Also a thank you to our families, who understood and supported the many hours invested in researching, writing, and editing this work.

# Preface

The transatlantic partnership that produced this book resulted from two writers quite separately wishing to tell the story of Middle Eastern military aviation from an Egyptian point of view. We began our research independently and started from different standpoints. Only when David Oliver, editor of *Air Forces Monthly* magazine, brought us together did our collaboration take off. Even today the five major wars and numerous skirmishes between Israel and its Arab neighbors are usually told from an Israeli angle. Clearly something was needed to provide more balance, but why did an American and an Englishman take up the challenge?

Lon Nordeen got to know the men of the EAF through his work at the McDonnell Aircraft Company. For him the faceless aircrews who flew those MiGs and Sukhois that briefly flitted across the West's TV screens, caught in the gun cameras of Israeli fighters, became real people—men of flesh and blood. After researching and writing a history of the Israeli Air Force, Lon decided that the story of these courageous Egyptian fliers should also be told. Thousands of miles away in England, David Nicolle had worked for BBC TV News during the 1960s and had become convinced that the Western media were biased in their coverage of the Arab-Israeli confrontation. During a visit to Cairo in May 1970, when the War of Attrition was reaching its climax, he was invited to the wedding of a complete stranger—the son of a shopkeeper—in a typical gesture of Egyptian hospitality. Unfortunately the reality of Egypt's war was suddenly brought home when this ceremony was called off three days before the wedding because the bride's brother, an Air Force pilot, had been shot down in combat over Sinai. From then on David Nicolle started collecting published and unpublished, English and Arabic, sources concerned with the history of the EAF. In 1990 Nordeen and Nicolle pooled their information and their enthusiasm for a neglected piece of aviation history; this book is the result.

# A Note on Transliteration

The authors have attempted to write the names of people and places consistently and accurately according to the Library of Congress system of Arabic and Hebrew transliteration. The only exceptions are internationally recognized names such as the city of Cairo or President Nasser. Nevertheless, problems arose when some original English language documents were quoted, as these often used other spellings. Similarly the authors hope that members of the EAF, and other persons with Arabic names, will forgive the fact that their names have been spelled in a way that is not the way in which they themselves spell their own names when writing in English! Whenever possible we have tried to explain such discrepancies in the Index.

# Ranks

Egyptian Army ranks, as used by the EAAF until 1937:

| | |
|---|---|
| Mulāzim Thānī | 2d Lieutenant (2d Lt.) |
| Mulāzim Awwal | 1st Lieutenant (1st Lt.) |
| Yūzbāshī | Captain (Capt.) |
| Bimbāshī | Major (Maj.) |
| Qā'im-maqām | Lieutenant Colonel (Lt. Col.) |
| Amīr Alāy | Colonel (Col.) |
| Liwā' | Brigadier (Brig.) |

REAF ranks, from 1937 until late 1950s:

| | |
|---|---|
| Ṭayyār Thānī | Pilot Officer (Pilot Off.) |
| Ṭayyār Awwal | Flight Officer (Flt. Off.) |
| Qā'id Sirb | Flight Lieutenant (Flt. Lt.) |
| Qā'id Asrāb | Squadron Leader (Sq. Ldr.) |
| Qā'id Janāḥ | Wing Commander (Wing Comdr.) |
| Qā'id Liwā' | Group Captain (Group Capt.) |
| Qā'id Firqah Jawwīyah | Air Commodore (Air Commod.) |
| Qā'id Usṭūl Jawwī | Air Vice Marshal |
| Qā'id Asāṭīl Jawwīyah | Air Marshal |

# Introduction

Napoleon once described Egypt as the most important country in the world, and he was speaking before the days of oil. Egypt certainly has a unique geographic position at the junction of Asia and Africa, the Red Sea and the Mediterranean. Culturally it also stands at the center of the Arab world, is largely Muslim but has a large and influential Christian minority—all these factors support Napoleon's original opinion. Today Egypt has the added importance not only of being an oil producer itself, but of lying close to the world's most important oil resources. Finally, it remains a leader of those largely underdeveloped countries known as the third world.

Following the collapse of the Soviet Union and the apparent end of the cold war, the world has not yet become a more peaceful place. On the contrary, ancient rivalries have resurfaced, often in a very bloody manner. Many such crises have erupted in the third world, and this is where western military planners expect the majority of wars to be fought in the foreseeable future. But accurate and reliable information on the armed forces of the third world is hard to come by, and this is particularly true of their air forces. In the past these have also been largely disregarded, if not by professional military planners, then at least by the general public.

One of the most important, powerful, and combat experienced third world air forces is that of Egypt, the EAF. Current developments in the Arab-Israeli dispute mean that the EAF can, perhaps, look forward to a period of greater peace than it has known for half a century. At the same time the EAF remains in many ways a typical third world air force. A greater understanding of its background, problems, potential, and limitations will be valuable as the West faces an uncertain twenty-first century.

This book is an attempt to tell the story of the EAF not only in terms of brave men facing appalling difficulties—though there is plenty of that—but to show how a third world air force developed within its own political, cultural, and military context. Third world air forces tend to be very different from those of the West, or indeed those of what used to be the Soviet bloc, and there are many reasons for this. Most were created for political rather than purely military reasons, and many continue to play important political as well as military roles. The study of third world air forces also poses its own peculiar problems for the aviation historian. Where the EAF is concerned these have been a lack of officially released information, a deeply engrained tradition of military secrecy, and a certain suspicion of westerners. As one British air attaché in Cairo pointed out:

> When the EAF imported Soviet equipment, it also imported Soviet doctrine and hence became almost paranoid about the security of all military information. This mental conditioning took place over a twenty year period, in fact a whole generation of military men. Access to military bases was not permitted, especially to Attachés who had ulterior motives, and EAF personnel were warned that communication with foreigners carried the threat of dismissal from the service. When I arrived as Air Attaché in Cairo in 1983, this secretiveness was still quite apparent, visits to bases did not occur and we were issued with a social list of EAF officers with whom we might converse. Needless to say these were all trusted generals who rarely granted interviews.[1]

Such attitudes have changed over the past decade. An easier atmosphere has developed, and the paranoia that crept into official documents during the 1950s to 1970s is now rarer. Similar problems would be faced by those researching the military history of many recently independent postcolonial countries, but where Egypt is concerned there are also different traditions in the use of language. In the West understatement and the downplaying of emotions tends to be admired in documents dealing with the carnage of war, whereas in the Arab world an element of exaggeration is acceptable. To westerners who have neither knowledge nor understanding of Arab-Islamic culture this can seem like a disregard for the truth. It is not. It merely reflects attitudes that were common even in Europe not so long ago. One only has to look at the way the English characterized the supposedly "emotional" French soldiers of the Napoleonic era, or what they saw as the "mercurial" Italians of the Second World War, to see how military prejudice can grow from cultural misunderstanding.

## A Leading Third World Air Force

Many third world air forces are plagued by budget, training, or leadership problems, and sometimes all three. Here the Egyptians have been no exception. Under such circumstances mere numbers of aircraft possessed—"bean counts," as the Americans say—can be meaningless. Infrastructure and peacetime preparation are important for all armed forces, but the EAF faced specific difficulties in these spheres.

To date, the only third world air forces to have received much outside attention are those of India and Pakistan. But most of their operations have been directed against each other rather than against a western or "top league" enemy. The EAF, along with several other Arab air forces, has had the dubious privilege of being pitted against one of the world's best—the Israelis—not once but on several occasions, each time under different politicomilitary circumstances. The EAF has also been in operation elsewhere in the Arab world and in sub-Saharan Africa. These facts alone make the history of the EAF important. At the same time the EAF remains a vital element in the military situation of the Middle East, perhaps the world's strategically and economically most sensitive area.

Most third world air forces exist in regions where concepts of western democracy are either absent or only recently introduced. Unlike some other Arab countries, Egypt has never been a true military dictatorship, though since 1952 its rulers have had military backgrounds. Egypt's armed forces have, however, played a leading role in the formation of national policy. Although the Egyptian Army remained the dominant military influence, the EAF and its leadership clearly influenced government policy on several occasions, for example, in the pre-1956 move toward the Soviet Union, in a gradual break with the Soviet Union before and after 1973, and in relations with the United States at the time of the Camp David Agreement. On all three occasions the Egyptian Air Force was reacting to constraints forced upon it by those who were at the time its main equipment suppliers: the British before 1956, the Soviets around 1973, and, during the Camp David negotiations, in anticipation of supplies of high-technology American aircraft. Egypt's own attempts to develop advanced technology, such as the jet fighter and surface-to-surface missile programs of the 1960s, reflected President Nasser's claim to leadership of the Arab world. They were also further examples of how the EAF and the Egyptian Air Defense Command were used as tools of the country's international aspirations.

The EAF's central position in the Arab-Israeli dispute has given it more combat experience than most other air forces. The international ramifications of this dispute also provided the EAF with unique experience of British, Soviet, and most recently American equipment and military doctrines. Italian, French, and Chinese aircraft and aircraft of various other origins have also served in the Egyptian Air Force. Such varied experience makes the EAF's own opinions well worth listening to, particularly as its long and eventful history has given it a clear awareness of its own typically third world problems.

## Modern Egypt's Military Heritage

In the West the popular image of Egypt's military history is, unfortunately, still a mixture of myth, ignorance, and prejudice. Even in Egypt itself the fact of the country's lack of control over its own destiny until the Revolution of 1952 has been exaggerated for political reasons. Unfortunately this gives the impression that the Egyptian people were somehow not involved in their own military history, and that for two thousand years the Egyptians did not fight their own battles. This again is a gross oversimplification of a complex and interesting slice of military history.

Military service has been theoretically compulsory in Egypt since the days of Muḥammad ʿAlī early in the nineteenth century. The selection process was arbitrary and service remained deeply unpopular, particularly among the *fallāḥūn* peasantry. Those who could afford to do so bought themselves exemptions, and this became an important source of government revenue. Nevertheless, Winston Churchill's oft-quoted and highly racist remark that "It appeared easier to draw sunbeams out of cucumbers than to put courage into the fellah" was not only disproved by military events but was also contradicted by those British officers who commanded the Egyptian Army between 1882 and 1936. British military reports from the late nineteenth century are naturally couched in the patronizing style typical of the British Empire at its peak, but they still shed a remarkably sympathetic light on the "native" Egyptian soldier. He was regarded as steady, tough, and uncomplaining but rather unimaginative. Though less aggressive than his Sudanese counterpart, the Egyptian was more disciplined, while the patience of the Egyptian soldier when wounded became proverbial among their British officers.

One thing that set Egyptian troops, noncommissioned officers (NCOs), and even Egyptian officers apart from their British opposite numbers was a passion for parade-ground drill and the fact that the Egyptians volunteered for extra whenever possible!

More serious were the deep divisions that separated the Egyptian common soldier from the Egyptian officer, a feature that to some extent persists to this day. At their worst during the nineteenth and early twentieth centuries, these reflected the often very different cultural and ethnic as well as economic and educational backgrounds of soldiers and officers. Egypt, though in reality virtually independent, theoretically remained part of the Ottoman Turkish Empire until 1914. Its khedives (*khidīw,* autonomous governor) were of Albanian origin, and their dynasty, next as independent sultans and then as kings, ruled the country until Colonel Nasser's Revolution in 1952. From 1882 to 1936 Egypt was also under British control, and British influence persisted in a more or less subtle manner until the 1952 Revolution.

Until then the country's ruling and military elites remained largely Turkish, Circassian, or Albanian, despite the fact that several rulers encouraged the recruitment of Arab-Egyptian officers. As in many countries, Egypt saw the emergence of military families, generations of whom tended to join the armed forces. At first these were largely Turkish or Albanian, but various Arab-Egyptian families gradually developed their own "military tradition." Military families continue to exist, of course. A remarkable number of the EAF's pilots are still drawn from families that, though now thoroughly Egyptian, have Arab, Turkish, Albanian, Bosnian, or Circassian ancestry and that have sent several generations of sons into the armed forces.

In the nineteenth century, rivalry between officers of differing ethnic origin was a serious problem and had undermined the Egyptian Army's effectiveness by the time of the British occupation in 1882. The old officer corps was then summarily retired, though many were reengaged when the Army was recreated in 1883. During the subsequent British occupation, great efforts were made to stamp out the old system of promotion by patronage and to ensure that officers rose solely by merit. Unfortunately the habits of patronage, so deeply engrained in Arab society, were never quite removed. Some more critical observers suggest that they persist to this day.

Under the British occupation, and indeed until the Revolution of 1952, a military career had low status in Egyptian society, and the military elite was in many ways separated from the rest of society. As Jacques Berque, one of the most perceptive historians of modern Egypt, pointed out, the army had been absent from the Egyptian political scene since the failure of Colonel ʿUrābī's military revolt in 1882 led to the British occupation of Egypt.[2] By the 1930s things were again beginning to change. A new Anglo-Egyptian treaty supposedly changed the relationship to one between allies rather than one of military occupation. King Fārūq's Speech from the Throne in April 1938 stressed the need to strengthen the Egyptian armed forces, which was a very popular theme in a country that had just regained a degree of independence from British domination. But it also fitted the ruler's own personal ambitions, and several observers noted that officers now took their oath of loyalty to the king, not to Egypt's new constitution. In government circles there was a widespread feeling that everything should be done to raise the prestige of the officer corps, and Egyptian pride in an army and air force recently released from British tutelage was genuine. There had, however, recently been a military coup in Iraq, and while King Fārūq did all he could to keep the army out of everyday politics, he also tried to make it an arm of royal authority. He cultivated the new officers, particularly those of Egypt's fledgling air force, with flattery and promises of royal patronage to promote their careers. This inevitably caused divisions, not only within the armed services themselves but in the country at large where many people no longer saw King Fārūq as truly representing the Egyptian nation.

During the Second World War the Egyptian armed forces suffered additional strains. Not only was the officer corps divided between loyalty to an undeserving monarch and loyalty to its country. It was also split on whether to support an Allied cause represented by Britain—the country still in military occupation of Egypt despite the Treaty of 1936—or to support an aggressive Fascist Italian-German Axis which proclaimed its support for Egyptian freedom while crushing that of every other country within reach. Fortunately for Egypt those who said help Britain defeat the common foe won the argument.

Under President Nasser there was a rapid change in public attitudes. The armed forces were once again seen as reflecting the nation's aspirations rather than those of a ruling clique. A military career gained higher status. As a result the quality of recruits improved and men were drawn from more varied backgrounds, but outside Egypt and the Middle East old attitudes and deep-seated prejudices remained.

Throughout most of the twentieth century the Arabs, their armed forces, and above all their air forces have been subject to appalling prejudice in Europe and America. The EAF was the target of almost consistently hostile propaganda which was seen at its most blatant from the 1950s to the 1970s. The worst period was probably in the 1960s when the EAF became the butt of overtly racist humor in the western media. This reinforced a growing sense of paranoia in an Arab world that all too often ascribed western prejudice to "Zionist plots" or a supposed "Jewish control of the media," which in turn made the Arabs themselves sound racist!

As a result the sheer courage of Egyptian pilots has not yet been fully understood, despite the West's oft proclaimed admiration for a "heroic underdog." The last and only time that Americans got a hammering in an air war was over the Philippines in 1941–42. The British experiences over the Western Front in 1917 and in the early years of the Second World War are also a long time ago, though comparable memories may be fresher in the minds of other European countries. As one reviewer of an early draft of this book wrote, with a little exaggeration: "being in the EAF is not like being in the USAF. These guys have all lost most of their friends. Most have been shot down. Yet they still go and do things like attack the Israelis with L-29s (jet trainers)."[3]

Those close to the realities of Egypt's military situation naturally have a clearer understanding of the sometimes brutal realities of the Arab-Israeli conflict. Many recognized that, even during the worst tensions between Egypt and the West, the majority of the Egyptian officer corps, particularly in the EAF, were never really "antiwestern." Despite the antiwestern rhetoric of the government and media during the dark days of 1967–71, many men of the EAF paradoxically identified themselves with the United States Army Air Force (USAAF) at Pearl Harbor and the Royal Air Force (RAF) during the Battle of Britain. In the past there had been a tendency, encouraged by the Egyptian government and media, to blame others for Egyptian setbacks. Defeat in Palestine in 1948 was blamed on the British. The militarily disastrous Suez War of 1956 was seen, quite justifiably, as a heroic failure against overwhelming odds. For this reason the Egyptian armed forces constantly underestimated their Israeli foes until the catastrophe of June 1967. But that disaster could not so easily be explained away, and it became a watershed in attitudes, not least in the EAF.

After 1967 greater realism was followed by gradually more success. Those years saw tense confrontation and occasional full-scale fighting along the Suez Canal. The War of Attrition and the October War of 1973 gave the EAF its first real chance to show its abilities, as well as its failings. They also saw the EAF fighting back from a position of total defeat by the Israel Defense Force/Air Force to one of near parity. Appalling casualties were suffered, in the air as on the ground, and Egypt's ability to take such losses perhaps reflects the deeply religious character of Egyptian society. Only the most committed friend of Egypt would claim that the story ended on a note of total triumph, but the EAF did earn the respect of its foes. More important, the EAF had demonstrated that Egypt was a country the West would do well to have as an ally rather than allowing to drift into resentful neglect.

# 1

## Background (1910–1927)

In 1910 a handful of famous pilots arrived in Egypt and attempted to fly from Heliopolis, northeast of Cairo, to the world-famous Giza Pyramids and back. Not all made it. Only two years later an Egyptian officer laid claim to one of the "firsts" in the history of military aviation. This took place during the Italian invasion of Libya in 1912. Libya was then the only remaining part of North Africa ruled by the Ottoman Empire, neighboring Egypt being nominally an Ottoman province but in reality under British military occupation. Italy's act of unprovoked aggression not only angered the Ottoman Turks and their Libyan Arab subjects but also led several Egyptian Army officers to volunteer to fight these invaders. The British turned a blind eye when they joined Ottoman and Libyan forces already waging a highly successful guerrilla war against the Italians. But there were already men of Egyptian origin serving in Libya, and one of the most senior was Col. ʿAzīz ʿAlī al-Miṣrī ("ʿAzīz ʿAlī the Egyptian") of the Ottoman Army. He was of Circassian family rather than Arab-Egyptian origin but was, nevertheless, an early Arab nationalist. ʿAzīz al-Miṣrī's claim to fame was that he commanded the first anti-aircraft battery to have shot down an enemy aircraft in war. This was said to have been the Italian Nieuport monoplane flown by Capt. Riccardo Moizo, brought down by an old Austrian mountain gun on a vertical mounting on 10 September 1912.[1] Col. ʿAzīz al-Miṣrī and his men continued to fight on, even after peace was officially reached between Italy and the Ottoman Empire a month later, but were finally forced to retreat into Egypt at the end of 1913. ʿAzīz al-Miṣrī would reappear in the history of the Egyptian Air Force just under thirty years later.

### Egypt and the First World War

In August 1914 Europe threw itself into what some historians have described as a continentwide civil war. Within a few weeks Germany and the Austro-Hungarian Empire, which together ruled most of central and much of eastern Europe, were fighting Russia in the East, France, Britain, and Belgium in the west, and Serbia and Montenegro in the south. The distant empires of Britain, France, and Belgium were drawn in at once and were soon joined by Japan in the Far East. Little Luxembourg was engulfed at the very start. Next the Ottoman Empire, Bulgaria, and eventually Italy, Albania, Romania, even Portugal, Greece, the United States, China, and Brazil were drawn in to make this the first truly world war.

When the world crisis erupted Egypt lay under British occupation. The Egyptian Army was a tiny force and had been reconstructed along British lines since the British takeover in 1882. ʿAbbās II, the current khedive or nominal ruler of the country, was actually visiting the Ottoman capital when the Ottoman Turkish Empire threw in its lot with Germany late in 1914. This, coupled with the khedive's supposed pro-Turkish sentiments, led the British to depose ʿAbbās II on 17 December 1914. They replaced him with his uncle Ḥusayn Kāmil, who was declared a sultan and became, theoretically, an independent ruler. This ended Egypt's link with the Ottoman Empire, and the country officially became a British protectorate—though with no more freedom than it had before.

Britain still doubted the loyalty of Egypt's armed forces, but the *jihād* or "religious struggle" that the Ottoman sultan-caliph now proclaimed against Britain had virtually no impact. The Egyptian officer corps, while naturally resenting British domination, were fiercely nationalistic and longed for their country to be free from both British and Ottoman-Turkish rule. A few officers also felt a broader sense of Arab loyalty, though in 1914 such "Arab nationalism" was a very new idea.

With remarkably little appreciation of these anti-Turkish feelings within Egypt's armed forces, the British only permitted the Egyptian Army, Coast Guard, and Frontier Police a modest role during the four years of the First World War. Instead it fell to the ever patient Egyptian peasantry to labor and to die in large numbers, manning the Egyptian Camel Transport and Egyptian Labour Battalions without whom British successes against the Turks in Palestine and Syria would have been impossible.

Compared with the quiet heroism of these Transport and Labour Battalions, the Egyptian Army's role was almost insignificant. Yet the army did participate and at the same time had its first direct experience of the use of air power. At first the Ottoman Turks made a halfhearted attempt to seize the Suez Canal in February 1915, and the Egyptian Army played a minor role in its defeat. Later that year Italy entered the war as Britain's ally, and thus found itself once again at war with the Ottoman Empire. Not surprisingly the Ottoman sultan encouraged Libyan tribes of the reformist Sanūsī Muslim sect to rise once more against the infidel invaders—by whom the sultan meant both the Italians and the British since the Sanūsī straddled the frontier of Egypt and Libya.

Egyptian forces were involved in the suppression of Sanūsī dissent; however, it was further south, in the Sudan, that aircraft and the Egyptian Army played their most notable roles. During the nineteenth-century Egypt had conquered huge territories in what are now the Republic of the Sudan, Ethiopia, Somalia, and Uganda, but, like Egypt itself, the Sudan now lay under British control. Here the Ottoman sultan's call for a *jihād* fell on more receptive ears. ᶜAlī Dīnār, the semi-autonomous Sultan of Darfur province in the far west of Sudan, refused to recognize the British deposition of Khedive ᶜAbbās II and rose in revolt. This time the British decided that Egypt should play the leading role in crushing ᶜAlī Dīnār's large but primitive tribal army. A small Darfur Field Force was drawn from the best Egyptian units and was sent against ᶜAlī Dīnār. It was supported by four BE 2Cs of No. 17 Squadron, Royal Flying Corps (RFC), which arrived at al-Nuhūd 563 kilometers southwest of Khartoum in May 1916. From here, and from forward bases at Ḥillah and Abyaḍ over 200 kilometers further into the bush, the tiny RFC contingent commanded by Captain Bannatyne cooperated with the Egyptian Darfur Field Force, carrying out reconnaissance, bombing, and strafing Sultan ᶜAlī Dīnār's warriors.

ᶜAlī Dīnār replied to these raids with leaflets that told the Darfur Field Force's commander that he did not care what the "iron horses that flew in the air" did. However, on 15 and 16 May 1916 Captain Bannatyne drove the sultan's forces from the village of Biʾr Malit with machine-gun fire and 21 lb. Hales bombs. The Egyptian Darfur Field Force met ᶜAlī Dīnār's main army at Birinjīyah a week later and defeated it. On 23 May 1916, Lt. J. C. Slessor bombed the enemy rear guard, killing a camel that ᶜAlī Dīnār was preparing to mount, and effectively dispersed what was left of the Sudanese sultan's army.[2] At the end of this little-known campaign, Sultan Ḥusayn Kāmil of Egypt sent a letter of congratulation to the British commander in chief which gave particular thanks to the RFC.[3]

The Egyptian Army gained more knowledge of air operations during the next stage of the First World War. Slowly, in a costly campaign, the

British reclaimed Sinai from the Ottoman Turks, drove the enemy out of Rafah on the Palestinian frontier, but were then effectively blocked by the Turks' tenacious defense of Gaza. Here the enemy was greatly helped by a German air unit under Ottoman command. This was the Fliegerabteilung 300, better known as the 300th "Pasha" Squadron, which arrived at al-ʿArīsh and Beersheba early in 1916. Something of a stalemate ensued, in the air as on the ground, with the outnumbered but modern German Rumpler C Is and Fokker monoplanes being used more effectively than their British counterparts. Meanwhile the Turkish squadrons of the Ottoman Air Force were being used on the Dardanelles, Caucasus, and Iraqi fronts and to protect the Turkish coast. When the Arab Revolt erupted in Arabia in June 1916, however, the Pfalz A II parasols of the 3d Bölük (squadron) of the Ottoman Air Force were transferred from the Caucasus to bases in southern Syria and the Hejaz.[4]

German, Turkish, and British squadrons gradually acquired more modern equipment over the following months, but when, a year later, Gen. Sir Edmund Allenby assumed overall command of Allied forces in Egypt he promptly set about reorganizing and enlarging his air component. By the end of 1917 the German "Pasha" Squadron actually consisted of five sections (Nos. 300–304), while the purely Turkish 3d Bölük was supported by the 4th Army Aircraft Park at Damascus. During this period the Egyptian Army's role remained largely noncombatant. On the other hand the Palestine Brigade of the Royal Flying Corps now included the kite balloons of No. 43 Balloon Section, No. 21 Balloon Company, and here men of the Egyptian Army's 1st Infantry Battalion served as ground handlers.[5] During the actual assault, balloons directed artillery fire and the attack was successful, the Ottoman Turks and Germans falling back in full retreat. Five enemy aircraft were abandoned and burned at ʿIrāq al-Manshīyah airfield—a name that would also feature prominently in the history of the EAF in 1948. The struggle along this front would go on for many more months, of course, with three Turkish squadrons (3d, 4th, and 14th Bölüks), six "Pasha" sections, and a strictly German fighter unit facing an even greater number of British RFC and Australian squadrons by the end of the war. Other Egyptian troops played a fully combatant role in support of the Arab Revolt in what are now Saudi Arabia and Jordan. But here their experience of air operations was largely at the receiving end of the remarkable and little-known activities of the Ottoman Turkish Air Force's 3d Bölük which, though largely based in what are now southern Syria and Jordan, had aircraft operating from the holy city of Medina in Saudi Arabia as late as 21 March 1918.[6]

Meanwhile at least one Egyptian officer was training as an air observer, but unfortunately he was doing so at the Ottoman Turkish Air Force's Flying School at San Stefano (now Yesilköy) outside Istanbul. The Turks recruited airmen not only in what is now Turkey, but also from refugees driven out of regions recently conquered by Greeks, Serbs, Bulgarians, and Italians. Others came from the remaining Arab provinces of the Ottoman Empire including Iraq, Syria, and the Yemen. There were even a handful of Iranian officers flying for the Turks by the end of the Great War. Cavalry 1st Lt. ʿAbd Allāh, however, was listed as an Egyptian but was still under training when the war ended in 1918. Nothing further is known of his life and career.[7]

Egypt's direct military participation in the First World War had necessarily been small. Yet when the war ended the leaders of the tiny, ill-equipped, and inadequately trained Egyptian Army were fully aware of the importance of air power.

## Dashed Hopes but New Pride

Egypt had cooperated with the Allies during the First World War, but the country did not benefit from the Versailles Peace Conference that followed. All across Europe nations regained their independence, and new names appeared on the map. Outside Europe it was a very different story. Here the victorious powers—Britain, France, Italy, and even Greece—tried to extend their authority in direct contradiction of the famous Fourteen Points put forward by U.S. President Woodrow Wilson in January 1918. These had proposed the creation of independent nation-states as the basis for a more peaceful world order.

Instead Britain's domination over Egypt was confirmed, at least until the massive Egyptian unrest that erupted in 1919 forced the British to modify their position. In 1922 Egypt was declared an independent sovereign kingdom under its new ruler, King Fuʾād, who had inherited the throne on the death of Ḥusayn Kāmil in 1917. Even so, Britain retained control of Egypt's military forces and of the strategically vital Suez Canal. Not only was there still to be a large British military presence in Egypt, but almost all the commanding officers of the Egyptian Army remained British.

In the same year that Egypt became a kingdom, 1922, the government in Cairo first raised the idea of creating an Egyptian air force. It was to be a tiny organization consisting of two flights, each of four small two-seater aircraft, based at Sallūm and Rafaḥ. These flights would help the Egyptian Frontier Forces in their increasingly difficult struggle against smugglers crossing the border from Libya and Palestine.[8] Nothing more, however, was heard of the idea for another two years.

Civil unrest continued despite the declaration of an independent kingdom, and in August 1924 the British *sirdār* or commander of the Egyptian Army, Sir Lee Stack, was assassinated. Britain's reaction was to insist on a further reduction in Egypt's armed forces and tighter control over all modern weapons.

General Allenby, victor of the campaign against the Ottoman Turks in Syria during the First World War, was now the British high commissioner in Egypt. This made him, rather than King Fuʾād, the most powerful man in the country. Even so, Allenby recognized the strength of nationalist feeling in Egypt and in the Egyptian Army. He made several small yet significant changes to improve relations, and in 1925 Egypt was allowed to form its own Army Council to control the appointment and promotion of officers. The idea of creating an air force had been raised again in 1924, but in the aftermath of the murder of Sir Lee Stack the British authorities in Egypt did all they could to delay any such possibility.[9] Nor were they pleased at the Egyptian government's secret agreement with the famous German aircraft manufacturer Junkers for the establishment of a regular air mail service between Alexandria and Trieste in Italy using Junkers F13a seaplanes.

Two years later the nationalist-dominated Egyptian Parliament officially proposed the creation of an Egyptian air force. This time the new British high commissioner, Lord Lloyd, reluctantly agreed but still insisted that this force include British officers and that it remain under British control.

King Fuʾād always took a keen interest in military matters and had encouraged the establishment of the Fārūqīyah Naval School in 1927. The first eighteen cadets of this new Naval School had gone for further training to *HMS Worcester* in England, which established a precedent soon to be followed by the air force. It was against this confused and often embittered background of Anglo-Egyptian relations that an Egyptian Army Air Force (EAAF) finally came into being.

2

# A Start (1928–1932)

The creation of an Egyptian Air Force in 1932 was more of a political than a military action on the part of the Egyptian government. At that time the country was still under British domination, although it had been declared an independent kingdom ten years earlier. Nothing could be done without British approval—certainly not the creation of an air force. Even so, the training of a handful of pilots to fly a single squadron of unarmed aircraft proved to be a complicated business bedeviled with political pitfalls.

## Egypt Demands an Air Force

A sense of Arab nationalism had deepened during and after the First World War in many Arab countries. At the same time there was a resurgence of Muslim confidence as well as deep-seated resentment at being pushed around by Christian foreigners. The old Ottoman caliphate in Constantinople (now renamed Istanbul) had been abolished by Kemal Ataturk, Turkey's new secular leader in 1924. This meant that for the first time in 1,292 years the world's Sunni Muslims[1] no longer had a single religious authority who could claim spiritual leadership as the caliph (*khalīfah* or "vicar" of the Prophet Muḥammad) had done. Many Arab Muslims felt that a new Arab caliphate should take the place once filled by the Ottoman caliphs, the kings of both Egypt and Iraq seeing themselves as possible candidates. Egypt and Iraq were at the same time competing for leadership of the Arab world even though both still lay under the firm control of the British Empire. Iraq had, in fact, recently created its own tiny air force, the first Arab country to do so. National pride demanded that Egypt have an air force as well.

As soon as the Egyptian Parliament proposed the creation of an air force, two hundred or so Egyptian officers volunteered for the new arm. From these, four candidates were finally chosen in October 1928 following extremely strict medical tests and technical examinations. Three of them would become Egypt's first military pilots. Their names were Muḥammad ʿAbd al-Munʿim Mīqātī, Aḥmad ʿAbd al-Rāziq, and Fuʾādʿ ʿAbd al-Ḥamīd Hajjāj. They were promptly sent to the British RAF No. 4 Flying Training School at Abū Ṣuwayr next to the Suez Canal. There the Egyptians trained on a variety of aircraft including Avro 504Ns and DH9s, both developed from First World War bombers, plus Armstrong Whitworth Atlases and Vickers Vimys—the type in which two RAF pilots, Alcock and Brown, had made the first nonstop flight across the Atlantic in 1919. At this stage there was no thought of specialization, and the three Egyptian pilots were expected to be able to fly almost any kind of aircraft, large or small.

Their preliminary training was not without incident. 2d Lt. Fuʾādʿ Ḥajjāj, a distant relative of the Egyptian royal family, was officially attached to No. 13 Squadron RAF, and he virtually demolished an Avro (serial J8779) on 9 December 1929, but went on to receive his wings on 31 January next year. After graduating at Abū Ṣuwayr the three Egyptian junior officers traveled to England for specialized training, Mīqātī going to RAF Eastchurch for an armaments course, while Rāziq and Ḥajjāj went to Calshot for navigation. They were still encouraged to get experience on as many different machines as possible, Mīqātī adding Bristol Fighters, Armstrong Whitworth Siskins, and Westland Wapatis to his logbook at Eastchurch. The Wapati apparently had to be taxied fast with the joystick hard back to avoid tipping the aircraft on to its nose. Sam Patch, a well-known rugby football player who later rose to the rank of air vice marshal, was then the safety officer at RAF Eastchurch. Mīqātī, who also rose to become an air vice marshal, recalled: "Sam Patch once blew up at me for taxiing too fast. This made me so angry that I told him to try it himself. He did, and within fifty yards the Wapati was on its nose."[2]

Meanwhile Britain was still resisting the establishment of an Egyptian Air Force, and when in 1928 a more pro-British government came to power in Cairo the three new pilots had to rejoin their army regiments. Around the same time Cairo newspapers reported that an Egyptian named Muḥammad Ṣidqī flew his light aircraft from Germany to Cairo, being the first Egyptian to make such a long flight. He was hailed as a national hero on his arrival, and many people felt that Egypt had finally entered the air age.[3]

## A "Political" Birth

Not until September 1930 did the Egyptian minister of war convince Spinks Pasha, the British inspector general of the Egyptian Army, to agree to the creation of an army air force. By that time five more Egyptian officers were under training by the RAF at Abū Ṣuwayr, and in September 1931 the British De Havilland Aircraft Company won a contract to supply Egypt with five DH60 Moth Trainers. These little biplanes, similar to other Moth variants that made several astonishing long-distance flights in the 1930s, were predecessors of the famous Tiger Moth aerobatic primary trainer. This first Egyptian squadron was to consist of one British staff officer, one British flight commander, three Egyptian flying officers, one Egyptian staff officer, five British civilian advisors, and forty-six Egyptian mechanics, tradesmen, clerks, and laborers. Air Commodore A. G. Board, the chief staff officer of RAF Middle East Command, was then made director of Egyptian Military Aviation.

While Egypt's first order for Moth Trainers was being built, a pair of Avro 618 "Tens" from a defunct airline were sold to the Egyptian government by the enterprising British Airworks representative in Cairo. These aircraft were license-built versions of the more famous Fokker Trimotor, comparable to America's Ford "Tin Goose" of 1928. These arrived in Egypt from England on 18 January 1932, but they had been flown by British pilots and consequently landed without any fanfare to be quietly parked at one side of Almāẓah airfield. They do not even seem to have been given Egyptian markings and their new serial numbers (G-AASP becoming F200, G-AASR becoming F201) until after the first "official" arrival of the EAAF's first "official" aircraft almost five months later. The reasons for this lack of publicity clearly illustrate the political character of the Egyptian Air Force's birth.

When Egypt's first three military pilots, Mīqātī, Rāziq, and Ḥajjāj, arrived in England to collect their DH Moths, they were virtually confined to the residence of Dr. Ḥāfiẓ ʿAfīfī Pasha, the Egyptian ambassador in London, because the British government insisted that there be no publicity. In fact Britain wanted Egypt's Moths to be shipped to Alexandria in crates. This idea was endorsed by Air Commodore Board, director of Egyptian Military Aviation, who doubted the wisdom of sending a flight of fragile Moths across Europe and the Mediterranean Sea in winter. Unfortunately the Royal Iraqi Air Force had recently been set up with considerable British encouragement. Worse still, Iraq's first aircraft had flown from England to Baghdad in the full glare of publicity. Clearly the Egyptian government could not permit its first planes to slink into Alexandria harbor in wooden crates. But Board had, on his own initiative, already put the crated Moths into a ship that was now sailing south. A flurry of diplomatic messages led to their being unloaded at Gibraltar and sent back to England in another ship. Board promptly resigned, while Egypt's five Moths were reassembled at the De Havilland factory in northwest London. This work was partly carried out by the three Egyptian pilots under the direction of Flt. Lt. S. J. Stocks, who had been appointed Egypt's chief flying instructor. Many of those British advisors who helped the Egyptian Air Force over the years did so after various mishaps or clashes with their British superiors. Stocks was no exception, having had the misfortune to run an Imperial Airways flying boat ashore at Marsá Maṭrūḥ in northern Egypt.[4]

On 23 May 1932 the five DH Moths at last took off from Hatfield, De Havilland's main airfield north of London, flown by Mīqātī, Rāziq, Ḥajjāj, a British sergeant pilot, and Stocks. Their journey would take them in easy stages from Lympne in southern England, Paris, Lyons, Marseilles in France, through Italy via Pisa, Rome, then a two-day engine check at Naples, and Catania. Next they flew to Malta, on through Italian-ruled Libya, landing at Tripoli, Surt, Banghāzī, and Ṭubruq. Their first port of call in Egypt would be Marsá Maṭrūḥ before arriving in Cairo. At Banghāzī a representative of the British Shell Oil Company rather spoiled the British government's hopes of no publicity by taking and publishing photographs of the Moths refueling with Shell aviation gasoline.

On 2 June these first five aircraft of the new Egyptian Army Air Force (EAAF) arrived safely at Almāẓah airport northeast of Cairo to be greeted by King Fuʾādᶜ, the young Crown Prince Fārūq, all available troops, the diplomatic corps, and the government. A march-past by the aircrew dressed in rather oversized RAF flight suits drew a huge cheer from a vast crowd of excited Egyptians. For the delivery flight and this arrival ceremony the aircraft had green and white Egyptian flags painted on their upper wing surfaces and fuselage sides. Such markings were strangely reminiscent of those used by the Ottoman Turkish Air Force in late 1914 or early 1915. Beneath their wings, however, the DH Moths wore the green and white "cockade" roundels that were to be carried by Egyptian military aircraft until 1958.

3

# Fledgling Years (1932–1934)

The creation of the EAAF had no real effect on the balance of power between Egypt and the occupying British forces. The political gulf between the two nations remained immense, but, paradoxically perhaps, the EAAF did become an area in which military personnel could work together amicably. In fact, these early years established a tradition of good professional relations between the British RAF and the EAF which survived the following difficult decades. At first things went very smoothly, with the EAAF gradually increasing in skill and, rather more slowly, in size. It was not permitted to arm its aircraft, but it was given difficult and sometimes dangerous duties that provided its personnel with experience and confidence. The most important of these was the fight against drug trafficking.

Fascism and Nazism were spreading across Europe, but Egyptian politics continued to be dominated by a struggle between the royal palace and those seeking constitutional change. On top of this were the complex rivalries of traditionalist, Islamist or "literalist" Muslim, secularist, pro-Fascists, and pro-Communist movements. Above all there was the continuing tense relationship between Egypt and Britain. Sometimes it seemed as if the tiny EAAF was little more than a fine flying club whose personnel were drawn from the social elite of Egypt and where such political tensions had no place. In reality, however, this was not the case, and many young officers in the EAAF were as concerned about their country's future direction as were those who rioted on the streets of Cairo.

## Laying Foundations

The new EAAF remained part of the Egyptian Army under the direct authority of the War Ministry in Cairo. Its first squadron was essentially a training unit intended to give experience in air operations, maintenance, and administration to officers who would then form the nucleus of a larger force. As it happened an air force man would not take overall command for another twenty years, even after the EAAF became a separate service as the REAF (Royal Egyptian Air Force) in 1937. The top post always went to an army officer. Initially this reflected a lack of suitably experienced air force personnel, but later it almost certainly had a political motive, reflecting the greater loyalty of the Egyptian Army to the existing regime.

In the beginning the British insisted that one of their officers take charge, but the Egyptian government was very unwilling to see a British officer in command of a force that was being portrayed as a symbol of Egypt's independence. Eventually both sides agreed that a Canadian officer, Squadron Leader (Sq. Ldr.) Victor Hubert Tait, be placed at the head of the EAAF holding the rank of qā'im-maqām (lieutenant colonel). Prior to this Tait had been the chief signals officer in RAF Middle East Command, and as commanding officer (CO) of the EAAF he still reported to the British *sirdār* or commander of the Egyptian Army.

Qā'im-maqām Tait's first task was to establish airfields, and those he selected are still used today. He remarked about this task: "I chose most of the equipment and airfields myself. Almaza and Dakhaylah had been civilian airfields. Otherwise I just chose a good flat bit of desert, had the stones cleared away and a wind-sock put up. We didn't have much trouble with equipment but the desert sun used to rot the fabric-covered aircraft. Really we did better than the RAF with this problem because there were so many good sail-makers from the Nile, so we could make repairs very easily."[1] The fledgling air force had more difficulty finding skilled engine mechanics and personnel capable of performing more complicated services in a climate so hard on machinery.

Qā'im-maqām Tait's next job was to choose a staff. One of the officers he selected was Flt. Lt. S. N. Webster, the prestigious winner of the famous Schneider Trophy air race and a personal friend. RAF NCOs were selected for secondment to the EAAF on the basis of both technical ability and athletic prowess. Tait had a traditional British belief in the merits of organized sports as a means of keeping the Egyptians fit and strengthening unit morale. Perhaps Tait's most valuable skill was supreme tact and sensibility, which enabled his Egyptian and British subordinates to work together without friction—an ability that is warmly remembered by EAF historians.[2] He was also able to pick officers from the very cream of the Egyptian Army, while many EAAF ground crew came from the Egyptian Police and Fire Service, bringing with them mechanical skills. As he recalled: "Lots of the men went back to the Police afterwards. By 1939 half of the Cairo Traffic Division seemed to be ex-Air Force men! . . . As for the officers, ninety-nine percent of the cadets volunteered for the Air Force. We took the best but we also made them do plenty of exercise. The Egyptians weren't too good at that, but they learned. In the end we won practically all the Armed Forces sports matches."[3] An apprentice school was also set up for civilian volunteers who would subsequently be employed as technical specialists.

Early Egyptian pilots were generally of a very high standard. ʿAbd al-Ḥamīd Dighaydī, for example, left the RAF's Central Flying School at Upavon in England with the rare Grade A1. He later became a war hero in Palestine in 1948, fighting in the Faluja Pocket alongside Colonel Nasser. By then, however, Dighaydī's tendency to have trouble with the military authorities found him back in the ranks of the army. Dighaydī's air force career effectively came to an end in 1968 when he was among those senior officers tried for dereliction of duty in the wake of the disastrous June 1967 War, even though he was acquitted. A certain lack of discipline seems to have been a problem with young officers all too aware that they were the cream of Egypt's armed forces. Another pilot named Khalīfah was only found to have been using his Avro 626 to visit his girlfriend when a mechanic reported that its propeller was slightly bent—caused by landing in a field near the young lady's home. Once again, however, this lack of discipline went together with more positive qualities and Khalīfah later served as a chairman of the UN Atomic Energy Commission.[4] Among others who joined the air force around this time were Ḥasan ʿĀkif, who became King Fārūq's aide-de-camp and personal pilot, and Ibrāhīm Ḥaqqī, who, as the De Havilland Aircraft Company's representative in Cairo, negotiated the purchase of Egypt's first commercial jet airliners, the Comets of United Arab Airlines.

## Fighting the Drug Smugglers

The EAAF's first duties were ceremonial, but the next were much more serious when the Moths and Avro "Tens" were called in to help the war against opium. The drug hashish had been part of the Egyptian way of life since at least the thirteenth century, but during the First World War foreign traders introduced hard drugs into the country. The notorious legal system of "Capitulations" forced upon Egypt by outside governments in the late nineteenth century meant that foreign drug smugglers were effectively immune from Egyptian law. As the scourge of heroin took hold of the country, opium poppies were also grown illegally by Egyptian *fallāḥūn* peasants. Russell Pasha, British head of the Egyptian Police Force, estimated that by the mid-1920s there were half a million heroin addicts out of a population of only fourteen million. But the very success of Egypt's Central Narcotics Intelligence Bureau made drug prices soar.[5]

The EAAF was called in to help with aerial surveillance. Egyptian peasants would plant small patches of opium poppies in the middle of their bean fields, and since beans grew to a greater height than the poppies it was almost impossible for police on the ground to spot the illegal crops. But from the air the bright purple flowers of the opium poppy were highly visible. In a typically Egyptian way, Qāʾim-maqām Tait and his pilots developed a technique that was both effective and merciful to the poverty-stricken *fallāḥūn*. As the DH Moths had neither cameras nor radios, their crews had to circle the poppy patches for some time to plot the position on maps before returning to base to inform the police. It then generally took half a day for ground patrols to reach the offending bean field, by which time the peasant would have summoned his friends to root out the illegal crop. Such a result also avoided expensive prosecutions. The EAAF's task got a bit harder when the farmers learned to bend down the poppy heads and cover their blooms with handfuls of soil if aircraft were operating in the area, but the pilots and observers soon learned to deal with this trick as well.

Meanwhile drug trafficking across the desert increased, and so the sturdy little Moth trainers were fitted with extra hand-pumped fuel tanks to patrol border areas. Suspect camel caravans were then reported to the Egyptian Frontier Forces. The large trimotor Avro "Tens" played a particularly important role here as they were equipped with radios and could communicate directly with ground patrols. Unlike the Nile Valley *fallāḥūn*, however, the Sinai bedouin who actually took the smuggled drugs into Egypt were well armed and often would shoot back. They also tended to be very good shots and often gave the scattered Egyptian Frontier Force patrols a hard time. Nevertheless, the frontier force usually picked up the smugglers' tracks where they crossed into Egypt from Palestine. Patrols would be sent in pursuit while the information was radioed to other outposts. Traveling only at night, and usually having a good start, the smugglers would release their camels to graze like any other bedouin herd once they reached the relative safety of the Sinai mountains some thirty-two kilometers east of the Suez Canal. Next night they would head for the canal, where swimmers would be waiting to carry the valuable cargo to the other side.

The Egyptian authorities clearly needed motorized transport and aircraft to combat the drug smugglers effectively. The first full-scale EAAF antismuggling operation was carried out in April 1933. The frontier police had detected a camel caravan on 23 April and soon caught two smugglers. There were clearly others, however, so an all-out effort was made to intercept them. On 25 April 1933, three EAAF DH60 Moths arrived at al-ʿArīsh from Cairo, flown by Bimbāshī (Major) Stocks, Mulāzim Awwal (1st Lt.) Ḥajjāj, and Mulāzim Thānī (2d Lt.) Nājī. The aircraft almost immediately picked up the smugglers' tracks, but it was too late in the day to follow them, so the planes returned to al-ʿArīsh. On the following morning the Moths found the smugglers, reported their position to the police, and returned to Cairo, their mission successfully accomplished.

At the end of April 1933 one of the EAAF's trimotor Avro "Tens" was used on a similar operation. With Stocks piloting, Mulāzim Awwal Rāziq as second pilot, Mulāzim Awwals Dighaydī and Mīqātī as observers, the aircraft found the smugglers' tracks very quickly and radioed an alert to the police. A long chase ensued right across Sinai to within a couple of kilometers of the canal. The band of smugglers then admitted defeat and scattered, leaving their drugs on the sand.[6] These Sinai patrols were not without danger, and casualties were suffered. A couple of years later a solitary Avro 626 biplane was flying across the desert when the Egyptian pilot spotted something on the ground below. He flew lower to investigate, but as he circled the aircraft stalled and crashed into a hillside. The pilot was killed, but the observer, Ḥasan Maḥmūd, was only injured. Later in his career Ḥasan Maḥmūd was the first air force officer to became commander of the Egyptian Air Force following the Revolution of 1952.

Other work that fell to the EAAF included aerial spraying during a nationwide antimalaria campaign and aerial photography for the Egyptian Antiquities Department. This was an urgent necessity as new irrigation schemes meant that archaeological sites were fast disappearing beneath expanding areas of cultivation. The EAAF's work was, in fact, so successful that the air force won a special commendation from the now aged King Fuʾād who had a particular interest in Egypt's antiquities. A huge photographic mosaic inside the entrance to the Cairo Museum remains an example of the EAAF's efforts.

## The First Expansion

With only two Avro trimotors and a single flight of DH Moths, the EAAF clearly needed more aircraft, and the versatile Avro 626 was selected by Qāʾim-maqām Tait himself.[7] It would remain Egypt's standard military aircraft until 1937, despite the fact that the EAAF's first such squadron was dogged by misfortune. Shortly after the Egyptian Frontier Force, aided by the EAAF, captured its biggest Sinai drug-smuggling caravan ever in July 1933, the government in Cairo announced that it was buying ten Avro 626s. This biplane was a modified version of the famous Avro 621 Tutor trainer. It was capable of undertaking initial flying training, bombing, photo reconnaissance, gunnery, radio and night flying training, as well as navigation and even seaplane instruction if fitted with floats. Though basically a two-seater, it could carry three crewmen since the rear gunner's cockpit also contained a prone bomb-aiming position. This versatile aircraft was eventually sold to no less than fourteen countries.

Fourteen Egyptian officers came to Avro's Woodford airfield near Manchester, England, to learn to fly their new aircraft under the command of Tait and Webster. This time the British press gave full publicity to the Egyptians' arrival and their inspection by the Egyptian ambassador, while every step taken by the pilots was eagerly reported in an enthusiastic Cairo press. After three weeks of training the Avro 626s were flown down to Lympne on the southern coast, ready for their long flight back to Egypt. The first problem came on 13 November when one of those sea fogs, so typical of England's Channel coast, rolled in while Mīqātī and a British pilot were both airborne. The British pilot got down with only minor damage, but Mīqātī landed in a nearby marsh and destroyed his plane. One was repaired and the other replaced under an insurance policy that Tait had arranged only a few days earlier with the British Aviation Insurance Company—but not before the Egyptian ambassador had made another formal inspection and noticed that the squadron was two planes short.

When at last everything was ready for the first leg of the flight home, the ten Avro 626s took off. They were flown by Qāʾim-maqām Tait, Bimbāshī Webster, British Warrant Officers Smith, Dingwall, Oldham, Whitlock, and Roberts, Egyptian Mulāzim Awwals Munʿim, Dighaydī, Mīqātī, and Ḥajjāj, plus Egyptian Mulāzim Thānīs Khalīfah, Nājī, and Ḥaqqī, together with three Egyptian NCOs and an Egyptian radio operator. The weather forecast had been favorable, and the English Channel was a sparkling blue, but the fliers found the French coast hidden beneath a blanket of cloud. Conditions grew rapidly worse as the flight made its way south before turning inland toward the Paris airfield at Le Bourget. Three aircraft soon got separated. One landed safely at the French Air Force base of Villacoubley. Another crash-landed with minor damage after developing engine trouble 16 kilometers south of Abbeville, the two British warrant officers aboard being shaken but unhurt. The third aircraft crewed by Mulāzim Awwal Fuʾād Ḥajjāj, and NCO Shuhdī Duss was not so lucky. Lost and alone in unfamiliar territory, Ḥajjāj appears to have carefully lost height, perhaps hoping to find his bearing from the ground beneath. But the cloud continued down to ground level as fog. Apparently Ḥajjāj found himself heading toward a high electrical power line and tried to dive beneath it, but the Avro struck the ground, bursting into flames. Both men died. They were the EAAF's first fatal casualties, and Fuʾād Ḥajjāj's name heads a list of twenty-one air force officers and men who lost their lives between 1933 and 1946.[8]

Tait and Webster collected the bodies of their young comrades at Blangy sur Bresle and accompanied them to Cairo. Here cultural differences between Egyptians and westerners finally emerged, for, as Tait recalled, he and Webster were not keen on what they saw as the emotionalism of an Arab funeral at the little cemetery outside Heliopolis, and so they went to a nightclub. There they met the representative of the British Aviation Insurance Company, who urged them to renew the EAAF's insurance policy. Back in Paris the Egyptian ambassador arranged for the shaken EAAF airmen to enjoy the hospitality of the French Air Force for several days before restarting their long journey to Cairo. Bad weather next held up the flight at a French air base outside Dijon, in the heart of France's Burgundy wine country. According to Mīqātī, a culture clash was avoided this time. Nevertheless, some Egyptian pilots were rather groggy as a result of French hospitality when they took off the following morning.

Landing for an overnight stop at Qaābis in Tunisia, another Avro bent an axle and a village blacksmith was summoned to straighten it out. Tait must have felt some relief when the remaining eight aircraft flew in formations of three and five over Cairo. There was to be no rest, however. The Royal Egyptian Aero Club had organized a "Tour of the Oases" air rally, and the Avro 626s had arrived home just in time for three of them to shepherd a large number of foreign and Egyptian light aircraft across formidable stretches of desert. The rally route lay between Asyūṭ, Khārijah, Dākhilah, Farāfirah, and Baḥrīyah in December 1933. The EAAF also prepared all the maps, provided most of the landing grounds, and also located several pilots who force-landed in the desert. The EAAF's first experience of air rallies did not pass without loss, however. Bimbāshī Stocks was carrying seven race officials in one of the EAAF's Avro "Ten" trimotors as an engine stopped dead on takeoff when the machine was less than a hundred feet up. Two engines were still running as Stocks turned back toward the landing ground. Then these engines also died, and, quite correctly, Stocks landed straight ahead but came down in rough ground and demolished the aircraft. No one was hurt, but the episode was seized upon by one of Egypt's most virulently anti-British magazines which accused Stocks of incompetence. Unable to accept criticism from such a source, a British aviation magazine rose to the challenge with an almost hysterical defense of the unfortunate Stocks.[9] While the Egyptian and British personnel of the EAAF worked together without friction, the tense political situation between their two countries was as bad as ever.

The Egyptian government was very eager to expand its air force, but the British Foreign Office retained doubts about arming Egypt with modern weapons. Popular feelings in Cairo were certainly deeply resentful of the continuing British military presence in Egypt and tended to boil over at the slightest provocation. The British authorities in Cairo had, in fact, become so sensitive to such demonstrations that they even objected when old King Fuʾād unveiled a statue of Muḥammad ʿAlī, founder of the Egyptian ruling dynasty, at his birthplace of Kavalla in what is today Greece! Muḥammad ʿAlī was, in fact, from a family of Muslim Albanian tobacco merchants with no previous royal pretentions.

Nevertheless, the destroyed Avro "Ten" was soon replaced by a similar Westland Wessex trimotor, a machine that landed—its pilots agreed—like a feather cushion. Meanwhile the Avro 626 had suited Egypt's needs so well that in June 1934 ten more were purchased. This second batch were still not fitted with guns, gun mountings, or bomb racks, their role being strictly limited to aerial reconnaissance and photography, although their cameras were Williamson F28s of the most modern type. These multipurpose aircraft soon performed important missions including an emergency survey of the annual Nile flood which had reached its highest level in fifty years.

The EAAF's pilots were growing in experience, and some had now logged more than seven hundred flying hours. Their formation flying was considered excellent even by the perfectionists of the RAF, just as the parade-ground drill of Egyptian soldiers had impressed the British Army forty years earlier. New Egyptian crews were being trained at the RAF's No. 4 Flying Training School (FTS) at Abū Ṣuwayr near the Suez Canal. Other EAAF pilots regularly attended specialist courses in England, Mulāzim Awwal Rāziq going on the very popular two-year engineering course at RAF Henlow.

As Egypt dragged itself into the modern world, aviation became a fashionable subject, and the EAAF's exploits were followed eagerly. Many Egyptians continued to agitate for political reform, rapid modernization, and real independence from the British. Among these was a militant student leader who got himself wounded by a police bullet while leading a demonstration in 1935; his name was Jamāl ʿAbd al-Nāṣir, the future President Nasser. Equally patriotic young men were now entering the air force straight from the Egyptian Army Cadet School. The first such cadets to transfer without having to serve in an army regiment were sent to train at Abū Ṣuwayr in 1934. Ibrāhīm Ḥasan Jazzārīn was among them. His generation had entered the Military College in the year the EAAF had been formed, but after two rather than the usual three years he and his colleagues, Muḥammad Faraj and Muḥammad Ibrāhīm Abū Rabīaʿh, were selected for Flying Training at the RAF's No. 4 FTS at Abū Ṣuwayr. There they were taught by a group of very experienced British instructors led by Thomas Pike, later Sir Thomas Pike, marshal of the RAF. The Egyptians flew the ever reliable Avro 504N and the less lovable Armstrong Whitworth Atlas which the students generally regarded as a very dangerous airplane. As Jazzārīn recalled: "We were the only three non-RAF students at Abu Suwair at the time. I remember the motto hanging in the mess—Better to be Mr. Late Pilot than Late Mr. Pilot. Abu Rabia was a really good pilot. He left No. 4 FTS with the very unusual 'exceptional credit' rating. Later he commanded our first proper fighter squadron."[10]

These first years had been difficult for the EAF, but it had laid good foundations and its men were growing in experience. The main problems had been external political ones rather than problems stemming from the EAAF itself. This would remain the case for many years to come.

# 4

# A New Treaty and a World Crisis (1935–1939)

Egypt and its new air force were not immune from the looming world crisis of the 1930s, although the country was largely preoccupied by its tense relationship with Britain. As totalitarianism spread in Europe and the Far East, Britain recognized the threat to world peace and set about improving Anglo-Egyptian relations. The result was a new Anglo-Egyptian treaty of 1936 which allowed Egypt much greater control over its own destiny, yet still the land of the Nile was not truly free. Next the EAAF became a separate service within the Egyptian armed forces as the Royal Egyptian Air Force (REAF). Its personnel and equipment were increased as, beyond Egypt's borders, the war clouds gathered. By the time the Second World War broke out in 1939 the REAF had grown into a small but quite efficient force which might have played a significant role in the struggle against Fascism and Nazism, had not the old problems that bedeviled British-Egyptian relations intervened once more.

## The Italian Threat

The British military presence in Egypt might have been unwelcome, but it was benign when compared to what was happening in neighboring Italian-ruled Libya. Here the Sanūsī Muslim sect had continued to resist Italian rule even after the First World War. It had done so very effectively by pinning Italian occupation armies to the coast and in a few inland oases. But Mussolini, the Fascist dictator who seized power in Rome in 1922, decided that such dissent would end, whatever the cost. The brutal war of genocide that followed was largely ignored by the outside world and has remained virtually forgotten to this day, even though its memories continue to shape Libyan attitudes to the West. The Arab peoples were, however, appalled. This was particularly true in Egypt where many people, including army officers, secretly helped the Sanūsīi by supplying arms and information.

The savage Fascist conquest of the Libyan interior was largely completed before the EAAF was created. The end had come in January 1931 when the oasis headquarters of the reformist Sanūsī Islamic sect at Kufrah finally fell to the Italians after a ferocious aerial bombardment. This bombardment, in fact, contributed to the Egyptians' desire for their own air force. The Italians then built a two-hundred-mile long wire fence from Jaghbūb in the deep Saharan desert to the Mediterranean coast. This was designed to prevent Sanūsī guerrillas from crossing into Libya from Egypt or escaping back westward. Backed up by regular patrols of Italian aircraft and armored cars, it successfully stopped traffic between the two countries. In 1932 Italian forces finally subdued the last Sanūsī guerrilla resistance, but by then thousands of Sanūsī men, women, and children had been shot or captured as they tried to flee across the wire into Egypt. This in turn left a legacy of bitterness which caused the Sanūsī to rally to the British cause when the North African desert campaigns of the Second World War broke out eight years later.

The slaughter of the Sanūsī was not forgotten in Egypt, and in 1935 money for a fourth squadron of aircraft was agreed without question in the Egyptian budget. That same year Mussolini's Fascist Italy also invaded Ethiopia, completely altering the strategic situation in northeastern Africa. British forces in the Sudan now faced Italian Fascist armies to both east and west. Recognizing these changes, Britain sought a new relationship with Egypt and started negotiations for a new Anglo-Egyptian treaty.

In May 1936 old King Fuʿād died, and was replaced by his youthful son Fārūq. The new king was an unknown quantity to both Britain and his own Egyptian subjects. He was fortunate in coming to the throne in the same year that the much-heralded new Anglo-Egyptian treaty caused an upsurge of Egyptian patriotic pride. Under this new agreement the British *sirdār* or commander of the Egyptian Army was replaced by an Egyptian and the many British officers in the army were rapidly superseded by a permanent British Advisory Mission. All remaining British officers now took their oaths of allegiance directly to the Egyptian crown. Even so, the mounting threat from Italy made a complete British withdrawal from Egypt and the Sudan inconceivable, at least in London.

The rapid expansion of both the Egyptian Army and Air Force[1] had the effect of broadening the social spectrum from which Egyptian officers were chosen. It was now easier for the sons of lower middle-class villagers, artisans, and shopkeepers to gain a commission. Many of the men who would seize power in 1952 entered the Egyptian officer corps at this time. Among them were Jamāl ᶜAbd al-Nāṣir, ᶜAbd al-Laṭīf Baghdādī, and Jamāl Muṣṭafá Salīm, the latter two of whom joined the air force upon graduating from the prestigious Military Academy.

Meanwhile an officially sponsored public fund raised more than a million Egyptian pounds to buy new aircraft for the EAAF, and in 1936 a contract was signed for half a dozen Hawker Audax biplane bombers, built under license by Avro.[2] These arrived the next year to supplement the EAAF's two general purposes squadrons, one trainer squadron, and single communications flight. The Avro Aircraft Company almost seemed to have cornered the market for military aircraft in Egypt at this time, perhaps as a result of the success of the tough little Avro 626s. The EAAF also acquired the high-speed twin-engined monoplane Avro 652 Mk. II prototype, which was tested for its suitability as a light bomber. A later version of this type earned fame in the RAF as the well-loved Avro Anson. Egypt also bought two Avro 641 Commodore five-seater biplanes, the first of which served as Qāʾim-maqām Tait's personal transport.

The Egyptian government had hoped to expand the country's armed forces, increasing the air force to forty-nine planes within a year and to one hundred by 1941. In 1936 Major-General Marshall-Cornwall, the British Military Mission commander, issued a very sobering report on the condition and needs of Egypt's armed forces which showed how unrealistic the government's plans were. The army had no machine guns, support, or antitank weapons. There were no tanks, radios, or antiaircraft guns, while training was, in the report's words, "based on tactical ideas of almost proto-dynastic antiquity." But the EAAF came out rather better, being described as a small but well-trained organization of thirty-eight aircraft—though the planes were not armed and their crews had no practice in bombing. The Advisory Mission suggested a three-year expansion program designed to develop one army cooperation and one bomber-transport squadron to be based at Cairo, plus a bomber and a fighter squadron at Alexandria.[3]

Meanwhile another Tour of the Oases Air Rally had been organized, originally to take place in March 1936. It was to have included parts of the Red Sea coast, but this leg had to be abandoned because of the Italian invasion of Ethiopia, and in fact the entire event was postponed until February 1937. Even so, it should have been a peaceful affair, supported as before by the EAAF which patched up damaged aircraft and found pilots who had come down in the desert. When it was all over the rumor in the EAAF was that the final winner of the rally, the German pilot Freiherr Speck von Sternberg in a large Junkers 86, had made a photographic survey of possible landing sites in the Western Desert. If so, then Von Sternburg's work may have come in very handy for the Luftwaffe a few years later.

All British officers in the Egyptian Army had now been withdrawn from command positions, but such a move was clearly impractical in such a technological service as the air force. The transfer of the EAAF from British to Egyptian control was delayed until April 1937, whereupon Spinks Pasha, the former British *sirdār,* told Tait that he would have to leave. Tait replied that he now took orders only from the Egyptian minister of war, and when a meeting with the minister was arranged the new position of air advisor to the EAAF was invented specifically for Victor Hubert Tait who was also promoted to the RAF rank of group captain. Nevertheless, the British Advisory Mission still made sure that the British played a major role in the operations of every EAAF squadron. Nos. 1, 2, and 3 Squadrons were still led by British officers, though Nos. 1 and 2 did have Egyptians as second in command. A number of British warrant officers and NCOs were also attached to every unit as instructors and supervisors.

Up till now EAAF officers had held army ranks, but in 1937 new ranks copied directly from those of the British RAF were introduced. Unfortunately the change had the effect of reducing air force officers by up to one full grade in a move to stop the Egyptians from taking over command of the squadrons and flights they had in reality been leading for months. Few if any air force men yet had the close contacts with the Egyptian palace enjoyed by some senior army officers, and although Tait argued against this transparently political move, his advice was ignored. At the same time the EAAF was renamed the Royal Egyptian Air Force (REAF) and a popular artillery officer, Liwāʾ (Brigadier) ʿAlī Islām Bey, was made its first director, having previously been commander of the Egyptian Military Academy. ʿAlī Islām then selected Squadron Leader Dixon of the RAF as his chief staff officer.

In April 1937 two more squadrons of British aircraft were purchased, eighteen additional Audaxes being delivered in 1938. In the late 1930s the British airplane manufacturing company Armstrong Siddeley also offered to set up an aircraft factory in Egypt. The proposal was studied by the Egyptian government, but no decisions were made before the Second World War broke out. The REAF's role was now becoming much more serious as international tensions rose. At the same time the REAF started to suffer from the kind of political interference that would continue to obstruct its development until quite recently. Some of its officers also resented finding themselves under an army officer, however amiable. Nevertheless Major General Marshall-Cornwall, head of the British Military Mission, could report that by May 1938 the REAF was making remarkable progress and should be regarded as a definite factor in those forces available for the defense of Egypt.[4] Such additional forces were certainly needed as Britain's RAF was appallingly overstretched and the threat of global war was growing acute, but even so the REAF's contribution to Imperial Defence was inevitably small.[5]

An Egyptian Flying Training School had been set up in 1937 using the remaining Moths for ab-initio training, and a year later the Egyptians received their first Miles Magister monoplane primary trainers. Unfortunately this delivery was rapidly followed by three fatal accidents when Magisters failed to pull out of dives. A board of inquiry, on which the Miles Company was also represented, found that the control surfaces were inadequate for the Egyptian climate, and after modifications had been made the machines were returned to service. The only REAF aircraft to possess any offensive capability, the Audaxes of No. 4 (Bomber) Squadron based at Dākhilah, were also experiencing severe hot weather problems with their fuel tanks. These aircraft were the first that the British military authorities permitted to carry guns—but not to fire them as certain vital components were still not delivered. Similarly the six Fairey Gordons sold as a target towing flight were denied any towing gear![6] Various schemes were meanwhile under way to replace the REAF's toothless Audaxes with twin-engined Bristol Blenheim bombers and the mixed equipment of No. 3 (Communications) Squadron with large Vickers Valentia transports, but such hopes were rapidly overtaken by events.

## The Last Days of Peace

During Italy's invasion of Ethiopia Egypt had been the only non-League of Nations state to apply full sanctions against the aggressor. Now, as the world crisis deepened, the REAF found itself face to face with an Italian Regia Aeronautica that posed a real and potent threat to Egyptian security. During the Munich Crisis of 1938, REAF Audaxes and Avro 626s were based on the Mediterranean beach at Marsá Maṭrūḥ and at Siwa Oasis further south in the Sahara Desert. From there they patrolled the Libyan frontier, often meeting Italian Savoia Marchetti trimotors doing the same thing on the other side of the barbed wire fence that the Italians had erected from Jaghbūb to the coast. Three aircraft would fly each patrol, one Avro equipped with a radio, two Audaxes with single Lewis guns. These were the only Egyptian machines to carry any workable armament, possible now that the Egyptians were no longer under complete British control, but their crews still had virtually no weapons training. Nevertheless, the three aircraft, flying at a steady four hundred feet, would follow the coast from Marsá Maṭrūḥ to the frontier and thence down the frontier to a landing ground on a plateau twelve miles beyond Siwa Oasis. Among those who led such patrols was Flt. Lt. Ibrāhīm Ḥasan Jazzārīn, who subsequently rose to the rank of air commodore in the REAF.[7] During this period the Hawker Hinds of No. 113 Squadron RAF were also based at Marsá Maṭrūḥ, making a photographic survey of the frontier zone. The REAF continued these patrols for another two years, until Italy finally entered the Second World War in 1940.

Meanwhile the rest of the REAF was busily exercising alongside the British Suez Canal Brigade. Following the severe fright provided by the Munich Crisis of 1938, Britain temporarily changed its attitude and declared an eagerness to see the REAF expand. The main effort seems to have gone into increasing the number of Egyptian pilots, of whom there had been only fifty at the start of 1938. But the REAF was only responsible for initial training, a one-year course resulting in a minimum of one hundred hours flying time. Flying officers were then sent for further training in England or India. At the same time existing airfields were expanded and new ones built in the Suez Canal Zone and the Western Desert.

The first modern military aircraft to be delivered to Egypt in substantial numbers were the Westland Lysanders that reequipped No. 1 (Army-Coop) Squadron early in 1939. At the same time the RAF's No. 208 Squadron received its Lysanders, both units converting together. Outsiders might, in fact, have been forgiven for regarding the REAF as a mere extension of the RAF. New uniforms had been introduced, exact copies of RAF apparel even down to a side-cap known locally as the *fārūqīyah,* which replaced the traditional but impractical *ṭarbūsh* or "fez," which was now only worn on ceremonial occasions. The REAF was also dependent on the RAF for almost all the sophisticated aspects of maintenance. Yet the REAF was independent enough to have its own peculiar problems. The most pressing was a lack of experienced senior officers for administrative duties. ʿAlī Islām, the air force director, was described as a "pleasant but totally ineffective officer" by one British report, while the Air HQ at the Cairo War Ministry was regarded as quite inadequate.[8]

During the winter of 1938–39 the REAF Flying Training School graduated its third completed class, while the first batch of trainees from the Air Mechanics School also joined their units. Egypt's air force was, in fact, distinctly bottom-heavy in skills. While REAF technicians could solve such problems as the six Gordans by manufacturing their own towing gear, the question of forming an effective fighter force was getting urgent. War was clearly imminent, and Britain eagerly supplied second-line Gloster Gladiators in such numbers that the Egyptian government actually asked for deliveries to be slowed down. The REAF simply could not absorb these aircraft fast enough, and instead many Gladiators went into an RAF storage unit at Abū Qīr for use by whomever needed them first. No. 2 (Fighter) Squadron with nine Gladiators but no spare parts was formed at Dākhīlah and took part in air defense exercises with the RAF in April 1939. No. 4 Squadron's Audaxes also played the part of the "enemy," while No. 1 Squadron's Lysanders worked both with British units in the Canal Zone and with Egyptian Frontier Force armored cars based in Baḥrīyah oasis.[9]

As soon as this exercise ended, Liwāʾa ʿAlī Islām was retired, much to the annoyance of some in the REAF where he had been popular. His replacement was another Egyptian Army brigadier, an artillery officer with staff experience named ʿAbd al-Wahhāb who was viewed with more favor by the British than by his own men.[10] In practice, effective command of the REAF fell to the commander of the Almāẓah air base, Sq. Ldr. ʿAbd al-Munʿim Mīqātī, who had been one of the country's first three military pilots.

Shortly before the outbreak of the Second World War, the Egyptian Army's first anti-aircraft regiment was established, this ancestor of today's highly effective Egyptian Air Defense Command being regarded in both the Egyptian and British headquarters as a priority. By August 1939 the British Military Mission reported that its shooting was as good as that of a regular British Army anti-aircraft regiment.[11] At that time the most vulnerable targets were the Alexandria naval base, the city of Cairo, the oil refinery and storage tanks at Port Said, the Suez Canal facilities, and the forward military base at Marsá Maṭrūḥ.

This then was the situation as far the Egyptian armed forces were concerned as the threat of a new world war approached. Italy was still seen as the most likely aggressor in the eastern Mediterranean, and there seemed no reason why the REAF would not play its full part in defending Egypt if it was attacked.

# 5

# Neutrals at War (1939–1942)

The first two years of the Second World War convinced many people, even some in America, that the Axis powers of Nazi Germany and Fascist Italy would win. These years were ones of almost unrelieved disaster for the Allies. France was defeated, and Britain stood alone with parts of her worldwide empire coming under increasing attack. During the first months of the Nazi invasion of Russia it also seemed as if the Soviet Union would collapse. Among peoples who felt themselves to have been oppressed by Britain, France, or Russia, pro-Axis groups appeared. In the Ukraine they fought alongside German troops; in Iraq they set up a short-lived pro-German government, and sympathy for the Axis powers could even be found in Egypt. It never affected the great majority of Egyptian civilians, nor of the Egyptian armed forces. Nevertheless, the British authorities were aware of such feelings and would not trust the Egyptian military with sensitive tasks. This in turn caused further resentment, particularly in the REAF. As a result the lowest point of Britain's fortunes in the Desert Campaign against Italy and Germany also saw the REAF grounded because of subversion by a handful of officers.

## Egypt's Phony War

In October 1939 the storm broke, but over Europe not over Egypt. This was fortunate for not only was Britain's military presence very limited but Egypt's own recently revived military strength was insufficient to defend the country from a determined attack. The Egyptian Army was still tiny, and the only REAF units yet operational were a fighter squadron, an army cooperation (AC) squadron, a communication

(Comm.) squadron, and the first flight of a second Gladiator squadron.[1] For its part the Egyptian Army could provide two heavy anti-aircraft regiments, one light anti-aircraft regiment, and two searchlight regiments. These were particularly valuable as the British suffered from a severe shortage of anti-aircraft defenses. An elaborate and highly effective Air Observation System also ringed the coasts and deserts.[2]

These forces went to their war stations on the outbreak of hostilities, as planned by the existing Anglo-Egyptian air defense agreement. Under a "Combined Plan for the Defence of Egypt" agreed by Cairo's Ministry of Defense of the British General Maitland-Wilson, Egyptian forces were responsible for patrolling the Libyan frontier, contributing to the Marsá Maṭrūh and Alexandria garrisons, protecting the Mediterranean coastal railway, providing coastal defenses for the Alexandria Naval Base, and assisting in the anti-aircraft defense of Marsá Maṭrūh, Alexandria, and the Suez Canal. They would also operate a small mobile force from Baḥrīyah Oasis in the Sahara southwest of Cairo and set up a line of observation posts deep in the Western Desert to warn of enemy aircraft.[3] By this means the Egyptian Frontier Forces could observe any hostile activity without the presence of British troops provoking the Italians. At this stage both Britain and Egypt particularly feared an Italian motorized strike across the desert, probably stemming from Kufrah in southern Libya and aimed at separating Egypt from Sudan by seizing the Nile between Aswan and Wādī Ḥalfāʾ. Meanwhile most REAF squadrons were based at Almāẓah and Suez with a station headquarters (HQ) at Dākhilah. The Air HQ remained part of the Ministry of War at ʿAbbāsīyah outside Cairo.

The Middle East's "Phony War" of 1939 to early 1940 was even more phony than that in France, at least until Italy's declaration of war in the summer of 1940. Misrair, the country's civil airline, was "militarized" to cooperate in anti-aircraft and searchlight training. No. 1 AC Squadron's Lysanders also went to work, one flight making a twice-daily reconnaissance of the Suez Gulf and the southern approaches to the Canal. One aircraft would fly from Suez to al-Ṭūr on the Sinai coast in the late afternoon, returning to Suez early the following morning, and communication with ships in the approaches to the Suez Canal was by Aldis lamp. A second flight of Lysanders stationed at Baḥrīyah Oasis and Maʿāṭin Baqqūsh cooperated with Egyptian Frontier Forces by making operational reconnaissance sorties 240 kilometers into the Western Desert. These Suez and western Oases units were rotated every three months.[4]

Though the war had yet to reach Egypt, the pattern of future Egyptian military activity in the Western Desert between the Nile Valley and Italian-occupied Libya was already emerging. All major operations would inevitably take place in the open terrain between the Qaṭṭārah Depression and the Mediterranean Sea to the north, but a potential if minor danger also loomed to the south. Here a narrow gap existed between the Qaṭṭārah Depression and the Great Sand Sea. The western end of this desert "pass" was guarded by the Siwa Oasis, while to the east a wide expanse of relatively firm ground led to the Baḥrīyah Oasis, the Nile Valley, and Cairo itself. A large army could not invade by such a route, but small raiding forces might do so. South of the Great Sand Sea there was another gap between the "sand seas," leading from the Kufrah Oasis in Libya to Egypt's Khārijah Oasis and the Nile Valley. Both these "desert passes" were later made famous by Britain's Long Range Desert Group and both were defended by the Egyptian Army, supported by the REAF, throughout the Second World War.

When no war actually erupted, the REAF returned to its peacetime stations to continue training and expansion, while the Flying Training School at Almāẓah increased its courses. Among those who graduated as pilot officers on 17 January 1940 was a twenty-two-year-old from Alexandria named ʿAbd al-Ḥamīd Abū Zayd, who immediately went to the newly formed No. 5 Squadron flying Gladiators. Eight years later, as an acting squadron leader, Abū Zayd would be Egypt's most successful fighter pilot in the skies over Palestine. For the more experienced Gladiator pilots of No. 2 Squadron, early 1940 meant further combined air defense exercises over the Delta with the British. For No. 1 Squadron's Lysanders and No. 3 Comm. Squadron's two newly delivered Avro Ansons, it meant an ambitious long-distance flight south to the Sudan to accompany a high-ranking Egyptian delegation. At the same time a pair of Percival Q6s arrived at No. 3 Squadron to form the nucleus of a Royal Flight. Various problems were immediately experienced with these aircraft, the worst being overheating in the desert climate, though this was eventually cured by Flt. Lt. Ibrāhīm Jazzārīn who fitted the Q6s with tropical oil filters.[5]

No. 4 Squadron's promised Blenheims failed to materialize, and the REAF's ambitious expansion program was soon facing serious difficulties. These ranged from the loss of Group Captain Tait as air advisor, as he returned to England one day before the Battle of France broke out, delays in aerobatic training because of limitations imposed by the

Magisters' carburetor systems, and the fact that a dozen ex-RAF Hawker Hart advanced trainers had not been modified for desert conditions. The REAF eventually made its own modifications to these particular aircraft. In a vain effort to diversify sources of supply, King Fārūq had purchased two MS and EM-24 light aircraft from Hungary shortly before that country finally joined the Axis alliance. They proved to have extremely poor performance but were sent to No. 3 (Comm.) Squadron anyway.

A lack of cooperation between Egypt's civil and military authorities also hampered the air force's director, ʿAbd al-Wahhāb, at every turn. On a more prosaic level the REAF was almost totally without spares for which, naturally enough in the circumstances, the RAF had priority. Unanimously complimentary reports from British, Indian, and Egyptian units concerning No. 1 Squadron's army cooperation work in the Canal Zone could not hide underlying weaknesses that were already sapping REAF morale.[6] A sense of complacency was also creeping back into British attitudes toward the Middle Eastern situation, a Foreign Office report of 15 December 1939 stating that "It is no real interest of ours that the Egyptian Army should reach the maximum of efficiency in training and equipment.... We should lose no sleep over inefficiency in Egyptian high places or the deterioration of army material through bad officership. So long as we can fairly place the blame on the shoulders of the Egyptians, these deficiencies are, in the long view, no loss to us."[7]

Meanwhile there were still British officers in key positions throughout the REAF. Sq. Ldr. V. A. Pope served as an instructor with No. 1 Squadron, Wing Comdr. P. B. Coote with both Nos. 2 and 5 Squadrons. Two British flying instructors, Wing Comdr. E. Britton, and Flt. Off. L. F. Humphrey, similarly played a vital role in the REAF's Flying Training School.

## The Bombing Starts

Italy's entry into the war altered the entire Mediterranean situation. Egypt was now in the front line, and its forces were soon actively engaged. Small-scale Italian air raids against the British naval base of Alexandria started almost at once, causing high civilian casualties. Egypt's political position was, however, very ambiguous. The Chamber of Deputies or Parliament resolved that the country would defend itself if attacked, but when members of the Saʿadist Party proposed a declaration of war on Italy, they were voted down by their cabinet colleagues and resigned. Instead an increase of military forces was proposed with martial law being imposed throughout Egypt. Technically this remained the situation until Egypt declared war on Germany early in 1945.

This left the REAF in a very strange situation. While the royal palace and many politicians remained studiously neutral if not occasionally hostile, the Egyptian Army, the REAF, civil service, and most responsible individuals assisted the Allies. In fact many Egyptian officers called for an immediate advance from Siwa Oasis to seize Jaghbūb, the neighboring oasis in Libya which had been Egyptian during the nineteenth century.[8] A Mobile Brigade under Prince Ismāʿīl Dāʾūd actually moved to Qaṣabah, south of Marsá Maṭrūḥ, to cover Siwa and, while Egyptian troops in Maṭrūḥ itself were effectively placed under British command, this Mobile Frontier Force as well as the Siwa garrison remained a purely Egyptian concern. Nevertheless, in August and September 1940 all Egyptian Army units were integrated into the defense plan for the Western Desert. Although the British seem to have had quite a high opinion of Prince Ismāʿīl Dāʾūd, his behavior did not always go down well with those young pilots who were still flying REAF Lysanders in cooperation with the Mobile Frontier Force. One young pilot, ʿAlī Muḥammad Labīb, recalled how Ismāʿīl Dāʾūd habitually wore a magnificent cavalry uniform resplendent with medals. On one occasion he made a surprise inspection visit to the Lysander Flight deep in the Western Desert, in company with the army commander in chief. The prince then invited the pilots to his tent for dinner, and young ʿAlī Labīb was astonished to find it full of costly Persian rugs, silk hangings, incense burners, and the sound of soft European music, presumably from a gramophone. Among the elaborate dishes produced for that night's feast was one made from a desert gazelle. Not surprisingly the REAF aircrew felt they had suddenly stepped into the world of Sheherazade's "One Thousand and One Nights"![9]

From the British point of view, the Egyptian anti-aircraft regiment was of particular value to the Allied war effort. Though it still had only 70 percent of its searchlights, its guns were now fully operational, the largest units being used to defend Alexandria and Marsá Maṭrūḥ. Other batteries were sited around Cairo, the Almāẓah air base, Abū Qīr, Ḍabʿah, Aswan, Ziftá, Suez, Binhah, Kafr al-Zayyāt, and the Muḥammad ʿAlī Barrage.[10] Like Prince Ismāʿīl Dāʾūd's Mobile Frontier Force, the REAF, though integrated into the overall defense plan, remained under separate Egyptian command. Suggestions that the British buy its equipment for RAF use were resisted, and Egyptian aircraft continued to fly defense sorties over both Cairo and Suez. In those early months the major threat came from Italian SM 81 bombers which, first by day and later by night, focused their attacks on British military installations in Alexandria. Here Egyptian as well as British soldiers were killed in such air raids. Closer to the frontier other Egyptian Army anti-aircraft batteries were also in regular action, bringing down three enemy machines and capturing an Italian pilot and a radio operator.[11]

After Italian land forces finally moved into Egypt to occupy Sīdī Barrānī, the entire air defense of Lower Egypt was reorganized. Protection of Alexandria, the Delta, and Canal Zone fell to No. 255 Fighter Wing based at al-Miks with Sector HQs at al-Marīyah and Ḥulwān. This force, though largely consisting of RAF and Commonwealth units, also included the two now operational REAF Gladiator squadrons. They had been undertaking affiliation exercises with a flight of No. 80 Squadron RAF's Hurricanes based at al-Marīyah, south of Alexandria. It seems to have been here that, in the summer 1940, the Egyptians had the repeated, and for them rather amusing, experience of nightly bombing raids during which Italian bombs always hit the British side of the airfield, never their own![12] The RAF's reaction to this phenomenon is unrecorded. On completing these affiliation exercises, No. 2 REAF then moved to Almāẓah to cover Cairo, No. 5 REAF moving to Suez.

On 28 November 1940 the Italians made their first air raid against the Suez Canal, but No. 5 Squadron's Gladiators were unable to make contact with the raiders. Generally speaking the pilots of No. 5 Squadron were inexperienced, averaging only 150 flying hours on all types. They also considered, rightly enough, that their Gladiators had little hope of intercepting the high-speed SM 79s that the Regia Aeronautica was now using in daylight operations.[13] Already Axis parachute mines in the canal were forcing Allied shipping to unload at Suez, thus causing severe congestion. But while these raids were of greater strategic importance, the REAF was still active over the Western Desert, in particular around Siwa Oasis. Here, a few miles from Italian-occupied Libya, an isolated detachment of the Egyptian Frontier Force kept in touch with the rest of Prince Ismāʿīl Dāʾūd's Mobile Brigade via a flight of six Lysanders.[14] These came from No. 1 Squadron, commanded by Flt. Lt. (acting Sq. Ldr.) Ṣāliḥ Maḥmūd Ṣāliḥ, and were based at Ḍabaʿah. Subsequently Ṣāliḥ Maḥmūd

Ṣāliḥ acted as the REAF's chief of air operations at al-ʿArīsh during the Palestine War of 1948–49. Egyptian ground forces in Siwa also had several skirmishes with Italian troops, capturing some. In August 1941, as Rommel advanced into Egypt, the British military authorities insisted that Siwa be handed over to a British garrison, who lost it just under a year later. Meanwhile the Egyptian Mobile Frontier Brigade, supported by REAF Lysanders, withdrew to Baḥrīyah Oasis from which it continued to patrol the desert "pass" between Baḥrīyah and Siwa.

## "A Secret from Our Government"

Meanwhile No. 1 Squadron's second flight of Lysanders, based at Suez, continued to fly dawn and dusk antisubmarine and mine-spotting patrols over the Suez Canal's southern approaches. In Egypt the RAF's spares situation was getting difficult, but that of the REAF was already critical, No. 1 Squadron as a whole being down to only six serviceable aircraft at one point. The situation was so serious that other aircraft were called in to help. So while No. 1 Squadron's Lysanders flew short-range patrols from Suez, two Ansons and the Avro 652 from No. 3 (Comm.) Squadron based at Almāẓah ranged further down the Gulf of Suez. The REAF's shortage of trained aircrew for these important missions apparently meant that men from No. 4 (Bomber) Squadron were also trained to fly twin-engined Ansons and to fly their share of patrols. They would fly as far south as Hurghada (al-Ghardaqah), escorting convoys on the final stage of their long journey around Africa. One Anson was lost during these missions, the crew being killed during an emergency landing in thick fog, while the Lysanders attacked suspected Italian submarines on several occasions.[15] For a while the danger from mines, air raids, and Italian submarines based in Eritrea was, in fact, so grave that President Roosevelt closed this area to U.S. shipping.[16]

Throughout these months the still "neutral" Egyptian government turned a blind eye, all REAF patrols from Suez and Almāẓah being undertaken without its official knowledge since Liwāʾ ʿAbd al-Wahhāb listed all such activity as training.[17] In reality, of course, the Egyptian government must have known what was going on. By December even Egypt's civil airline, Misrair, was involved servicing the REAF's long-range Ansons after such patrols. But at the very end of 1940 Liwāʾ ʿAbd al-Wahhāb resigned as head of the REAF, not for political reasons but because he could not get on with a new army chief of staff. The British were sorry to see him go; nevertheless, his successor, another experienced army rather than air force man, Liwā᾽ (Brigadier) ᶜAlī Muᶜāfī, made a good impression as soon as he arrived.

Italy had conquered British Somaliland, and the winter of 1940 and spring of 1941 remained a dangerous period in the Red Sea. Since the fall of France this route was vital, and its convoys came under frequent air attack right up to 4 November 1941. Hence the combined RAF-REAF Red Sea patrols played an important if unglamorous role in keeping Britain's Middle Eastern forces supplied. Among the No. 4 Squadron pilots who flew such patrols were Shafīq Ḥasīb, who later flew Dakota "bombers" and transports against the Israelis in 1948, and Muḥammad ᶜAdlī Kafāfī who, receiving his wings in 1939, rose to command No. 4 Squadron after the Second World War.

Raids on the Suez Canal were also growing more intensive, air-dropped mines sinking a number of ships and causing even more serious delays. Against the Italians' SM 82s of the 41st Gruppo operating from Libya and the Cant 1007bis of the 172nd Squadriglia in the Aegean,[18] the REAF's Gladiators could do little. Suez itself was, in fact, bombed five times in October and November 1941. The threat against Allied communications at the northern end of the Red Sea continued even after the fall of Mussolini's East African Empire, and the Egyptian fleet suffered its first serious casualty when a small Coastguard and Fisheries Administration transport, *HMES Amīrah Fawzīyah,* was sunk in shallow water off Suez during an air raid on 28 July 1942. This ship was subsequently raised and repaired; it took part in the Palestine War of 1948, during which it was again claimed destroyed by the Israelis, and continued in the coastguard service into the 1950s.

The zone from the Suez Gulf to Ismāᶜīlīyah was now a purely Egyptian responsibility by day, though it became a British responsibility by night. Egypt's second Gladiator squadron remained responsible for the defense of Cairo. Throughout this period and long after, Egyptian Army anti-aircraft gunners and searchlight crews were also in constant action in Alexandria, Cairo, and along the Suez Canal. There they earned the unstinting praise of their British and Commonwealth comrades, not only for their steadiness under fire but also for their accurate shooting. In addition, the Egyptian Army had two fully trained bomb disposal squads in Cairo and Alexandria by October 1941. Another anti-aircraft battery, far away in Port Sudan, had a quieter time after the fall of

Mussolini's East African Empire and was one of the few, though by no means only, unit of the Egyptian armed forces to serve outside Egypt proper during the Second World War.[19] It is also interesting to note that several of Egypt's leading soldiers in the highly successful Air Defence Command during the October 1973 War first saw action with these Egyptian anti-aircraft batteries.

In October 1941 the REAF received its first four Hurricanes which went to the Flying Training School as a conversion Flight. Seven Mk. Is were subsequently formed into a new No. 6 (Fighter) Squadron for the night defense of Cairo. From early 1942, this squadron also covered Alexandria and remained theoretically operational until January 1943. One aircraft was lost on 29 July 1942, and by December only five of No. 6 Squadron's Hurricanes were still airworthy.[20]

## Subversion and Disgrace

Raids by Britain's famous Long Range Desert Group deep into the south of Italian-occupied Libya were countered by the seizure of Siwa Oasis by a force of Italian armored cars in September 1942. To the British this was a minor military reverse, but it had far-reaching reverberations within the Egyptian armed forces. Siwa, so long successfully defended by Egyptian troops, had become something of a symbol. Now it had been lost by British troops, and the consequent blow to British prestige in the Egyptian Army and the REAF was out of all proportion to the real military significance of the fall of Siwa.

Unfortunately there was already a cloud of doubt hanging over the loyalty of certain sections of Egypt's armed forces, including several individuals in the REAF. Trouble had really begun when Winston Churchill and the very anti-Egyptian nationalist British ambassador in Cairo, Sir Miles Lampson, forced Gen. ʿAzīz ʿAlī al-Miṣrī to retire as inspector general of the Egyptian Armed Forces. ʿAzīz al-Miṣrī, the Circassian-Egyptian officer who claimed to have shot down the first enemy aircraft ever brought down in war, had never disguised the fact that he was an Arab patriot who wanted to see the end of the British military occupation of Egypt. Yet he felt even less love for the Italian Fascist occupiers of Libya and had no record of sympathy for Germany. In fact the fuss that the British made about al-Miṣrī having a German secretary ended up as farce because she proved to be Jewish and fiercely anti-Nazi!

In Egypt the general was widely respected, not only as a hero of the Libyan resistance to Italy but also for the role he played in the Arab Revolt against the Ottoman Turks during the First World War. There ʿAzīz al-Miṣrī had organized the first regular Arab rebel forces in what is now Saudi Arabia, months before the famous Lawrence of Arabia arrived on the scene. On the other hand, despite his position as inspector general, ʿAzīz al-Miṣrī was little more than a figurehead, having little contact with the army and being out of touch with modern military ideas. In some respects he was more popular in the Egyptian palace than in the Egyptian Army. Yet he did have staunch friends in the officer corps, and the tactless, offensive way in which the British forced him to take extended leave upset even those who had previously regarded him as an outdated interloper.

General al-Miṣrī's retirement was finally confirmed in August 1940, and for almost two years he kept out of the limelight. Then, in May 1942, the general attempted to make an unauthorized flight to Syria which was still ruled by pro-German, Vichy-French, colonial authorities. Two REAF airmen went with him; Pilot Off. Ḥusayn Dhū al-Fike Ṣabrī, a member of the "Free Officer Movement" led by Jamāl ʿAbd al-Nāṣir (Nasser) and subsequently Egypt's ambassador to Switzerland, and Pilot Off. Munʿim ʿAbd al-Raʾūf, who was to later involved in an attempt on President Nasser's life. After ʿAzīz al-Miṣrī made an unsuccessful attempt to rendezvous with a German aircraft at Jabal Ruzzah, the two Egyptian pilots took an REAF Anson from Almāẓah, but were almost immediately forced to land in a Delta cotton field when the aircraft's engines seized up, probably because their oil pumps had been tampered with. At his subsequent trial al-Miṣrī insisted that he had actually been on a secret mission for Colonel Thornhill of British Intelligence, an idea that even the now rather paranoid Sir Miles Lampson considered possible.[21]

Rommel's Afrika Korps was now knocking on the gates of the Nile Valley, and many responsible British civil and military officials in Cairo thought that Egypt would soon fall to the Nazis. Fear and confusion at the British embassy infected many Egyptian officials, and several Egyptian officers, including the late President Sadat, felt that it was time to do whatever they could to save their own country from the anticipated collapse of the British Empire.[22]

According to secret British reports of May and June 1942, "subversives" in the Egyptian Army and REAF were few in number and mostly came from what was regarded as a staunchly royalist or "pro-palace" group of officers.[23] These royalist officers were said to have formed a "secret organization" with its own banner consisting of an Egyptian flag, a revolver, and a lanyard or bowstring—an object used throughout Ottoman Turkish history to execute corrupt or traitorous officials. British intelligence officials thought it was caused by intensive German propaganda targeting the useful cooperation between Egyptian and British armed forces. But these officials also recognized that the organization was more anti-British than pro-German, while it also had close links with Islamic "anti-imperialist" groups in other countries, particularly Turkey. These men were, in reality, straightforward Egyptian patriots who thought of their own country first. King Fārūq's Palace officials played on this fact and also flattered their professional military pride. The inclusion of some of the best, most-qualified, and most-respected men in the REAF in this "secret organization" also caused concern to British officials in Cairo.[24] The late ʿAbd al-Raḥmān Zakī, a respected historian and first director of the Egyptian Military Museum, was included in these British Intelligence reports. He later made it quite clear that the "secret organization" was in no way pro-Axis but, fearing that the British would be unable to defend Egypt from Rommel's Afrika Korps, wanted to save their country from as much damage and humiliation as possible.[25] Identical views were expressed in the 1970s by the late Air Commod. Ibrāhīm Jazzārīn who was also included in these British reports.[26]

Whatever the truth behind General al-Miṣrī's intended flight to Vichy-held Syria, the subsequent defection of two REAF pilots in July 1942 confirmed British fears. Anwar Sadat, then a relatively junior army officer, was involved in this extraordinary episode when he and his associates drafted a "treaty" which they proposed to offer to the Germans in expectation of Rommel's conquest of Egypt. Four REAF pilots then took small hand-held aerial photographs of some British military positions inside Egypt which would be offered as tokens of good faith to the Germans. Copies were given to Pilot Off. Aḥmad Saʿūdī Ḥusaynm, who then flew another man's Gladiator across the Axis lines on 7 July 1942. At 0702 hours, over Ḍabʿah, Saʿūdī Ḥusayn was intercepted by two Bf 109s of the Luftwaffe's I/JG27 and fell to the guns of Lieutenant Stahlschmidt, a fighter pilot who ended up as Germany's third highest-scoring ace in the Western Desert.[27] Apparently the recognition signals previously arranged with the Germans were not seen or not understood. A second REAF pilot, Warrant Off. Muḥammad Riḍwān Salīm, did manage to reach the German lines in a second Gladiator the following day, subsequently being captured by the advancing Allies in Germany in 1945 and returned to Egypt for trial.[28] It did not say a great deal for British Intelligence reports that neither of these men, nor Anwar Sadat of the army, had been included in lists of supposed "subversives" within the Egyptian armed forces. The Egyptian government thereupon grounded the entire REAF on its own initiative by having all aircraft magnetos removed, while the REAF High Command also offered to hand over its Almāẓah facilities to the RAF. This was the nadir of the REAF's fortunes in the Second World War, and the Egyptian government was so disillusioned as to suggest disbanding the entire air force!

The final page of this sad story came in the autumn of 1942 when Anwar Sadat and another REAF man were finally arrested and imprisoned for their involvement with a German spy ring. Meanwhile British and Allied forces first halted and then reversed Rommel's advance in the two battles of al-ᶜAlamayn. As they regained their nerve, the British even complimented the REAF for "remaining calm" during the previous near disastrous weeks and strenuously opposed any punishment of the air force as a whole, being convinced that active disloyalty was limited to very few personnel. Indeed the British Advisory Mission reported a general atmosphere of shame and resentment in the REAF against those who had defected, while many of those who had previously sympathized with the "secret organization" felt that these defectors had overstepped the mark.[29] In general, however, the crisis sparked relatively little resentment in the rest of the army and REAF where there was widespread feeling that the British accusations contained more than a little truth, though the Egyptian officer corps was upset by the necessity of washing its dirty linen in public.

Liwāʾ (Brig.) ᶜAlī Muᶜāfī insisted on bearing the final responsibility and resigned. He was replaced by another army man, Amīr Alāy (Col.) Ḥusnī Ṭāhir, as no REAF officers were yet senior enough for the post. ᶜAbd al-Wahhāb and ᶜAlī Muᶜāfī were both awarded the CBE (Commander of the Order of the British Empire) by Britain in 1943. Meanwhile three senior REAF officers, fourteen officer pilots, and seventeen NCO pilots were temporarily transferred to the army as a disciplinary measure. Clearly dislike of Britain's military occupation was strong enough to drive some younger and less experienced officers into the arms of a far more cruel "imperialist" power. Even today this 1942 crisis remains a hot political issue, and many of the relevant British government documents have yet to be released for historical scrutiny.

The Egyptian Air Force had entered 1939 with high hopes and considerable confidence. For over two years it had contributed a great deal to the Allied war effort, despite the fact that Egypt remained theoretically neutral. Yet the end of 1942 saw the REAF in a sorry state. Only the most optimistic observer could have anticipated how the REAF's fortunes, and its contributions in the war against Fascism and Nazism, would improve during the last three years of the Second World War.

# 6

# Forgotten Allies (1942–1945)

From late 1942 until the end of the Second World War, the REAF dragged itself back from being grounded and in disgrace to a position in which its personnel were helping the British RAF in numerous ways. Most such roles were noncombatant, but one squadron played a fully operational role under direct Allied command.

## Changing Fortunes

The autumn of 1942 saw the fortunes of the North African campaign at last turning in Britain's favor at the two battles of al-ʿAlamayn. This period also saw the REAF at its lowest ebb—in morale, the confidence of its allies, and its appallingly antiquated equipment. But the end of 1942 also saw the REAF begin a long and sometimes painful process of reconstruction, the first in the history of an air force that has overcome many hard knocks. Faith in an Allied victory also revived, and when, in 1943, Axis forces were finally driven out of North Africa this was genuinely welcomed by Egyptian officers who, by and large, felt that they should have been given a more active role to play.

The task facing the British Advisory Mission to the REAF was a tough one. Indeed an official report of 21 December 1942 pointed to an almost total lack of modern equipment and spares. The work of the British advisory officers was described as "depressing" and the Egyptians' lack of enthusiasm as "understandable."[1] A complete overhaul of the disciplinary structure was the first priority. Almāẓah was also reorganized as one unit, instead of two as previously, while RAF Middle East Command agreed to transfer equipment from its own stocks instead of the REAF relying on supplies directly from Britain. By this means dozens of problems were overcome, some of which had been as minor as the REAF's acute shortage of red dope for fabric-covered aircraft.

Morale now rose rapidly, and the REAF's senior men found that they had a sympathetic ally in the person of Group Capt. Johnny Chick, chief of the RAF section of the British Advisory Mission. Training, however, remained a fundamental problem that could not be solved so easily, even though much of the REAF's equipment was now only suitable for a training role. The thirty-four serviceable Miles Magisters and sixteen available Avro 626s of the REAF's Flying Training School had been transferred from overcrowded Almāẓah to Khānkah in the Delta, together with one surviving DH Moth. Yet these were used for refresher courses rather than preparing new aircrews, as no new officers were recruited into the REAF until the end of the Second World War. In December 1942 the REAF had, in fact, only 121 officer and 17 NCO pilots. Technical training schools in armaments and wireless were also established at Almāẓah. Meanwhile many aircraft, including the Wessex and the Gordans, were grounded after having been flogged too hard with inadequate spares. A detailed survey of units illustrated just how bad the situation had become.[2]

This then was the sorry state of the REAF at the close of 1942, but by July 1943 Group Captain Chick could report excellent relations between the REAF and the British.[3] Amīr Alāy Ḥusnī Ṭāhir, the new REAF commander, imposed a highly centralized authority, proving to be a strong disciplinarian and very pro-British. All but one of the officers and men who had been sent back to the army in 1942 returned to REAF duties, and all that was needed to revitalize morale was to give it what the British Advisory Mission described as a "real role" in the war. Problems, of course, remained. These included acute congestion at Almāẓah and a continuing shortage of spares for the REAF's varied aircraft. The accident rate had, however, dropped sharply. Morale and discipline continued to improve throughout 1943 with the arrival of modern aircraft and increased flying time. But one shortcoming was to survive the war and far beyond, the inadequacy of administration in the Air HQ at the War Ministry, Cairo. So the start of 1944 not only saw the REAF back in a position to cooperate actively with the RAF, but it also saw the British eager for the Egyptians to do just that. Even so, some British reports of the time read with a certain irony when viewed with the benefit of hindsight.[4]

## The Fighter Squadrons

Among the more up-to-date aircraft now being offered to the REAF were a number of Curtiss P-40 Tomahawks. The idea of supplying these was first mentioned in March 1943 against a background of concern that it might lure Egypt away from "buying British." In April 1943, six ex-RAF P-40s were handed over after a good deal of fuss as to whether they had originally been sold to Britain by the United States or were Lend Lease aircraft, in which case the RAF had no right to pass them on to a third party. In the end the P-40s proved to have been bought, so no obligation was left to inform the United States of the transfer. Six more followed later, and between May and October 1943 No. 6 Squadron, now renumbered No. 17 (Fighter) Squadron under its new CO, Flt. Lt. Ṣalāḥ Farīd, was reequipped with P-40 Tomahawks, while its Hurricanes were put in storage. No. 17 Squadron was then placed under the control of the RAF Air Officer Commanding Air Defence East Mediterranean for the protection of the Nile Delta, along No. 2 Squadron.

Problems emerged at once. On these ex-RAF machines, British magnetos had replaced the original American versions. Unfortunately they were too heavy for the coupling and tended to break, one Tomahawk soon being lost because of engine failure during takeoff. This REAF squadron had originally been intended for Mediterranean convoy protection work, but the Tomahawks' unreliability meant that they could not operate over the sea. The British then offered to lend further Hurricanes so that No. 17 Squadron could cooperate fully with the RAF, but the Egyptian government regarded Hurricanes as outdated by 1944 and offered to buy Spitfire Vs. This the RAF was not prepared to accept, though later it did lend some Spitfires.

Throughout 1944 No. 17 Squadron, based at Almāẓah, had to be content with training sorties over land until the Tomahawks were finally grounded and returned to salvage in November. From then until February 1945, when ten Hurricanes arrived on loan from the RAF, No. 17 Squadron had nothing to do, though its personnel were fully trained.

Meanwhile the REAF's ever-reliable Gladiators soldiered on as best they could. The aircraft of both No. 2 and No. 5 Squadrons were given new fabric, No. 2 Squadron's fifteen serviceable aircraft remaining at Suez for the air defense of the southern Canal Zone. Here one section stood at readiness all the time, but their effectiveness was limited as they lacked the VHF radio now standard on modern RAF machines. Officially the old Radio-Telegraphy contact with the RAF sector operations room at Sanhūr (Qārūn) had been discontinued but, refusing to be denied a combat role, the pilots of No. 2 Squadron arranged to have a watch kept specially for them on the old frequencies.

By this means No. 2 Squadron was able to enter combat on two confirmed occasions in the early summer of 1943 and had a number of other unconfirmed brushes with high-flying German Junkers. The results are unknown and, given the disparity in equipment between the REAF's aging Gladiators and the Luftwaffe's latest raiders, are unlikely to have been satisfactory. They remain, however, among the last occasions when Gladiators entered air combat. At least one Egyptian pilot ʿAbd al-Ḥamīd Abū Zayd who had once been included in a British list of suspected "subversives," was proposed for a decoration following his efforts on these occasions.[5]

Meanwhile No. 5 Squadron retained its fourteen Gladiators to the very end of the Second World War. They were at first based at Almāẓah to defend Cairo, while at the same time training in night interception. No. 5 Squadron then moved to Suez in the summer of 1943 and was still flying anti-aircraft cooperation sorties in Gladiators over the Canal Zone in June 1945.[6]

## Met Flight and Humble Helpers

More directly under British control was No. 1411 Meteorological (Met) Flight at Almāẓah, which had been set up in January 1942. Its Gladiators, and subsequently Hurricanes, carried Egyptian markings, although the unit was still officially part of the British RAF and under RAF command. It was, however, entirely operated by REAF pilots and maintenance crews from August 1943 to early 1947 as part of a substitution scheme designed to release RAF pilots for active duty. Not that Met flights were without their dangers: one REAF pilot was killed following a bad weather crash landing in his Gladiator near Khānkah on 30 October 1944. At first twice-daily flights were carried out at 1100 and 2300 hours, but toward the end of 1944 these were replaced by a single climbing sortie at 0100 hours, by which time several Egyptian pilots were becoming very competent at night flying.[7]

At a rather humbler level the REAF also helped the Allied war effort with balloons and radios. In May 1943 the British had approved the setting up of two REAF balloon squadrons of about five hundred men,[8] and in February 1944 REAF radio operators were made available to take over nonoperational duties in the Delta area, thus again releasing RAF personnel for more urgent tasks.[9] Meanwhile Egyptian Army anti-aircraft units continued to defend the main Egyptian cities as well as other locations, and by July 1943 the British considered these to be the most effective of all Egyptian Army units. Egyptian troops also defended Allied airfields in the Egyptian Delta.

## Shipping Protection Patrols

The business of reequipping the REAF with Hurricanes, and thus enabling it to cooperate with the RAF over the Mediterranean, proved to be a complicated business. At the start of 1943, No. 17 (ex-No. 6) Squadron's single flight of Hurricanes Mk. Is were all grounded, partly as a result of the political crisis, but largely because they were barely airworthy. Yet by the summer of that year REAF technicians had got them all flying once again alongside the unsatisfactory Tomahawks. This was another tribute to the ingenuity of the REAF's engineering division which could call upon the skills of, among others, Sq. Ldr. Rāziq, who had done a two-year intensive engineering course at RAF Henlow.

Meanwhile the REAF had offered to take on extended convoy protection work far beyond Egyptian territorial waters. The British government was not yet willing to accept this as Egypt was still officially neutral. The REAF responded by offering to take over a complete RAF Hurricane squadron's equipment and so ease the RAF's manpower shortage. Britain countered by suggesting that a completely new REAF Hurricane squadron be formed.[10] Egypt agreed but wanted financial assistance, so Britain offered eighteen Hurricane IICs at £5,000 each, to be placed under the RAF's Air Defence Middle East Command for the static defense of Egypt. The Egyptians then requested that these planes be loaned free of charge, there already being an established precedent for lending aircraft to Egypt when the REAF took over RAF duties. Unfortunately the British Treasury balked at replacing any resulting active service losses. Then came the Italian surrender which altered the whole military situation in the Mediterranean. The idea of the REAF replacing an RAF squadron was dropped, and Egypt agreed to buy eighteen Hurricanes for shipping protection work, while the British finally agreed to make good any losses.

In the event No. 2 Squadron was reequipped with these aircraft, rather than No. 17, eighteen ex-RAF Hurricane Mark IICs having been transferred to the REAF in September 1943. As early as May 1942 the British appear to have selected the twenty-five Egyptian pilots they considered best suited for the REAF's proposed Hurricane squadron, and some of these men may even have been given some training over the next year.[11] No. 2 Squadron was also the first REAF unit to get proper VHF radio, and its Hurricanes began by training alongside No. 208 Squadron RAF which had itself recently converted from Hurricanes to Spitfires. During their time at the RAF Air Gunnery School the pilots of No. 2 Squadron were reported to have achieved "exceptionally good" results.

No. 2 Squadron was now under Flt. Lt. (acting Sq. Ldr.) Muḥammad Ḥāfiẓ, whom British records described as "very competent." He would later serve as director of the REAF Office in the Egyptian Ministry of War in 1949 and become one of the most respected officers in the Egyptian Air Force. ʿAlī Muḥammad Labīb, who had earlier been astonished at the luxury of Prince Ismāʿīl Dāʾūd's desert banquet, served as his second in command responsible for communications. On 2 January another well-known Egyptian pilot, ʿAbd al-Ḥamīd Abū Zayd, transferred to No. 2 Squadron as a flight lieutenant, and exactly a month later this unit reached full RAF operational standards, Sq. Ldr. N. P. Hancock of the RAF having helped in operational training.

On 16 January No. 2's Hurricanes were placed under the Allied Air Officer Commanding Air Defence Eastern Mediterranean, again in close cooperation with No. 208 Squadron RAF.[12] Unfortunately Egyptian personnel as a whole still suffered from the patronizing attitudes of their British comrades, as betrayed even in complimentary British reports, such as: "the bearing and standard of REAF personnel came as a surprise to the RAF who now welcome their presence."[13] While two flights of the 208's Spitfires were based at Barsis, No. 2 Squadron operated from Idkū near Alexandria, but this arrangement only lasted a short while before 208 Squadron was sent to Italy in March. No. 2 Squadron's Hurricanes then took over full responsibility for shipping protection duties from 1 April 1944 under the overall control of No. 219 Group RAF. As such it was the only REAF squadron to be placed fully—and officially—under RAF command during the war. Shipping in the eastern Mediterranean was, of course, far from secure as German forces had succeeded in taking over the entire Aegean Sea from where they made regular sorties.

On 15 June the Egyptian squadron extended its range of operations by moving from Idkū to Marsá Maṭrūh. Three fatalities were suffered during these operations, two Hurricane IICs disappearing without trace over the Mediterranean on 8 June, and a third colliding with an RAF Beaufighter on 11 September. The Hurricane pilot, Midḥat Muḥammad Qaṣdī, who had again been on a British Intelligence list of potential "subversives," and the British crew all being killed. A subsequent RAF inquiry decided that both pilots had been at fault, but the Egyptians were of the opinion that neither airman was to blame and that the accident resulted from an RAF ground controller instructing the Beaufighter to make his landing approach right through the path of a flight of REAF Hurricanes which were already airborne.[14] No. 2 Squadron's duties were to protect naval convoys sailing between Alexandria and Sallūm, two Hurricanes constantly flying over such ships until they passed out of the Egyptians' patrol area. In November the squadron returned to Idkū where it continued to cooperate with the Royal Navy and Alexandria anti-aircraft units in air defenses. In February 1945 it returned to full REAF control, spending the rest of the war cooperating with the British Royal Navy and the Alexandria anti-aircraft defenses in various training exercises. Even so, it was soon virtually grounded as the supply of RAF spares had promptly dried up.

The men of No. 2 Squadron had been eager to prove themselves in RAF eyes, and they also regarded their service under British command as very useful experience, even if politically delicate. For their part the British clearly also regarded their efforts as valuable and wanted to decorate the men, but unfortunately the Egyptian government found this politically unacceptable and vetoed the idea.[15] It is worth noting that No. 2 Squadron became the REAF's elite unit, bearing the brunt of air operations over Palestine in 1948–49, by which time it was, however, flying Spitfire LF9s. February 1945 also saw the purchase of twenty Spitfire Vs for No. 2 Squadron, and in June 1945 No. 2's Hurricanes were transferred to No. 6 Squadron.

## Convoy Leaders

More dramatic were the activities of twenty or so REAF pilots who volunteered for duty with the RAF's No. 216 Transport and Ferry Group. The first six volunteers[16] were selected in July 1943 to join a transport and ferry group responsible for the entire Middle East and North African war zone. By February 1944 twelve REAF officer pilots, all of whom had passed through the RAF's Middle Eastern Officer Training Units (OTU), were with No. 216 Group. The director of the REAF wanted them to return to their units after a six-month tour of duty, but they were proving so useful that the RAF insisted on keeping them longer.

Their original duties were to deliver new fighters—Spitfires V, VIII, and IX, Mustangs and Kittihawks plus a few Martinets—to collection centers in North Africa and southern Italy. Frequently, when the pressure was on, they also volunteered to fly new machines to front-line airfields in Italy, Sicily, and Sardinia, returning with battle-scarred aircraft. More than 350 were delivered to the Italian front in this way, and approximately 200 brought back for maintenance or repair. The opening of a new front in southern France broadened the ferry pilots' field of operations still further, and by the end of 1944 the Egyptian pilots alone—not including their colleagues from other countries—had transported more than 850 aircraft to and from various fronts.[17]

Such second-line duties were not without hazard. Apart from periodic bad weather and sometimes barely serviceable aircraft, these delivery pilots sometimes found themselves under fire. Pilot Off. Saᶜd al-Dīn Sharīf's Spitfire was, for example, severely damaged by ground fire during a delivery flight from Tunis to Catania in Sicily.[18] In fact, three Egyptian ferry pilots lost their lives during these operations. Flt. Lt. Zaytūn, who had also been listed by British Intelligence as a possible "subversive," died in a flying accident near Abū Qīr. Flt. Lt. Sᶜaīd Sābit was killed at Tunis, and Pilot Officer Rifāᶜī died when his Mustang exploded during a thunderstorm 95 kilometers east of Algiers on 24 September 1944. Four of the first dozen such pilots were withdrawn to be replaced by four new volunteers early in 1945, while some others converted to the delivery of multi-engined aircraft.

The survivors were now logging up a great deal of experience, and by the spring of 1945 Ḥashshād, Bakīr, and Sharīf were made convoy leaders from their base in Morocco. It was at this time that Pilot Officer Sharīf had another lucky escape. While leading a convoy of three aircraft from Algiers to Sicily, Sharīf was asked by one of the British ferry pilots to swap aircraft. He agreed, and they exchanged planes during a refueling stop and "teabreak" at al-ᶜUwaynah near Tunis. But during the subsequent Mediterranean crossing the aircraft that Sharīf had first been flying lost power, ditched into the sea, and sank.[19] Unreliable engines were the greatest problem for these ferry pilots. On yet another occasion an Egyptian was flying a damaged Mustang from a front-line base in Corsica for repair in Morocco when his engine cut out. He prepared to ditch in the Mediterranean and was about to hit the water when the engine restarted of its own accord. Nevertheless, it cut out a second time over the Algerian coast, and the pilot was obliged to land on a beach.[20]

## The Training Program

The REAF was, throughout this period, attempting to upgrade the capabilities of its senior men. Britain now permitted some senior REAF officers to attend advanced training courses, and a first group of REAF staff officers attended an RAF Staff College at Haifa in Palestine in 1943. They were soon followed by others, including Flt. Lt. Ibrāhīm Jazzārīn who, an Advisory Mission report stated: "was one of the Egyptian students on No. 11 Staff Course and was one of the five officers to obtain a B category. The influence of these officers is considerable and they have rendered valuable service to the Air Advisor."[21] Other REAF officers were also passing through the RAF's Nos. 71, 73, and 74 Officer Training Units, and, according to the RAF, the "results have been surprisingly good."[22] Competition for these fighter, fighter-recce, junior commanders, and signals courses was very keen, as was competition with RAF personnel on the courses themselves. One REAF pilot graduated second at No. 74 OTU, ahead of fellow pupils all of whom came from fully operational RAF units.

Ab-initio and intermediate flying training remained a problem, although all REAF instructors had qualified at the RAF's Central Flying School. No replacement pilots had reached Egyptian squadrons for the past three years, and not until February 1944 did the British authorities allow the REAF's Flying Training School (FTS) to admit a new course of twenty-two cadets. By then the REAF was short by some fifty pilots. At the same time there was a revival in ground training. The REAF's Miles Magister primary trainers were still adequate, but the Avro 626s and Hawker Audaxes were now completely out of date and were generally reserved for anti-aircraft sighting exercises in the Canal Zone. In July 1943 the RAF's No. 71 OTU lent two Harvards to the REAF's FTS, but these soon had to be passed on to the RAF's No. 70 OTU, whose need was thought to be greater. Other British aircraft were loaned in ones or twos, and not until 1946 was Egypt allowed to purchase other surplus Harvards.

The REAF tried to remedy broken training schedules by using ex-RAF Miles Masters, twenty-five of which were formed into an "intermediate training squadron" at the FTS. The wooden Masters, originally destined for Turkey, proved unsuitable for Egypt's desiccated climate. They eventually developed such severe wing cracks that they had to be returned to the RAF.[23] Four ex-RAF Miles Martinets destined for the Target Towing Flight early in 1945 also failed to arrive. The REAF's Flying Training School later obtained a handful of hand-me-down Hurricanes, at first Mk. Is, then probably some Mk. IICs after the war ended. RAF Middle East Command was, at the same time, debating whether to issue an ultimatum to the REAF, ordering it to transfer the intermediate and advanced training units from overcrowded Almāẓah to Ḥulwān south of Cairo. Fortunately this threat to the fragile relationship between Egypt and the occupying British authorities never materialized.

## The Neglected Squadrons

Almost all the REAF's energies were focused on its two "modern" fighter squadrons, Nos. 2 and 17, and to a lesser extent on the training program. As a result the remaining units tended to be neglected. The fourteen Lysanders of No. 1 (Army Cooperation) Squadron carried out anti-aircraft searchlight cooperation with the Egyptian Army throughout 1943, though by the end of this year the squadron was also familiarizing itself with a handful of overhauled Hurricane Mark Is. These were the old Mk. Is previously flown by No. 6 Squadron, and they still lacked VHF radio. On 16 January 1944, No. 1 Squadron's mixed force of Lysanders and Hurricanes cooperated with the RAF's No. 208 Squadron during a training exercise in which the Egyptians flew as No. 208's third flight. Code-named Operation Tussle, this took place in appalling weather between Cairo and Suez, pitting "Alexandrians" again "Kabritians."

The Hurricanes of what was now designated as No. 1 (Fighter-Recce) Squadron were then integrated into the Egyptian Army's new Air Support Control Organization. Two were set aside for photo-reconnaissance duties, having been fitted with twin cameras and oxygen equipment by the REAF's own workshops, while two officers were trained in photo-interpretation by the RAF. In June 1945, No. 1 was, however, completely reequipped with ex-RAF Hurricane IICs. Thereafter it maintained a very high degree of serviceability and was generally regarded as being well up to an RAF level of efficiency. Some of No. 1 Squadron's Lysanders were sent to Upper Egypt to carry out that most traditional of REAF roles, anti-drug smuggling patrols over the desert. Thereafter these Lysanders were retained for communications duties but were proving increasingly difficult to maintain. In fact, two of these aircraft got lost in the desert late in 1944. RAF and USAAF air searches failed to find them, but six days later the two pilots concerned arrived at Baḥrīyah Oasis on camelback, quite uninjured having force-landed on a limestone plateau 30 miles to the east.

In the summer of 1943, No. 4 (Bomber) Squadron—the Cinderella squadron of REAF—was still equipped with thirteen Hawker Audaxes. These still lacked bombsights, nor was there a proper bombing range available. Many of its pilots were, in fact, with the RAF Transport and Ferry Group or serving in other squadrons. Plans to purchase De Havilland Mosquito bombers were blocked by the RAF's unwillingness to release these aircraft, and when a few more Avro Ansons arrived late in 1944 for training purposes, these proved to have various parts missing. They were also severely corroded, but during the early months of 1945 the Ansons were made airworthy by REAF mechanics and were fitted with VHF radios so that they could cooperate with Egyptian Frontier Force patrols in the Western Desert. Despite all these efforts, the REAF's lack of an effective bombing force was made humiliatingly clear in the summer of 1945. On this occasion a serious clash erupted between drug smugglers and a frontier force patrol near Baḥrīyah during which the Egyptians had to call upon the British RAF for help.[24]

No. 3 Squadron's communications duties, flying a motley collection of Avro Ansons, Percival Q6s, the solitary Avro Commodore, one Hawker Hart, and a surviving DH Moth, kept this unit very busy. It ferried government officials around the eastern Mediterranean, as well as senior officers to and from the Middle East Staff College in Palestine. Early in 1945 a Royal Flight was established as part of No. 3 Squadron based at Inshāṣ, near one of the royal palaces.

## The Fārūq Flight

The REAF's new Royal Flight at first used of some of the nine relatively modern ex-RAF Ansons supplied to Egypt at the end of 1944.

Unfortunately all nine were grounded early the next year after corrosion was found in their main wing spars. In fact, a depressingly high proportion of the secondhand aircraft transferred from the RAF to the REAF proved to be structurally or mechanically unsound, and this heritage certainly contributed to an atmosphere of near paranoia during a "Defective Arms Scandal" that rocked Egypt following the Palestine War of 1948.

King Fārūq's Royal Flight also proved a monumental headache for all concerned—British and Egyptians. In December 1944 Britain decided to give the king an Avro Anson XIX. When the Iraqis heard of this proposal they were incensed because their ruler had been obliged to pay £10,000 for his Anson. On 24 January the British government sent an official letter to the Egyptian prime minister thanking him for the REAF's contribution to the war effort. On 7 February King Fārūq paid an official visit to his air force and the RAF seized this opportunity to get Air Marshal Sir Keith Park to hand over the Anson XIX complete with Egyptian royal insignia, air conditioning, and "luxurious appointments."[25]

The Americans then trumped Britain's small Anson with their own gift of a Dakota—larger, more modern, and distinctly faster than the British offering. Of even greater concern to the British embassy in Cairo was the king's aide-de-camp (ADC) and personal pilot, Sq. Ldr. Ḥasan ᶜĀkif. He had been the only air force officer among those originally suspended for suspected subversion whom the British refused to allow to return to their REAF squadrons. Three months later King Fārūq, almost certainly out of spite because he had himself been so humiliated by the British during the 1942 crisis, promoted ᶜĀkif and appointed him as his ADC. Since ᶜĀkif's anti-British sentiments were well known, the RAF's insistence that he was not sufficiently experienced to fly the royal Dakota—on which he had been trained by the USAAF at the "Aerial Harbour Cairo" at Payne Field (later known as Cairo West)—may itself be questionable.[26] Even so, ᶜĀkif's periodic visits to No. 3 Squadron were clearly so disruptive that the CO, Sq. Ldr. Ibrāhīm Abū Rabīᶜah who was one of the REAF's most experienced multi-engine pilots, asked to be relieved of his command, stating that he would no longer tolerate ᶜĀkif's interference.[27] This might indeed have reflected what the RAF described as the too easygoing attitude of the REAF's new director, Colonel Mitwallī, who had taken over on the retirement of Ḥusnī Ṭāhir in November 1944. The REAF itself may have agreed with this harsh judgment, for Mitwallī was retired shortly before the outbreak of the Palestine War with Israel in May 1948.

On 1 March 1945, along with many sideliners of the world conflict, Egypt officially declared war on Germany. The British Advisory Mission then suggested that Egypt pay £160,000 for RAF training courses during the war, which Egypt countered by suggesting that the RAF might care to pay an even greater sum for various services the REAF had rendered at a time when Egypt was, strictly speaking, neutral. Fortunately the matter was quietly dropped. In February twenty Spitfire VCs were purchased for No. 2 Squadron, whose Hurricanes went to the still largely earthbound No. 17 Squadron. This unit is also said to have received ten more Hurricanes on loan from the RAF in February 1945, though these might actually have been the same aircraft already loaned to No. 2 Squadron.

A representative of the American Curtiss company had met the REAF's new director late in 1944, reviving fears that Britain might lose a captive market for military hardware. Even as late as March 1945, however, the REAF was more keen to buy ex-RAF OTU aircraft including Hurricanes, Spitfires, and Wellingtons, the latter being intended to combat a continued menace of large-scale drug trafficking across the desert. RAF Middle East Command was keen to sell, but not so the British government, which was already under Zionist pressure in the face of the mounting crisis in Palestine. The attitude of the British government also caused considerable irritation to the British embassy in Cairo, where a new and more sympathetic ambassador expressed the opinion that "The position seems to me entirely senseless. Why are there no modern aircraft available, now that they are not required for bombing or gunning Germans or Japs? On the other hand one reads that 300 aircraft are being thrown into the sea."[28]

Shortly before the end of the Second World War special boards were established to investigate Allied war material still stored in Egypt; that for the REAF being under Group Capt. ʿAbd al-Munʿim Mīqātī. Mīqātī's group found about two hundred suitable aircraft, but, largely because the Egyptian government's current policy was to buy new machines whenever possible, only twenty-four various types were officially purchased, though a large number of other machines remained as abandoned scrap and thus a useful source of spares. These included the 101 ex-USAAF C-47s and C-46s at Payne Field (Cairo West). Although the British Advisory Mission opposed this scheme, twenty were bought from the Americans. They were then reconditioned to "zero hours" standard by REAF technicians for No. 3 Comm. Squadron's intended military mail service between Cairo and Khartoum.[29]

The last two years of the Second World War had seen the affairs of the Egyptian Army and the REAF dominated by a continuing quarrel between King Fārūq and his minister of war, between a palace clique deeply resentful of the high-handed way it had been treated by the British back in 1942 and a government that was doing as much as it could to support the Allied cause. In particular the king continually interfered with senior military appointments and promotions. Meanwhile, as the Second World War drew to its bitter close, far away in central Europe and the Pacific, the REAF's activities focused more and more on the increasingly tense Middle Eastern political situation. Britain was still a very unwelcome occupying power, and the Egyptian government was determined to build up the country's own ability to defend itself, thus dispensing with British tutelage once and for all.

## Operational Strength of the REAF in December 1942[30]

| Unit | Aircraft | Numbers | Comments |
|---|---|---|---|
| No. 1 (Army Cooperation) | Lysanders | — | Suffering fabric deterioration but still involved in training with gun cameras and formation flying |
| No. 2 (Fighter) | Gladiators | — | Standing in the open for two years, all needing new fabric, pilots "eager but frustrated"; currently flying refresher courses on Magisters |
| No. 3 (Communi-cations) | "Motley collection" | — | Limited to local test flying, Avrò Anson and Avro 652 suffering from shortage of spare plywood |
| No. 4 (Bomber) | Hawker Audaxes and Hawker Harts | — | Limited to local test flying by fabric deterioration and low serviceability |
| No. 5 (Fighter) | Gladiator | — | In best condition, with relatively new a/c, pilots keen, and CO highly regarded by British |
| No. 6 (Fighter) | Hurricane | 5 | Soon renumbered as No. 17 Squadron, perhaps because subversion had been particularly apparent in this unit, newest unit; training since end of 1941, pilots experienced but low morale, needing new CO |

7

# Threatening Horizons (1945–1948)

Even before the end of the Second World War tensions between Arabs and Zionist settlers flared up again in Palestine. Britain, exhausted by war and in a mood to start the process of withdrawal from empire, was no longer able to control the situation. The surrounding Arab governments, and those groups that would soon be governing Arab countries that were not yet independent, were well aware of the deep feelings Palestine aroused among their own people. For some years it looked as if Egypt could avoid being dragged into the deepening Arab-Zionist dispute, but by 1948 this was clearly no longer politically possible. Meanwhile the Egyptian government did all it could to increase the strength of the country's armed forces in case the issue came to war. Particular attention was given to the REAF, but once again the tense relationship between Egypt and Britain stifled any large-scale improvement in the REAF's capabilities.

## Continuing Anglo-Egyptian Tension

The end of the Second World War saw very different visions of the REAF's future role in Cairo and London. To the British the Egyptian Air Force was to be concerned solely with local defense, particularly of Egyptian airfields, while all other aspects of Middle Eastern defense would remain the responsibility of the British RAF. But to the Egyptians themselves, who had a clearer understanding of the looming realities of the Middle Eastern situation, the REAF needed to be a more effective force.

Back in February 1945, before the Second World War had ended, the Egyptian prime minister announced that in future his country would no longer lease any of its airfields to a foreign power. Meanwhile the REAF was already taking over several bases evacuated by the British. A meeting between the REAF's commander and the Curtiss Aircraft Company's representative in Cairo also sent a shudder through the British Foreign Office, which feared that America might break into the Middle East arms market. But despite such contacts the REAF still hoped to buy British, having its eyes on Spitfire LF9 fighters and Vickers Wellington bombers. RAF Middle East Command was eager to sell, but the Foreign Office would only permit limited sales of Spitfire LF9s, and no Wellington bombers. A British Parliamentary statement that twenty-one Spitfires, a smaller number of North American Harvard trainers, and an assortment of medium transports were supplied to Egypt between May 1945 and June 1947 was, however, inaccurate. This figure may have represented the aircraft allocated to REAF squadrons, but a secret report by the British Advisory Mission indicated that a much larger number were actually transferred to Egypt, even though many ended up in storage.[1] Nevertheless, these were hardly enough to tilt the military balance in Egypt's favor. The only combat aircraft to be included were the Spitfires, which, according to the British Foreign Office, were in flying serviceable condition only. The British also believed that the Egyptians were in no position to do further overhaul work.[2]

The new director of the REAF, Liwāʾ Muḥammad Mitwallī, had agreed to a complete overhaul of the Air HQ in Cairo, and this long overdue reform did a great deal to improve morale and discipline. Nevertheless, the REAF's operation flexibility was still limited by the fact that almost the entire force was concentrated at Almāẓah outside Cairo.[3] No. 3 (Comm.) Squadron was in greatest demand during the immediate postwar years. This included the Royal Flight whose commander, Flt. Lt. ʿAbd al-Salīm, had great difficulty maintaining discipline because King Fārūq's personal pilot, Ḥasan ʿĀkif, frequently contradicted ʿAbd al-Salīm's orders. Eventually the unfortunate ʿAbd al-Salīm was removed from command, just as ʿĀkif had previously forced Sq. Ldr. Abū Rabīʿah to resign.[4] Rumors that King Fārūq was using this Royal Flight to smuggle gold out of Egypt to secret Swiss bank accounts did little for royal prestige. At the same time the king also interfered in the REAF's day-to-day planning, blocking a visit by a delegation of senior officers to London late in 1946. King Fārūq did not, however, stop the REAF from sending some of its best men on specialist RAF training courses in 1946. Sq. Ldr. H. M. Tawfīq, Flt. Lt. ʿAlī Labīb and Pilot Off. Saʿd Allāh Hārūn went on a signals and radar course, while Sq. Ldr. Maḥmūd Ṣidqī went on the prestigious Bulstrode staff course. Several of these men later rose to prominence in the EAF, most notably the unfortunate Maḥmūd Ṣidqī Maḥmūd, who commanded the Egyptian Air Force during the disastrous 1967 June War.

When Group Captain Tait returned to Egypt shortly after the end of the war to visit his old friends in the REAF, he found that standards and morale were, however, rapidly returning to their prewar levels.[5] Meanwhile relations with the RAF remained genuinely cordial, although some members of the British Advisory Mission were now less experienced than the Egyptians they were supposedly advising. This did not deter the British War Office from declaring that "The REAF is only in a qualified sense a fighting force. It could, however, carry out simple tasks where no air opposition was encountered."[6] Then in March 1946 the Egyptian government decided not to renew the contracts of the officers and men of the British Advisory Mission. A few exceptions were made where technical advisors to the REAF were concerned, but otherwise the mission was to be closed at the end of 1947. Two months later Britain announced that it was withdrawing all its troops from Egypt, but not from the Suez Canal Zone. Britain would seek to negotiate a new Anglo-Egyptian treaty and held out the possibility of substantial sales of aircraft to Egypt, though all of these were now considered out of date by the Egyptians who submitted a list of much more modern types.[7] The Egyptian prime minister, Nuqrāshī Pasha, also appealed to the United States for assistance in training and modernizing Egypt's armed forces, and an REAF delegation visited America in November 1946. However, U.S. leaders in Washington did not respond favorably.

Late in 1947 Liwāʾ Mitwallī retired, and Liwāʾ al-Shaʿrāwī took over as the new commander of the REAF. Once again he was an army rather than an air force officer. On 31 December 1947, the British Advisory Mission was closed down after ten years of existence. In its final reports, drawn up in 1947, the Mission made it clear that although the REAF faced serious problems, it was a growing force that should not be ignored.[8] Morale was generally excellent, particularly in No. 2 and No. 6 (Fighter) Squadrons. However, the Spitfire pilots of these units lacked air gunnery training as the REAF still had no air gunnery targets. Four ex-RAF Boulton Paul Defiant target-towing aircraft had been given to Egypt as a gift by the British Air Ministry, but were in such poor condition that they were unsafe to fly. On the other hand, Egyptian aircrews "tended to be excellent" at air photography.[9] In general there were insufficient fighter pilots for them to train as full squadrons. Formation flying concentrated too much on display and was "operationally useless." Very few Egyptian Spitfire pilots had been given an opportunity to even fire their 20 mm cannon, and their aircraft had no oxygen equipment. The twenty ex-USAAF scrap C-47 transports that the British had expected to be a waste of time were, in fact, proving to be among the most useful aircraft the REAF possessed. Nevertheless, the REAF lacked its own repair depot and still had a manpower strength of only 200 officers and 2,000 men because so little training had been possible during the war years. The shortage of qualified ground technicians was even more acute and remained so for many years.

On paper the REAF seemed to have more aircraft than this small corps of pilots could use, and the situation described in the final reports drawn up by the British Advisory Mission in 1947[10] is unlikely to have improved much before the first Arab-Israeli war broke out in Palestine a year later. In reality the shortage of spares meant that by late 1947 Egypt's operational and serviceable front-line strength stood at no more than four antiquated Lysanders, twelve very tired Hurricanes, eighteen Spitfires in a better condition, and the twenty virtually rebuilt Dakotas and Commandos.[11] British ex-RAF technicians also worked under contract in the REAF's engineering and engine repair workshops in 1947, but it is not clear whether any of these civilian experts continued to work for the REAF during the subsequent Palestine War of 1948.

The withdrawal of the British Advisory Mission would, in the opinion of its members, mean that the REAF would soon degenerate into a "flying club" and that inadequate maintenance facilities "precluded the possibility of any sustained air operations."[12] Furthermore, this expected collapse in the REAF's capabilities meant that, in the opinion of the British Foreign Office, there was "a more urgent need than ever to dominate that area (the Middle East) for strategic and commercial reasons."[13]

While the British were expressing doubts about how the Egyptians would get on without their help, the Egyptian government was drawing up very ambitious expansion plans for the REAF. At one point this envisaged sixteen jet fighter squadrons, three coastal defense squadrons with Bristol Brigands, a communications squadron with an air-sea rescue flight, a short-range communications squadron with Percival Proctors, two medium bomber squadrons with North American Mitchells, and a tactical reconnaissance squadron with multi-engined jet bombers.[14] By the early months of 1948 this ambitious expansion plan had been quietly dropped. With no overall procurement plan left to follow, each department at REAF headquarters operated on a day-to-day basis, ordering equipment when needed, a procedure that led to inefficiency. Aircraft were still being ordered, and a few reached Egypt before the Palestine War broke out in May 1948. For example, in late 1947 Egypt announced its intention to buy six new De Havilland Devon transports, but the order appears to have been changed into one for the smaller twin-engined De Havilland Dove. Only three of these reached the REAF before the Palestine War, though three more were delivered later. Egypt also wanted to buy De Havilland Vampire jet fighters in late 1947, but Britain resisted this move. Yet in March 1948 the British Foreign Office allowed the Gloster company to sign a contract with Egypt for the sale of advanced Meteor jet fighters. Neither Meteors nor Vampires would reach Egypt until long after the Palestine War, though both types would fight against the British in the Suez Crisis of 1956. In January 1948 the Egyptian government started negotiations with the Taylorcraft company for a factory to be set up in Egypt where Auster VII airframes could be assembled, but nothing came of these plans. They were, however, the start of the Egyptian aviation industry.

## The REAF Looks East

The end of 1947 saw the British government adopting a slightly more helpful position toward Egypt as the crisis worsened in Palestine. Anti-U.S. feeling was meanwhile growing in Egypt as a result of the sympathy and support Americans gave to the Zionists in Palestine. Problems arising from Egypt's wish to modify the terms of the 1936 Anglo-Egyptian treaty and to subject British aircraft flying over Egypt to the same regulations as the aircraft of other countries were resolved by compromise on both sides. Even so, the British government still believed that Egypt was unlikely to take part in any fighting in Palestine, a belief apparently shared by many if not most in the REAF.[15]

The British had pinpointed the REAF's problems very accurately, but they had gravely underestimated the Egyptians' ability to cope with them. Despite these difficulties, the Egyptian fighter force consisted of reasonably experienced and confident personnel. Led by Wing Comdr. Muḥammad al-Janzūrī and Sq. Ldr. ʿAbd al-Ḥamīd Abū Zayd of No. 2 Squadron, the REAF's force of Spitfires was the second most powerful air component in the Middle East. The British still fielded by far the largest and most experienced air force in the eastern Mediterranean, but Egypt's Spitfires were probably more effective than Turkey's very mixed air arm. Iran had only just started to build an effective air force, while Britain had so far made limited progress in expanding the Iraqi Air Force.

Relations between Britain and Egypt were not helped by incidents such as that on 6 January 1948 when an REAF Spitfire fired on an RAF Anson that had strayed into an area of anti-aircraft exercises around al-ʿArīsh, close to the tense frontier with Palestine. Unfortunately another RAF aircraft was also fired on and hit ten days later, after which all British aircraft flying from Palestine to Egypt received fighter escorts. The Egyptians explained the somewhat trigger-happy behavior of their pilots by pointing out that they were, "on alert against attacks by Jewish aircraft" from Palestine, while the British admitted—within the privacy of Foreign Office documents—that the Egyptians were technically within their rights when opening fire.[16]

There were numerous unauthorized flights across the Sinai frontier during the months immediately preceding the Palestine War, and several British or British Commonwealth registered aircraft were, in fact, impounded by the Egyptian authorities for smuggling arms, or the aircraft themselves, to Zionist forces in Palestine.[17] Many of those that got through joined a fast-expanding Zionist "secret air force." Sometimes civil registered Egyptian aircraft were involved, though not necessarily making illegal flights. In April an Egyptian airline pilot, Capt. Muḥyī al-Dīn Sūsah, strayed over a British military area in southern Palestine while flying his private plane from Transjordan to Egypt and was forced down. But he landed near a Jewish settlement and was captured. Sūsah was eventually returned home via the Red Cross, but the Zionists incorporated his aircraft into what would soon become the Israeli Air Force.

Anglo-Egyptian relations were particularly strained by Egypt's commandeering of the Hawker Fury prototype during a sales demonstration tour. Egypt had expressed an interest in this new British fighter, and so, in April 1948, Bill Humble, a Hawker chief test pilot, flew the first prototype to Cairo for the forthcoming Heliopolis Air Display. On this occasion it carried the civil registration G-AKRY instead of its original RAF serial number NX798. Humble arrived on 21 April and three days later put on a demonstration of aerobatics and high-speed passes for the Royal Egyptian Aero Club at Almāẓah.[18] This display so impressed the Egyptians that an excited crowd swarmed over the aircraft as soon as Humble landed. Ḥalīm Ṭāhir Zakī, who was soon to fly C-47 "bombers" over Palestine, then tried out the machine, after which he sent a glowing flight test report to the REAF.

On 27 April Bill Humble gave the machine a ten-minute test flight preparatory to his homeward journey. Just before his intended departure the Hawker test pilot discovered that the Fury had been moved from the civil to the military side of Almāẓah and that the aircraft had been impounded by the Egyptian authorities who turned it over to the REAF. So Bill Humble had to fly home aboard a C-47 of BOAC to face unfair and unfounded allegations of "gun running."[19] The Hawker company eventually negotiated an official sale on 30 December 1948 as part of a deal to supply further Fury fighters. By then, however, the Hawker Fury prototype had already been in combat over Palestine, had shot down at least one enemy aircraft, and had itself eventually crashed in the Mediterranean Sea.

The Egyptians' seizure of the Hawker Fury demonstrated their frustration with what had become a semi-official British arms embargo. After Ṭāhir Zakī made a few more test flights, the Fury was handed over to Muḥammad Ibrāhīm ʿUbayd of the REAF's engineering division. Despite having no service manuals for this type of aircraft, he and his highly experienced maintenance crews worked on the machine, installing four cannon from a Spitfire plus a gun camera, four universal bomb carriers, and armor plating for the pilot.[20] In the first week of May 1948 the Fury prototype was flown to al-ʿArīsh in northeastern Sinai to join the squadron of Egyptian Spitfires led by Wing Comdr. Muḥammad al-Janzūrī. Together they would form the cutting edge of a small REAF component in support of the Egyptian Army's Expeditionary Force in Palestine. Not surprisingly the British protested at the Egyptian seizure of the Hawker Fury, the next few months seeing a flurry a diplomatic notes between the British embassy and the Egyptian government. It seems that, at first, the aircraft was flown in combat by the REAF while still wearing its British civilian registration. At a meeting with the British ambassador the Egyptian prime minister agreed that this was quite unacceptable and promised to investigate the matter.[21] On the original commandeering of the Fury, however, the Egyptians remained unrepentant, pointing out that they were forced to take such drastic action because the British refused to supply the REAF with the aircraft and supplies it needed. The British markings were soon removed, and the

Hawker Fury prototype subsequently carried the REAF serial number 701.

On 15 May, exactly three weeks after the seizure of the Fury, the British mandate over Palestine came to an end. The Zionist settlers declared their independence, making a point of doing so several hours before the mandate officially expired, and announced the founding of the state of Israel. The Arab League, which brought together the governments of most of the independent Arab countries, declared that Palestine was and would remain an Arab country in which Jews could live as equals but not as conquerors. In Cairo the Egyptian government ordered its armed forces to advance into Palestine the following day, and the first Arab-Israeli War began.

The period between the end of the Second World War and the start of the Palestine War had seen Egypt go from one international political crisis to another. Relations between Egypt and Britain seemed to improve, but the underlying tensions remained. For the REAF this was a time of hope and expansion, but the hope was not fulfilled and the expansion proved to be something of an illusion, as was soon to be demonstrated in the skies over Palestine. Courage and determination would not be enough when the REAF was faced by a far better trained, more experienced, and eventually larger foe.

## Operational Strength of the REAF Early in 1947[22]

| Unit | Aircraft | Numbers | Comments |
|---|---|---|---|
| No. 1 (Fighter-reconnaissance and fighter-bomber) | Spitfire LF9 1 | 6 | Low-level fighter-bomber version, lack of training on Spitfires, CO "not good," very short of pilots, average of 6 |
| No. 2 (Day fighter) | Spitfire LF9 | 11 | Medium-altitude fighter version, very short of pilots, average of 5, CO "very good and keen" |
| No. 3 (Communications) | Anson, Dakota, Percival Q6, Magister, Lysander | 15 | Very busy but only 1 qualified C-47 captain |

*Continued on next page*

*(Continued)*

| | | | |
|---|---|---|---|
| No. 4 (General reconnaissance) | Anson | 6 | Training hampered by used communications unit |
| No. 5 (Fighter) | — | — | Exists only on paper but is planned to have Spitfires |
| No. 6 (Day fighter) | Spitfire V | 16 + 3 unserviceable | Very short of spares, more pilots than other fighter squadrons and more experienced |
| Royal flight | Lysander, Anson, Dakota, and Magister | 9 | |
| Meteorological flight | Hurricane | 4 | Still operating efficiently |
| Elementary flying training squadron | Magister | 25 + 3 unserviceable | |
| Intermediate and advanced flying training squadrons | Harvard | 16 | |
| Navigation flight | — | — | Planned to have Ansons |
| Anti-aircraft and target-towing flight | Defiant | 0 + 4 unserviceable | |

# 8

# The First Offensive (1948)

The Palestine War of 1948 was the first time that the REAF carried out sustained operations without considerable help from the British RAF. It also saw the REAF go on the offensive for the first time in its history. Initial results appeared satisfactory, but as soon as the Israeli enemy acquired modern fighter aircraft Egyptian weaknesses became all too apparent. On the other hand, the REAF managed to sustain its operations, contrary to what the British and most other outside observers had expected. Nor did the rise of an Israeli Air Force stop the REAF from supporting the Egyptian Army in the field, which was its primary task. Operations got more difficult and more dangerous, with mounting casualties, but the REAF's most serious losses during this first stage of the Palestine War resulted from a mistaken clash with the RAF, not from Israeli action. When the UN managed to impose its first truce in June, the Egyptian Air Force could look back on the previous month's operations with considerable satisfaction.

## War in Palestine

Arabs and Israelis still argue over who started the Palestine War, or Israeli War of Independence. Arabs point out that the Zionist settlers were already fighting the Palestinian Arabs and claim that regular Arab armies moved in to save the retreating Palestinians from being pushed into the desert. Israelis claim that the Arab invasion of 15 May 1948 was unprovoked aggression intended to drive the Jews into the sea. The Arab League never seems to have formulated a coherent policy, and the Arab governments used Palestine as an arena in which to carry on their political rivalries. Military cooperation between the Arab armies was at best haphazard and was often nonexistent. Even as late as June 1948 the Egyptian government refused to consider that Egypt was actually at war, regarding the operations of the Egyptian Army as a "police operation" to protect the Palestinian Arabs from Zionist attack.[1] Egyptian forces were, in the eyes of the Egyptian authorities, merely taking over from the British who had failed to maintain peace in Palestine. One thing is certain: the Arab governments gravely underestimated the military strength of the Zionists and the huge sympathy that their new state of Israel would receive from the outside world.

For several years neighboring Arab countries had made clear their commitment to stop what they saw as a creeping Zionist seizure of Palestine. They, like other Arabs further afield, regarded Palestine as a land that had been Arab for at least seventeen centuries. While declaring that they felt no enmity toward the Jewish people as such, they made it plain that they were prepared to fight in defense of an Arab Palestine where, they said, Jews could live as equals but not as masters. For their part most Zionist settlers and their sympathizers felt no confidence in such declarations and believed that the Arabs planned to drive all the Jewish colonizers out of Palestine entirely.

While Britain governed Palestine under the mandate, originally established by the old League of Nations and now confirmed by the new United Nations Organization, the rhetoric of both sides meant little. The only fighting was the small-scale but undoubtedly savage struggle between Zionist settlers and indigenous Arab inhabitants within mandated Palestine, with the British attempting to control an ever more volatile situation. As the situation deteriorated, British troops found themselves fighting a guerrilla war against both Zionist Jews and Arabs, suffering particularly heavy casualties to Zionist terrorism. Once Britain announced its intention of pulling out, Arabs and Zionists fought savagely for control of villages and important communications centers. The greater social cohesion and superior weaponry of the Jewish settlers soon gave them the advantage, and Zionist military groups took control of many villages. The Arab inhabitants were sometimes forced to flee and sometimes chose to do so in the hope of finding refuge with neighbors or relatives. In all cases, however, they believed that peace would enable them to go home.

While the Palestinian Arabs seem to have been operating without any overall strategy, the main Zionist "underground army," the Haganah, had drawn up its final detailed plan for military operations in Palestine.

This was Haganah Plan D of March 1948 which, two months before the state of Israel was declared, replaced all previous Haganah plans. It was designed "To gain control of the area allotted to the Jewish state and defend its borders, and those of the blocs of Jewish settlements and such Jewish populations as were outside these borders." Plan D was intended to begin after the British were out of the way, but problems at the UN, where support for the Zionist point of view seemed to be slipping, meant that the plan was brought forward to late March and April. As a result the Haganah Army had captured a great deal of additional territory well before the regular Arab armies moved into Palestine on 15 May. These operations were also supported by a large number of light aircraft belonging to the Haganah's Sherut-Aṿir, or Air Service, which had been set up in November 1947.

The situation for the Palestinian Arabs was very different. They had neither aircraft nor air support, no means of evacuating casualties by air, and virtually no hope of any such help unless the surrounding Arab states intervened. Constant harassment by Zionist light aircraft also eroded the morale of men who were little more than an ill-armed, untrained, and generally uneducated peasant militia. Unlike the nomadic Arab bedouin, most of these Palestinian fighters were farmers or artisans with little experience of using rifles, and they were unaware of how often their light weapons actually hit the Sherut-Aṿir's little aircraft. They only knew that the enemy planes did not crash. They also saw RAF aircraft sharing the same skies with their enemies, further fueling a Palestinian belief that the British were now acting in collusion with the Zionist settlers.

Meanwhile the destruction of a railway train and the large number of resulting deaths added to a growing feeling among the Palestinian Arabs that the Jews were bent on their destruction. Such fears were fueled by stories of killings elsewhere—some true, some merely exaggerated rumors—that culminated an appalling massacre in the village of Dayr Yāsin by Zionist terrorists two weeks later. Here, in a small Arab village south of the Laṭrūn-Jerusalem road, the Irgun Gang killed 254 unarmed men, women, and children. The effect of this act of "ethnic cleansing" was dramatic. Fears that had been spreading throughout the Palestinian Arab population turned into a panic, and entire communities began to abandon their ancestral villages, joining others who had fled earlier. The Palestinian refugee problem had been born.

The reverse side of such massacres was that they encouraged ever more Muslim volunteers to come from Asia, Africa, and even Europe to fight in defense of Palestine. The Gaza area was, for example, already garrisoned by volunteers of the "literalist" Muslim Brotherhood, mostly men from Egypt and the Sudan, commanded by a Sudanese professional soldier named Ṭāriq al-Afrīqī.

Throughout these weeks the British steadily pulled their forces out of Palestine. The last area to be abandoned would be the "Haifa Enclave," a 10 by 20 mile area around the main Palestinian port of Haifa. This evacuation area was to be protected by a declining number of British troops and RAF fighter units based at Ramat Dawid, an airfield 15 miles southeast of Haifa.

The situation was further complicated by deep-seated rivalries between various Arab governments, most notably between Egypt and Iraq. Egypt already hoped to become the leader of the Arab world once British domination had been removed. Clearly the country would now have to take a lead in dealing with the main issue of the day, and for most Arabs that issue was Palestine. Yet popular fervor and political will alone are not enough, as Egypt's military leaders realized. The Zionist settlers, most of whom at that time came from Europe, had developed an impressive and very large paramilitary force obviously capable of overcoming local Palestinian Arab militias.[2] However, Egyptian leaders believed that these Jewish forces could not withstand an attack by the Egyptian and other Arab professional armies. This was a huge miscalculation as the Jewish community in Palestine was highly educated and well prepared to fight for its own state. As a result of the Nazi Holocaust and the appeal of Zionist nationalism, the Jews of Palestine were also supported morally and materially by million of sympathizers in Europe, South Africa, and North America.

## The REAF Prepares

Britain's self-imposed embargo on the sale of military aircraft to the Middle East came into effect just a month before the British mandate over Palestine was due to end. Its effects were soon felt by the REAF, particularly when the embargo became total on 4 June 1948. From then on the large amount of equipment, including aircraft, due for delivery to Egypt remained under lock and key at the RAF's No. 107 Maintenance

Unit at Kasfarit—in Egypt, but inaccessible for the Egyptians. The Egyptian Army, which in 1948 included a theoretical total of 50,000 men, found it difficult to raise the 5,000 trained troops required for it Expeditionary Force.[3] During the course of the Palestine War the Egyptian Expeditionary Force eventually rose to 40,000 troops, but this would include further volunteers from the Sudan, Saudi Arabia, Libya, Tunisia, and Morocco. The soldiers were told not to expect serious resistance from "civilian" Jewish settlers, while some Egyptian officers were led to believe that their role was to save southern Palestine from falling under the control of King ᶜAbd Allāh of Transjordan.

Compared with the difficulties the Egyptian Army faced in raising its Expeditionary Force, the REAF was well able to send several flights of Spitfire LF9s, flown by Egypt's most experienced pilots, to the forward air base at al-ᶜArīsh. There was, however, a great difference in experience between those of squadron leader rank and the rest. Squadron leaders and above had all been commissioned before 1941 and had operational experience under British guidance during the Second World War. The REAF's inability to train any additional aircrew during the war meant that the flight lieutenants had received their training no earlier than 1945, though some had been on RAF specialist courses since. Pilot officers were very much "new boys," fresh from flying school with a minimum of squadron experience and no operational experience whatsoever. The REAF's shortage of experienced pilots was amply demonstrated by the fact that men with valuable technical expertise, such as Flight Lieutenant Jibrāʾīl, one of the REAF's best maintenance officers, Group Capt. Maḥmūd Ṣāliḥ, the REAF's senior equipment procurement officer, and Wing Comdr. al-Malayjī, the REAF's senior training officer, all had to be used in front-line operations.

However, little aerial opposition was expected as the Zionists were known to possess only light aircraft capable of carrying only small bomb loads and small arms. The REAF's own primary task was close tactical support of the Egyptian Expeditionary Force as it advanced to secure the UN-designated Arab zones of Palestine. A detailed operational account of the REAF's operations over Palestine in 1948–49 has yet to be published,[4] and one of the very few Egyptian books on this air war was written by Munīrah Kafāfī, the daughter of a pilot who was killed in the conflict. After conducting interviews with many of her father's colleagues, she summarized the role expected of the REAF at the start of the war:

> The tasks of the air force were clearly going to be linked to, and in no way separate from, those of the army. On the first day of the war Egypt's Royal Armed Forces received orders to cooperate on the battlefield. At that time the air force's role was to provide front-line assistance to ground and naval forces. The work of the aircraft would therefore involve aerial transport carrying men and their equipment, and carrying out battlefield support. This led to a dispersal of effort, particularly by the long-range aircraft. As to why the air force's aircraft were not all put into one task, it would seem that the Egyptian technicians were hampered by their lack of spares and the need to put all they had into modifying transport aircraft into "bombers." . . . In spite of some doubts about these unproven modifications, the Egyptian Air Force certainly made them carry out their role quite well, even though the Egyptian aircraft in question could only carry approximately 675 kgs of bombs.[5]

By now the British Advisory Mission had ceased to exist, and consequently the British government no longer received accurate assessments of the REAF's current status. But after the end of the Palestine War, the British air attaché in Cairo did draw up what he believed, on the best available evidence, to have been the REAF's Order of Battle on 15 May 1948, at the very start of this first Arab-Israeli war. The only aircraft that would have any real combat role were the Spitfire LFs, of which no more than ten to twelve were serviceable.[6] It is not entirely clear whether a "joint squadron" using the available Spitfire LF9s and the REAF's most experienced pilots from Nos. 1 and 2 Squadrons was assembled at al-ʿArīsh, or if No. 2 alone was sent to this forward air base. The remaining serviceable Spitfire Mk. Vs of No. 6 Squadron were not to take part in operations for many months. Five C-47s had been modified so that bombs could be rolled manually out of their side-loading doors, and these now formed the nucleus of a new No. 8 Bomber Squadron under Wing Commander Abū Rabīʿah. The unconverted C-47s, and perhaps also some C-46s, of No. 3 Squadron under ʿUmar Shakīb, were mainly involved in moving men and material across Sinai since the road and rail link from the Nile Delta to the Palestine frontier were severely overburdened. Reconnaissance and communications tasks apparently fell to the REAF's antiquated Ansons and Lysanders. By limiting their forward force at al-ʿArīsh to no more than thirty aircraft, including both combat and noncombat types, Egypt's resources were not excessively strained. This force was commanded by Wing Comdr. Maḥmūd Ṣidqī al-Malayjī, with Group Capt. Ṣāliḥ Maḥmūd Ṣāliḥ, though senior in rank, as his chief of operations.[7]

Egypt was not, of course, the only country to have military aircraft in the vicinity. The RAF still had two Spitfire squadrons, Nos. 32 and 208 based at Ramat Dawid, to cover the final evacuation of British personnel from the Haifa Enclave, plus various transport and communications aircraft. But the bulk of the Palestine-based RAF units had already been relocated to Cyprus and the Suez Canal Zone in Egypt. Syria had a few armed Harvard trainers plus some light aircraft. These were probably the aircraft mentioned in Cairo's English-language *Egyptian Gazette* newspaper which, on 12 April 1948, reported a Palestinian Arab Liberation Army claim that its ground forces "will soon be supplemented by an air unit now being assembled outside Damascus." Iraq could supply a small number of Hawker Furies, though it never committed more than a single flight of these modern fighters to the war effort, as well as several Avro Anson light bombers which would play a more significant role.

The Zionists of Palestine were also not without air power. They had secretly built up a sizable flying organization that already totaled some forty pilots and three squadrons of medium transports and light aircraft. Many foreign "Mahal" volunteers or mercenaries were being recruited to aid the Zionist cause, particular effort being made to enlist men with combat experience in high-performance aircraft during the Second World War. The Jewish air arm, or Sherut-Aỵir, now seized many of the military airfields and smaller airstrips in Palestine as the British withdrew. The REAF knew that the Zionists had pilots and light planes, but the Arabs were not aware of the ease with which such a "private flying organization," supported by foreign volunteers and a highly experienced worldwide embargo-busting organization, could be rapidly turned into an effective air force. Nor did the Arabs realize that combat-experienced pilots from many countries, by no means all of them Jewish, would rally to the Zionists so quickly, nor that illegal sources of fighter aircraft could be found so easily.

On the ground the earlier British estimate of Jewish military manpower in Palestine proved to be inflated. Even so, the newly created state of Israel quickly fielded an army of 30,000 Jewish troops and paramilitary personnel. Though it was as yet short of weapons, this first Israeli Army immediately outnumbered the 20,000 Arab soldiers committed to battle, less than 15,000 of whom were trained regular troops from the armies of surrounding Arab states.

## Into the Attack

As the end of Britain's mandate over Palestine approached, the REAF at al-ʿArīsh awaited the order to attack. Beyond the frontier, 30 miles to the east, thousands of Palestinian Arab families were already being driven from their villages by intense fighting, while Zionist settlers and the new Jewish army dug in to await the promised invasion by the armies of Egypt, Transjordan, Iraq, Syria, and Lebanon.

Some REAF officers were not prepared to wait. Flt. Lt. ʿAbd al-Laṭīf Baghdādī and Flt. Lt. Ḥasan Ibrāhīm, both members of Colonel Nasser's secret "Free Officers" movement, had earlier applied for extended leave so that they could help organize a privately funded Palestinian air force to help the hard-pressed Palestinian Arab Liberation Army led by Fawzī al-Qāwuqjī. These efforts apparently came to nothing, unless ʿAbd al-Laṭīf Baghdādi and Ḥasan Ibrāhīm were involved in setting up the Syrian Air Force's first operational squadron outside Damascus, and Baghdādī was reported flying REAF C-47s back in Egypt later in the Palestine War.

In and around al-ʿArīsh, the Egyptian Expeditionary Force was making final preparations for its march into Palestine on 15 May 1948. Late in the afternoon of the fourteenth, the Jews of Palestine declared their independence and the establishment of the state of Israel. Israel's first leader, David Ben-Gurion, proclaimed his friendship for the Arab peoples and called on everyone to live in brotherhood for the common good. Unfortunately decades of mistrust, fear, national pride, mutual suffering, and occasional massacres meant that the Arab inhabitants of Palestine had no faith whatsoever in Ben-Gurion's words of friendship. Too much blood had already been spilled, too many fields and orchards lost, too many villages destroyed. Nothing was left but war, a war in which the Zionist settlers now called themselves Israelis, while the Palestinian Arabs still saw themselves as part of a wider Arab community rather than forming a distinct or separate Palestinian Arab "nation." Almost half a century of bitter conflict would follow before even a glimmer of mutual understanding, even a grudging recognition that the enemy had a valid case, would appear on either side of the Arab-Israeli divide.

At midnight on 14 May the British mandate finally came to an end. Dawn was spreading as Squadron Leader. Abū Zayd and his fellow pilots strapped themselves into the cockpits of their Spitfire LF9s. Their final briefing had reminded everyone that their tasks were limited to ground support, and now final instrument checks were made as the glow of the aircrafts' exhausts faded in the light of dawn. Each aircraft took off in a pall of dust, and the REAF's first offensive war missions had begun.[8]

Abū Zayd and his wingman headed northeast over the Mediterranean parallel with the coast of Palestine, while two other flights turned toward the sun's glare and headed due east across the frontier. After 75 miles Abū Zayd turned abruptly east and dived toward the coast. Straight ahead lay the city of Tel Aviv, its streets still in shadow but the rooftops bright in the early morning sun. The two Spitfires flew over at 1,000 feet. Abū Zayd and his wingman swept across the Yarqon River, almost touching the roof of Reading power station. The two pilots then dropped their 250 lb. bombs near the hangers of Tel Aviv's Śedeh Dov airfield which lay just beyond. The attack destroyed a C-47 that had just landed. Abū Zayd then circled round the field, his wingman following, then made three further firing passes. Other attacks by Egyptian Spitfires on Śedeh Dov airfield were reported on this first day of the war, by the end of which the REAF had destroyed the C-47 hit by Abū Zayd, an Israeli Republic RC-3 Sea Bee, an RWD-13, a Beech Bonanza, and a DH Rapide.

According to Israeli sources, Abū Zayd's wingman, a pilot named Maḥmūd Barakah, was brought down by small-arms fire that damaged his aircraft and wounded him in the head during the very first REAF attack on Śedeh Dov. Barakah is said to have then flown north, away from Egypt and other Arab-held territory, before making a forced landing on the beach near Herzliyya. His aircraft was then reportedly towed away by a jeep and handed over to the newly formed Israeli Air Force which restored it and used it against the REAF. The Egyptians, however, deny having had a pilot named Barakah.[9]

Further south the other REAF flights destroyed or damaged several Israeli aircraft at Nirʿam, the "Negev Squadron's" base, which also suffered one man killed and several wounded. Unfortunately one of the REAF strikes hit a remaining British outpost at al-Burayj near Gaza where the 651st Air Observation Post Squadron was still based. Damage was caused, one soldier being killed and eight wounded, but Britain made no complaint as the presence of this unit was supposedly secret and was certainly not known to the Egyptians.[10]

The REAF task force operating over Palestine remained concentrated at al-ʿArīsh until the final weeks of the war. During the first phase of fighting the Egyptians won several victories, and the al-ʿArīsh squadrons provided ground support and harassed the Israelis whenever possible. Convoys carrying supplies to fortified kibbutz settlements in the Negev were attacked, and determined efforts were made to undermine morale in Tel Aviv. It should, however, be pointed out that at this early stage of the war the REAF attacked only the defended airport and aircraft found there, not the city of Tel Aviv itself. Elsewhere it was also trying to restrict itself to specifically military targets such as fortified kibbutz settlements. However, all the Arab armies were finding their gains slow and costly. Israeli settlements were strongly fortified with entrenchments and well supplied with ammunition before the war began. They were proving far harder to overcome than the Egyptian Army had expected.

The REAF was often called in to help the ground troops, but its Spitfires could carry only two small bombs and were vulnerable to ground fire. The modified C-47s, with their out-of-the-door bombing techniques, were very inaccurate and completely unsuitable for the close support of Egyptian forces on the ground. Nevertheless, pressure was maintained. On 16 May the abandoned RAF air base at ʿĀqir was bombed, and on 17 May the Israeli airfield at Śedeh Dov was again attacked, as were the facilities at Kefar Sirkin airfield. On 18 May, REAF C-47 bombers and Spitfires struck at Tel Aviv itself but again attempted to limit themselves to military or strategic targets. The central transport depot was damaged and a fuel dump destroyed, but there were many civilians among the resulting forty-one Israeli deaths.

This was almost certainly the first mission by the REAF's converted C-47 "bombers." Many others would follow, though the same pilots and sometimes the same aircraft would also have to carry out normal transports duties. Air Maj. Gen. Yaḥyá al-Shannāwī explained the background and primary role of the REAF's few multi-engine pilots to Munīra Kafāfī, whose father then flew Dakotas:

> I was then the general staff officer at the al-ʿArīsh air base and I was also a comrade of your father (ʿAdlī Kafāfī) in the war of 1948. . . . Your father served in No. 4 Squadron and flew its Anson aircraft until 1945. Then he moved to No. 3 Squadron. The main work of No. 3 Squadron (during the Palestine War) was to maintain a constant air bridge, carrying men and equipment from Egyptian air bases up to the battlefield. The aircraft used by No. 3 Squadron at that time were Dakotas, and among our pilots in this unit were ʿAbd al-Laṭīf Baghdādī (later a close confidant of President Nasser), Yaḥyá al-Shabāwa, ʿAbd al-Moneim ʿAtā Allāh, ʿAmr al-Jamāl, and ʿAmr Shakīb. The

commander of your father's comrades in No. 3 Squadron was ʿAmr Shakīb (later air major general).[11]

On 18 May the fortified kibbutzim outside Dayr Sunayd and at Negba were also attacked. One Egyptian plane made a forced landing after being hit by ground fire during these operations. It was described in an REAF communiqué as a "heavy bomber" that had attacked the Tel Aviv area but crash-landed within Egyptian lines, the pilot returning to his base safely. But Israeli sources suggest that it was a Spitfire that crash-landed near Rishon LeZion, Flying Off. Jamāl ʿIrfān Ṣāfī al-Nāṣir being captured.[12]

Squadron Leader Abū Zayd spared no efforts in giving a good example to his men, averaging two sorties per day during the first weeks of fighting. He flew both Spitfires and the single Hawker Fury which he alone was permitted to fly in combat.[13] By day the skies belonged to the REAF, although by night the Israeli Air Force (IAF) carried out harassment bombing and supply and medical evacuation missions. Israeli light aircraft bombed concentrations of Transjordanian Arab Legion vehicles on many nights, and on 17 May the Syrians claimed to have shot down an Israeli light aircraft near Tiberius.

Initially the Egyptian Army advanced successfully, while other Arab armies moved into Palestine from the east and north. Israeli units in the south were tied up in the defense of besieged settlements, including those of Gat and Gal-On east of Faluja. The Egyptian high command particularly targeted the sprawling Israeli settlement of Negba, realizing that unless it was captured the Israelis could use it as a base from which to attack the Arabs' east-west vital supply lines through Faluja. The UN Partition Plan had proposed two states in Palestine, one Arab and one Jewish. Each was divided into sections, and one of the "points of no frontier" where two Arab and two Jewish zones met was near Faluja, adding still further to the strategic importance of this area. On 21 May 1948 an Egyptian tank shell hit the water tower at Negba, causing serious damage, and the same day a pair of REAF Spitfires strafed the settlement, killing its commander who was directing anti-aircraft fire. The following day the Egyptians again bombarded Negba using artillery, tank fire, and aircraft.[14]

Meanwhile the IAF was growing into a real modern force. Large numbers of pilots, Jews and non-Jews, arrived in Palestine to fly for Israel as volunteers. These ex-USAF, ex-RAF, and Commonwealth pilots became known as the Mahal ("volunteers from abroad"). Compared to the REAF aircrews, they had recent and often considerable combat experience. In addition to volunteer pilots, many planes, including fighters, were now arriving in Israel. Back on 23 April, before the establishment of the state of Israel, the Czechoslovakian government had agreed to supply the Jews with surplus Czech-built versions of the famous German Messerschmitt Bf 109. These were known as Avia C210s, the first being flown to Israel aboard large transport aircraft on 20 May, five days after the creation of the state. But it still took the Israelis two weeks to assemble and test their Avia C210s before sending them into action as the IAF's new No. 101 Fighter Squadron. This meant that the pilots had little time to familiarize themselves with these machines. Most of the pilots were foreigners, though there were some Israelis in No. 101 Squadron. Nor were the Egyptians ignorant of this dangerous development. The REAF put a great deal of effort into locating and attempting to destroy the Israelis' new fighter base before the Avias became operational.

Britain was still in the process of withdrawing its forces from Palestine and those remaining still controlled the Haifa Enclave plus several airfields. The RAF base at Ramat Dawid housed Nos. 32 and 208 Fighter Squadrons flying the latest Griffon-powered Spitfire FR 18s. These units were due to evacuate Ramat Dawid on 22 May and to move to a new home in Cyprus, from which they would cover the final British evacuation of Haifa due on 30 May. At 6:00 A.M., on the very day the RAF were to leave, two Spitfire LF9s appeared over Ramat Dawid and circled a few times. Then, to the astonishment of the curious RAF ground crews, the Spitfires attacked. The first aircraft dropped two 250 lb. bombs and the second strafed the field, leaving two RAF Spitfires burning on the ground. The British then put up a standing patrol of Spitfires from No. 32 Squadron, and the anti-aircraft defenses were fully manned.

The Egyptian pilots thought that they had finally found the IAF's fighter base, since it appears that the REAF had not been informed that British Squadrons would still be in Palestine on the 22 May. But they had also made a significant navigational error, believing that they had attacked the airfield at Megiddo, which lay five miles to the southeast. Since the REAF possessed no proper flying maps of Palestine, relying on a mixture of tourist and general ordnance survey maps, and were operating at maximum range deep inside enemy territory, their error was understandable. But the results were catastrophic. The REAF staff at al-ʿArīsh now determined to make an all-out effort to destroy the Israelis' fighter force before it could become a threat. At 7:45 A.M., two or three

Spitfire LF9s struck Ramat Dawid again, hitting the main hanger, destroying an unserviceable RAF Spitfire FR 18 and a C-47 transport. The standing patrol of British Spitfires was, meanwhile, away investigating the reported interception of an RAF Proctor communications aircraft, so at first the Egyptians were opposed only by anti-aircraft fire. This damaged one REAF Spitfire. The Egyptians were then caught by several furious RAF Spitfire pilots from No. 208 Squadron who had taken off during the attack, one REAF Spitfire being shot down. At 9:15 A.M., the REAF made a third attack using two Spitfires, but these were intercepted by four Griffon-powered RAF Spitfires from No. 208 Squadron. The combat was brief and one-sided, the more powerful British aircraft flown by better trained pilots quickly overwhelming the two Egyptian planes, one of which was shot down while the other was seriously damaged, crashing as it attempted an emergency landing.[15] In all three, the Egyptian pilots were killed, including some of the most experienced airmen in the REAF: Sq. Ldr. Muḥammad Naṣr al-Dīn, Flt. Lt. Saᶜd Ṣādiq al-Duwaynī; and Flt. Lt. Tutmus Kāmīl Ibrāhīm, a Coptic Christian; with Flt. Off. ᶜAbd al-Raḥmān ᶜInān being taken prisoner.[16] Four Spitfires were lost, another possibly being seriously damaged, out of a total of only eight based at al-ᶜArīsh. It was a devastating blow.

Once the truth was known back in Egypt, reactions varied. Senior men realized that the REAF could not afford to make an enemy of the RAF. Air Commodore Mīqātī offered to come to Ramat Dawid to apologize in person but was warned by a senior British officer, whom he had known for many years, that his reception would be very hostile.[17] The ferocity of the RAF's reaction was understandable as the men of Nos. 32 and 208 Squadrons had been due to leave Palestine that day, finally getting out of an unrewarding situation that they had come to hate. Two British officers and one airman had been killed, three men being wounded, three Spitfires and a C-47 destroyed, three more Spitfires being damaged—all by those they thought to be their friends. Some more junior REAF officers felt that the British should share responsibility for the extent of the tragedy, believing that the RAF could have warned the Egyptians of their error during the almost two hours that elapsed between the REAF's first and second raids. This view was expressed in its most extreme form by Ṣāliḥ Maḥmūd Ṣāliḥ, the REAF's director of operations at al-ᶜArīsh and once regarded as a staunch friend of the RAF:

> We learned that while two (Egyptian) Spitfires were providing air support to our land forces in Palestine, their pilots spotted a small number of aircraft

> parked at Ramat Dawid airfield. Since British sovereignty had ended and they had departed from Palestine on 15 May (or so the Egyptians believed), the pilots had no doubt that these aircraft were the new equipment of the Jews, and they attacked them. Reporting to the al-ᶜArīsh base commander on their return, he sent a further five Spitfires—out of the total of eight then based at al-ᶜArīsh—to go and follow up the attack on Ramat Dawid. Once there they fell into an ambush planned by the British who made an unexpected attack on the Egyptian aircraft and shot them down. Only ᶜAbd al-Raḥmān ᶜInān survived, being captured and then handed over to the Jews by the British. . . . When he returned home he recounted in detail the story of his sad and regrettable battle. It would have been easy for the British to warn the Egyptians of their mistake, thus avoiding the damage. But the British, it seemed, preferred the bloody loss of life which assisted the Jews and weakened Egyptian air power.[18]

There was, of course, no "planned ambush" on the part of the British, but the incident undermined still further the once excellent professional relationship between the REAF and RAF. Meanwhile only the Israelis profited from a tragedy that cost seven brave lives, learning a lesson but not having to pay the price.

The British now told the REAF that it must keep at least 20 miles away from the Haifa Enclave or risk being shot down by the RAF. After the Ramat Dawid tragedy the RAF brought back more fighters from Cyprus and delayed its evacuation of Ramat Dawid, the last British soldiers leaving Haifa on 30 May 1948. British aircraft were, in fact, fired on at least three times by both Arabs and Israelis during the Palestine War, excluding the major clashes between British and Israeli aircraft during the final week's fighting. On each of these earlier occasions the RAF admitted that its pilots had been slightly off course and that almost all incidents were a result of mistaken identity. Meanwhile the Egyptian government tried to impose serious restrictions on RAF units based in the Suez Canal Zone, particularly at night, but these were vigorously resisted by the British who continued to fly as before.

## Air Combat over Palestine

Naturally the REAF's remaining Spitfires continued to search for the Israelis' real fighter base and eventually located it at the ex-RAF airfield of ᶜĀqir. Egyptian air raids then destroyed one hanger and caused several casualties, but the Avias were not hit. Even so, the IAF's new 101 Fighter Squadron had to move north, into a new home near the sheltering orange groves of Herzliyya. The first combat sortie by the IAF's new Avia fighters took place on 29 May but was a costly failure. Four C210s attacked an Egyptian column south of Isdūd, one being shot down and its pilot killed, while another was damaged. Next day a lone Avia attacked Egyptian troops holding the ʿIrāq Suwaydan police fort but caused little damage. This important position overlooked the main crossroads east of al-Majdal. It would soon become the scene of some of the most intensive and strategically important fighting of the Palestine War.

The appearance of modern Israeli fighters did lead to a temporary cutback in operations by Egypt's unarmed C-47 "bombers." The threat of confusion between REAF Spitfires and IAF Avias, particularly by the Egyptian anti-aircraft gunners, also led the REAF to add "invasion stripes" around the wingtips of its front-line aircraft. These would remain a feature of Egyptian fighters until after the June 1967 War. REAF aircraft meanwhile bombed and strafed Israeli positions at Gan Yavne, Gedera, Negba, Nirʿam, Kefar Am, and Dorot on 30 May. On 2 June Wing Comdr. al-Janzūrī and his surviving Spitfire pilots carried out a major air bombardment of Negba to clear the way for a ground assault, but in the end rain and low cloud bogged down the Egyptian attack. The next morning Egyptian Spitfires were in action again, bombing Israeli bases around Sarafand and Rishon LeẒion in support of Transjordan's Arab Legion.

On the night of 1 June, Israeli aircraft bombed Amman in Transjordan with high explosives and incendiaries, the first such attack on an Arab capital. Two days later the REAF tried to use extensive cloud cover to strike back. Several C-47 "bombers" flew to Tel Aviv, but as they arrived the weather cleared and, on the afternoon of 3 June, two Egyptian C-47s came face to face with an Israeli Avia C210 flown by Modi Alon, the commander of the IAF's No. 101 Fighter Squadron. The "bombers" tried to escape out to sea, but Alon claimed hits on both. According to Israeli sources, one crash-landed on the beach 2 miles south of Bat Yam, while the other flew off trailing smoke, eventually to make a forced landing at Wādī Sukhrīr. The REAF crews escaped with their lives from this, the sixteenth and final Egyptian air raid on Tel Aviv before the first UN truce was imposed.

As the UN representatives in Palestine struggled to establish a ceasefire, air activity increased on all sides. The Israelis bombed the Syrian capital of Damascus on the night of 3 June and lost a plane during an attack on Egyptian vessels off Isdūd on 4 June. This was almost certainly the Fairchild Argus shot down by Sq. Ldr. ʿAbd al-Ḥamīd Abū Zayd while flying the REAF's one and only Hawker Fury. It was the first enemy aircraft to be claimed as a confirmed victory in the history of the Egyptian Air Force.[19]

Intense fighting continued between Egyptian and Israeli ground forces along a line of abandoned Arab villages near Isdūd. The Israelis had retreated to this point on 7 June, and here the advancing Egyptians stopped to regroup. On 7 June an IAF Auster J-1 Autocrat was also shot down near Lydda by Egyptian ground fire. On the night of 10–11 June, with the first UN truce due to come into effect the following day, Israel's Negev Brigade made a determined attack on the village of ʿIrāq Suwaydan, but they came up against Egyptian regular troops in well-prepared positions. At dawn, units of the Negev Brigade were still pinned down by Egyptian cross fire in barbed-wire entanglements. They were then strafed by a flight of REAF Spitfires from al-ʿArīsh, turning their retreat into a rout. Egyptian Spitfires also hit the isolated Negev settlement of Ruḥāmah seven times during a day of heavy fighting. The first UN truce then came into effect as a welcome relief to both the REAF and the hard-pressed Israelis.

As the fighting came to a temporary end the Egyptians could look back with some satisfaction on their first operations. The Arab armies had, however, only managed to take control of most of those areas already allocated by the UN to the proposed Arab state in Palestine. Almost nowhere had they significantly crossed into the UN proposed Jewish state of Israel. Despite the sometimes ridiculous rhetoric now coming from various Arab capitals, it seems that their true mission was to defend the designated Arab zones and perhaps retake Arab-designated areas that had already fallen to Jewish forces before the British mandate ended.

Only the catastrophic error at Ramat Dawid marred the picture for the REAF, which had done much better than its British critics anticipated, perhaps better than some of its own personnel had expected. Egyptian Spitfires had carried out reconnaissance and close-support missions with considerable determination, though their results had been limited by the Egyptian Army's inability to follow up with the same success. Longer range bombing by the C-47s in the Tel Aviv area had hit several very important targets but had also played into the hands of enemy propagandists who portrayed such missions as near genocidal attempts to carry on where the Nazis left off. For their part, the Arabs signally failed to make the same use of the even more imprecise Israeli bombing of Amman, Damascus, and various towns in Palestine. The REAF also had strict orders not to attack Haifa, the main Israeli port. This was not only the case while the port fell within the British Haifa Enclave, but apparently throughout the rest of the war as well.

The most significant result of this first round of fighting was that the Arab air forces had lost their first and only realistic chance of defeating the new and vulnerable Israeli Air Force. This was probably not realized at the time, but henceforth the IAF would be operating from a position of considerable qualitative, and occasionally even quantitative, superiority. In future the Egyptians would be facing a hydra-like enemy that seemed to grow stronger every time it was struck.

## Operational Strength of the REAF in April 1948[20]

| Type[a] | Numbers |
|---|---|
| Spitfire LF9 | 10–12 + 10–13 unserviceable |
| Spitfire V | 8 + 4 unserviceable |
| Anson | 9 + 3 unserviceable |
| C-47 | 6–7 + 2–3 unserviceable |
| Dove | 2 |
| Harvard | 12 |
| Magister | 10 + 10 unserviceable |

[a]This list excludes the REAF's small number of C-46s and its handful of other assorted communications or transport aircraft.

9

# Fighting the Hydra (1948)

A UN-brokered truce between Israeli and Arab forces came into effect on 11 June 1948. The following month was a period of frantic activity for both sides, who tried to mobilize political support at the UN while at the same time strengthening their military positions. Yet the success of the Israelis far outstripped that of the Arabs, primarily because Britain, so recently the occupier and mentor of Egypt, Jordan, and Iraq, was unwilling to get more involved in the already horrifying Arab-Israeli situation. Consequently when fighting broke out again the military balance had shifted significantly in Israel's favor. A second round of fighting erupted on 8 July, and although it lasted only ten days, the air war intensified considerably. A second UN truce lasted much longer, although it was frequently broken both in the air and on the ground. During this second truce all sides worked hard to increase the strength of their air forces, but in this the Israelis were much more successful than the Arabs. A third relatively brief bout of fighting flared up in mid-October, this time engineered by the Israelis who felt that they were now in a position to deliver a decisive blow against the Arabs. Despite making considerable advances, the Israeli attack was far from decisive, and the third UN truce was accepted on 22 October.

## The First UN Truce

Israel continued to use every means at its disposal to acquire weapons, stockpile supplies, and recruit volunteers for its fledgling army and air force during the UN truce, while the Egyptians found it much more difficult to acquire new equipment, and quite impossible to recruit additional trained aircrew. The REAF was particularly hampered by the British and American arms embargoes. Back on 27 May 1948, Col. Frank

Ryan of the Association of British Aircraft manufacturers in Egypt had expressed the confidential opinion that the REAF would come to a standstill within two months because of this shortage of pilots, spares, and proper maintenance facilities.[1] Colonel Ryan proved to be wrong, but Britain was certainly still unwilling to provide arms to fuel a conflict that could, the British government feared, draw in its forces stationed in the Suez Canal Zone.

However, Italy was more sympathetic and agreed to supply Egypt with fighter aircraft. On 23 June the REAF ordered its first batch of Macchi fighters, eight MC205Vs and sixteen MC202s reconditioned to MC205V standard by the addition of more powerful DB 605 engines. The deal was negotiated by ʿAbd al-Raḥmān Ḥaqqī, an REAF officer well known for his ability in the business world. The first four Macchis were shipped from Italy on 26 September, followed by other deliveries in October and November to be reassembled at Heliopolis. Contrary to widely believed reports, a sabotage attack on the Macchi airfield at Venegono by Zionist terrorists from the Stern Gang did not delay deliveries. In fact one Italian Air Force MC205V was destroyed, another slightly damaged along with an MB308 light aircraft. The Egyptian planes were already at another Macchi airfield at Valle Olona ready for crating and delivery.[2]

The REAF's fighter pilots were, however, dubious about the Macchi MC205Vs which they knew to be reconditioned rather than new machines. So on 24 October Aer Macchi's chief test pilot, Comandante Guido Casetiato, who had flown as a fighter pilot during the Spanish Civil War, put on a remarkable aerobatic and low-level display in front of ʿAbd al-Munʿim Mīqātī, Ibrāhīm Abū Rabīʿah who was now in charge of all REAF fighter squadrons, the squadron commanders, and a fair number of operational pilots. By the end of that month the first four MC205Vs to be assembled and given desert camouflage were fully operational at al-ʿArīsh. A total of fifteen would eventually enter REAF service before the end of the Palestine War in 1949.

Even before the truce, some crates of REAF spares had ended up in Haifa accidentally. The captain of the cargo ship *Sonja* admitted that they fell into Israeli hands. Meanwhile Britain unknowingly supported REAF operations through weapons spares and ammunition taken from the many abandoned Second World War weapons depots in Egypt. The REAF had to rely on such dumps and RAF cast-offs since the arms embargo cut off official deliveries of practically everything else. In complete contrast, the IAF sometimes found it hard to digest the volume of assorted equipment that now flooded into Palestine. However, it was able to continue a rapid pace of expansion due to the assistance of foreign volunteers and continued embargo-busting operations.

Despite these difficulties the Egyptians were able to maintain their operational strength, particular attention being given to the repair of damaged Spitfires. A subsequent report by Air Commod. G. W. Hayes, the British air attaché in Cairo, portrayed these REAF efforts in almost heroic terms.[3] Toward the end of the war the Egyptians were renovating German 250 and 500 kg bombs that had been abandoned in the Western Desert and were using their own locally manufactured "canister type liquid fire bombs" or Napalm. In fact the REAF was never starved of bombs in quite the same way as the Egyptian Army was starved of shells.

The old Spitfire Vs of No. 6 Squadron were also sent to al-ᶜArīsh in July. This was probably to enable some of the front-line Spitfire LF9s to be sent back to Cairo for a complete overhaul. Meanwhile the Egyptian Army's Expeditionary Force in Palestine was enlarged to 18,000 men. Nevertheless, the Egyptians, and indeed the Arab regular armies taken as a whole, were still unable to achieve numerical tactical superiority over the Israelis.

On many occasions during the truce the REAF was called out in response to Israeli air and ground actions. In mid-June it was discovered that Israel was trying to smuggle arms and ammunition through the Egyptian lines to their isolated outposts in the Negev. On 25 June the REAF retaliated by bombing two positions on Hill 113 near Negba which had been lost to the enemy the day before the UN truce came into effect. The same day two Egyptian Spitfires, perhaps engaged in the same attack, also intercepted an unidentified white-painted Auster which they circled, then fired upon and forced to land. This Auster was actually a UN aircraft flown by Lt. Col. Maurice Martin of the USAF who was on his way to Haifa. It suffered only superficial damage, being able to take off again and complete its journey after the Egyptian Spitfires departed. The REAF later apologized to the UN Truce Monitor Corps and explained that the Spitfire pilots had believed the Auster to be one of the many such aircraft flown by the Israelis.

On 7 July King Fārūq turned up near the front-line on one of his periodic inspection tours and visited the REAF base at al-ᶜArīsh. This morale-boosting royal visit was part of a wider tour during which King Fārūq inspected several Egyptian front-line units in Palestine. On one occasion shots were fired from Israeli-held territory in the direction of the king's car. An unidentified aircraft, also believed to be Israeli, then overflew al-ʿArīsh during the king's visit, though this time no shots were fired.[4] While the Egyptian king still dreamed of being the ruler who recovered Palestine, the true situation was becoming grimly apparent to Fārūq's general staff. These senior officers could view the position as a whole and recognized that Israel's military power was growing faster than Egypt and the other Arab forces could ever hope to trim it down.

## Ten Days of Fighting

As the agreed truce period neared an end both sides were eager to renew the fight. Egyptian forces fiercely attacked Israeli troops at Negba at 6:00 A.M. on 8 July with the objective of driving southward into the Israeli-held Negev Desert. The IAF struck back aggressively by bombing Egyptian targets at Bayt Jibrīn, although a planned strafing raid on the REAF's al-ʿArīsh base by four Israeli Avias was held up by maintenance difficulties. Early on 9 June the IAF tried again, but one Avia crashed on takeoff. The others failed to find al-ʿArīsh which was covered by clouds, and so machine-gunned targets near Gaza and shipping moored off the coast. Egyptian Spitfire pilots still regularly carried the fight to the enemy, and on 9 July a solitary aircraft came in low to strafe the IAF's new Avia fighter base near Herzliyya, claiming to have damaged five aircraft on the ground. July 9th also saw a strange and one-sided confrontation near al-ʿArīsh that had a most unexpected outcome. Air Commodore Mīqātī, the most senior man in the REAF (which was still under the overall command of Army Major General al-Shaʿrāwī) spent much of his time commuting between al-ʿArīsh and Cairo. Fellow officers described him as an experienced pilot who had bent every type of aircraft the REAF ever operated. He was certainly not a man to lose his nerve.[5] Mīqātī normally flew a Lysander on such journeys, and on the morning of 9 July he once again took off from al-ʿArīsh. The base was, at the time, at "condition red" in expectation of an Israeli raid. Mīqātī continued:

> I had been advised to keep the radio on, but I was still nervous as I set out across the Mediterranean. Fortunately my gunner—I don't remember his name—was a keen-sighted man, and he spotted an Israeli Messerschmitt (Avia C210) as it maneuvered into a position to attack. Of course my Lysander

> was a very old kind of airplane, but I'd flown these for a long time. Still we were at a big disadvantage, and you'd expect that such a contest could only end one way. The pilot of the Israeli aircraft came up behind us. I told my gunner to open fire just before the Messerschmitt came into range, and I went down to about a hundred feet. Then the gunner fired, and I throttled right back. You know the Lysander can drop like a stone to land in a field—like they did when the RAF took spies in and out of France. The Israeli must have been concentrating on keeping me in his sights because he dropped his nose to follow. He overshot and flew right in, almost level with me. I honestly felt sick in my stomach and—I don't know why—I saluted him, with my hand like this (at this point the air vice marshal gave a snappy RAF style salute). Then we flew straight back to Cairo.[6]

The Israeli pilot is believed to have been Robert Weeckman, an experienced combat pilot who had clocked up sixty five missions and two hundred hours active service flying with the USAF in the Second World War. He is likely to have stalled his Avia C210, a notably difficult machine to control, while trying not to overshoot Mīqātī's Lysander.

The heat and dust of al-ᶜArīsh were hard on both men and machines, but despite maintenance difficulties and the rigors of combat, the REAF maintained a high sortie rate. Abū Zayd in particular flew numerous air patrol missions in defense of al-ᶜArīsh as well as ground attack sorties in support of the Egyptian Army. Between 9 and 11 July the REAF's bombing and ground support missions grew in ferocity, with a very large number of targets being mentioned in official communiqués. An Israeli aircraft that tried to intervene was claimed shot down, though this was denied by the Israelis. The REAF admitted the loss of one aircraft during these numerous sorties. The plane apparently force-landed within Israeli lines, but the pilot reportedly escaped into Egyptian-held territory after a shoot-out with a pursuing Israeli soldier.[7]

The IAF fought back, but its few fighters were now generally concentrating their sorties on the northern front against the Syrians, Iraqis, and Lebanese who had no fighters capable of opposing the Israeli Avia C210s. On 12 July the Egyptian Army made another assault on the key position of Negba with the 9th Infantry Battalion supported by tanks and artillery. Every available Spitfire from al-ᶜArīsh supported the attack carrying maximum bomb loads. These air raids were particularly troublesome on this occasion, as the Israeli brigade commander in Negba pointed out in the following message to GHQ: "We are heavily attacked along the entire front from the air and by artillery and infantry assault. It is vital that their planes be brought down and that concentrations of their vehicles between Bayt Affa and Iraq Suwaidan be attacked."[8]

Apart from the increasing danger of interception by Israeli fighters, ground attack work posed other problems and hazards. ʿAbd al-Majīd al-Rifāʿī recalled that on the day after he was promoted he led a flight of Spitfires against one fortified Israeli settlement that was giving the Egyptian Army particular trouble. Munīrah Kafāfī tells the story:

> Egyptian aircraft became well known for their dexterity in low-level flying. But this also meant the pilots were highly visible (to the enemy) and faced greater dangers. Yet they increased their raids in close support of the front-line garrisons even though they were constantly in danger of flying into the hills. The Egyptian high command got worried about this and eventually sent orders banning such low flying, and the men obeyed. But Lt. General ʿAbd al-Majīd al-Rifāʿī smiled and told me the following strange story: "Naturally my comrades and I were surprised by this order to fly at increased altitude. But then came the day when I was promoted to flying officer, on the same day that we went to attack a settlement. The Egyptian aircraft hit the position but had no luck with the results. I was about to return as the position seemed to contain no enemy soldiers, when an element of doubt crept into my heart. So I went down low, despite the warnings of our leaders. In fact the propeller of my aircraft almost hit a haystack, but what do you think I found? The whole garrison of the settlement had gone into shelters, and their shelters were hidden between sheaves of grain. So now the Egyptian aircraft returned to demolish the place and complete the mission assigned to us.[9]

On 14 July the air war took another turn when three Israeli B-17 Flying Fortresses, being smuggled out of Europe, flew to Israel via Cairo and dropped a small bomb load on the way. At 7:55 P.M. four bombs from the B-17s fell on a residential area near the royal palace where wounded soldiers had assembled for a feast arranged by King Fārūq to mark the *Ifṭār* religious festival.[10] Many civilian casualties were caused. The Fortresses flew again that same night and dropped seven tons of bombs on the al-ʿArīsh and Rafaḥ areas in by far the biggest air raid of the war so far, though they failed to hit the airfield. Most other reported air raids on Cairo were probably "faked" by Egyptian anti-aircraft defenses to show that the capital was well defended or to cover the movement of wounded from the front by imposing a blackout.[11]

The REAF was now assembling a proper bomber force of its own at Almāẓah. Initially it consisted of C-46 transports from No. 4 Squadron which Egyptian mechanics had converted to roll-out-the-door "bombers," larger versions of the modifications already carried out on several REAF C-47s. The C-46 pilots managed to make quite effective night raids, the first of which seems to have been against the Israeli outpost at Beʾerot Yiẓḥaq at 5:30 A.M. on 15 July. Raids by flights of four Spitfires then continued against this target until the light of dawn brought a full-scale ground assault by the Egyptian Army.

Air-to-air confrontations between Egyptian and Israeli fighters gradually grew more frequent. Early in the evening of 18 July two REAF Spitfires, Mk. VCs, and three Israeli Avias met near al-Majdal as each was returning from a ground attack mission, one of the Spitfires being flown by Wing Comdr. Saʿīd ʿAfīfī al-Janzūrī. The Avias found themselves in a good position behind the Egyptians and shot down the wing commander before the Spitfires were aware of the Israelis' presence, al-Janzūrī being killed. This was claimed by Modi Alon as his third victory, and ʿAbd al-Ḥamīd Abū Zayd now became leader of the REAF fighters at al-ʿArīsh.

By the end of July, Israel had seized control of almost all the UN-designated Jewish state in Palestine, plus a large part of the UN-designated Palestinian Arab state.[12] The inhabitants of most, though not all, of the Arab villages that had fallen to the Israelis then became refugees. Some were forcibly expelled, while some fled before the Israelis arrived. The Arab armies had in turn captured several Jewish settlements within the UN-designated Arab state and in the UN-designated international zone around Jerusalem. The one Jewish settlement within the UN-designated Jewish state to have fallen was Mishmar HaYarden, which the Syrians captured early in the war. In the deep south, the Negev Desert had also been allotted to the Jews but was now cut off from the rest of Israel by the Egyptian Army. The Egyptians, however, made no real attempt to take control of the Negev.

## The Second UN Truce

On 18 July the UN once again managed to impose a cease-fire in Palestine, ending what came to be known as the "Ten Days Fighting." The subsequent truce lasted for almost three months and was a period during which both sides tried hard to strengthen their air forces, acquiring new aircraft and much-needed spares. Wing Comdr. al-Malayjī's main concern at al-ʿArīsh was to disperse his forces and avoid another, more successful, IAF attack on the airfield. Anti-aircraft defenses were increased, and various satellite airstrips were cleared further south in the Sinai Desert.

The most controversial of Egypt's arms purchases during the Palestine War were Stirling Vs, intended as heavy bombers for a new No. 9 Squadron. These civilianized versions of one of Britain's biggest Second World War bombers reached Egypt by a devious route and were, in fact, among the first weapons to be "smuggled" to the country. When the purchase had first been discussed, however, there had been no official arms embargo operating in the Middle East.

The unarmed Stirling V civilian transport was originally designed for the invasion of Japan, and Britain built some 160 of these aircraft. A detachment of No. 48 Squadron RAF Transport Command flew them from their base at Cairo West until they were replaced by Avro Yorks in 1946. In May 1947 a dozen of the most serviceable Stirling Vs were then bought from the RAF by the British firm Airtech Ltd. and, by one means or another, four ended up in Egypt, the first arriving on 27 July 1948. Four others were acquired at a later date.[13]

Two days later the REAF's first Stirling was doing circuits and bumps at Almāẓah with an Egyptian crew supervised by foreign instructors. But these aircraft were trouble from the very start, both in their operations and in their servicing, and never fulfilled REAF hopes. Bomb bays and bomb doors remained on the transport versions, but bomb racks, release mechanisms, and any form of defensive armament were lacking. The problem of self-defense was never solved, although one Stirling was later fitted with a turret from an old REAF Anson. Only two received makeshift bomb racks built up from ex-RAF scrap for entirely different aircraft, and it seems that these Stirlings did make a number of operational bombing missions. Sq. Ldr. ʿAdlī Kafāfī, one of the REAF's most experienced and highly qualified multi-engine pilots, was responsible for getting these aircraft into operational condition, as Brig. Shafīq Ḥasīb later explained to Kafāfī's daughter: "It was typical of ʿAdli that he selected and completely overhauled the first of these aircraft which he then used as the leading bomber pilot. And it was on this type of aircraft that he carried out by far the largest number of night raids on Israeli settlements, all of which were successfully carried out. But the main problem was that these operations all had to be carried out from Cairo West airfield on the desert road from Cairo to Alexandria."[14]

The REAF was also desperate to acquire more fighter planes to support the overworked Spitfires and replace combat losses. Orders had already been placed for Macchi planes, and now the Fiat company agreed to supply others. The REAF ordered a first batch of seventeen Fiat G55/A1 fighters and two G55/B fighter-trainers, with an option on twenty of the more advanced G59s. In the event no G59s were purchased, although a further eight G55/A1s and two more G55/Bs were acquired. Most of the first batch were aircraft retrieved from dumps in northern Italy at the end of the Second World War and restored by Fiat. Some of the G55/A1s were newly built, while the G/55Bs came from the Italian Air Force Flying School at Galatina.[15] These Macchi and Fiat fighters were the finest Italian designs and could compete on equal terms with the Israelis' Avia and Spitfire fighters.

Unfortunately Fiat deliveries did not start until September 1948 and were so slow that the last of the first batch of G55s did not arrive until 1949. Seven ex-RAF Sea Otters had also arrived before the end of the year as naval reconnaissance aircraft for No. 4 Squadron, much to the consternation of the British Foreign Office which, until May 1949, seemed unable to find out whether the RAF had sold or merely loaned them to the REAF.[16]

The only other source of supply open to the REAF was the Egyptian civil fleet, and soon several privately owned aircraft were being used to support the war effort. One such plane, a De Havilland Rapide, was destroyed on the ground near Gaza on 22 September during one of the IAF's "unofficial" air strikes during the second UN truce. The same day an REAF Spitfire that challenged the Israeli attacks was shot down, though the pilot was unharmed. Israel was also suffering losses, and on 14 August an IAF light reconnaissance aircraft was forced down near Gaza. Its crew of two were uninjured, but a crowd of enraged Palestinian refugees would not have allowed them to remain that way if two Egyptian officers had not appeared on the scene. Despite the UN truce, the IAF also claimed to have damaged two Spitfires and a C-47 at al-ʿArīsh during an air strike by Beaufighters on 2 October.

During the second UN truce the IAF received more fighters including several Avia C-210s, six Spitfires, and four P-51 Mustangs. Bristol Beaufighters supposedly intended for a film company turned up in Israel, as did a continuing stream of other assorted aircraft, a supply that continued even after the fighting flared up once more. The supply of new aircraft and other weapons now tipped the military balance heavily in Israel's favor. Of course not all the aircraft being smuggled in from Europe reached Israel safely. During the war two Noorduyn Norsemen flying from southern Italy lost their way and arrived near Gaza at night. Thinking they were near Tel Aviv, they called for guidance. No response came over the radio, but runway lights were suddenly lit some way to the south. They landed—to find that this was al-ʿArīsh and they were now POWs in Egyptian hands.

On the ground, Israel built up its garrisons in the Negev, the deep southern area cut off by Egypt's control of the main east-west road between the coast and the Judean hills through Faluja. Since Egyptian forces held the roads, a full-scale airlift seemed to be the only answer. Israeli and volunteer aircrews flew a nightly shuttle service using almost all the IAF's C-46 and C-47 transports. Code-named "Operation Dust," the airlift moved troops and supplies into the Negev, starting on 22 August.

After a few nights one of the C-46s developed engine trouble and had to remain at the desert airstrip near Ruḥāamah during the hours of daylight. It did not take long for a pair of REAF Spitfires on their morning patrol to spot the stranded transport. The truce was still in effect, so they restrained from attacking, but the IAF's secret was out, and in retaliation the Egyptian Army barred ground convoys carrying nonmilitary supplies to the Negev.

Nevertheless, Operation Dust continued, and the new Israeli troops in the Negev were soon involved in some severe fighting, despite the supposed truce. The two sides fought to control a series of *talls*[17] at Khirbat al-Maghaz just east of the main road leading south from Faluja. In this struggle, fought out by Egyptians, Israelis, and local Palestinian militias, both sides suffered heavy losses. Low flying ground attack Spitfires of the REAF were also involved, but in the end the Israelis retained control of these strategic mounds.

## The Third Round of Fighting

By mid-October the Israelis were ready to attack again. The road south was still blocked, and they had not tried to send any convoys into the Negev for some weeks. This time, however, the Israelis tried a new tactic, and the date selected for this new offensive was carefully chosen. Not only was a UN truce in force, but the thirteenth and fourteenth of October were the important *ʿId al-Aḍḥá* Muslim religious festival. As a result there were several days of national holiday in Egypt with no one working, no newspapers being published, and with many Egyptian officers on leave.

Late in the afternoon on 15 October three groups of planes took off and headed toward prearranged targets at the REAF's main base at al-ʿArīsh, the Egyptian Army HQ at Gaza, and a forward Brigade HQ at al-Majdal—all to be hit simultaneously at 5:40 P.M. As the sun dipped low, a convoy of Israeli trucks moved south toward the crossroads west of the ʿIrāq Suwaydan police fort. When challenged they refused to halt and, as hoped, the Egyptians opened fire. One truck burst into flames, probably ignited intentionally, while the others withdrew. Their task was done—Egypt had technically broken the UN truce. Messages crackled over the radio to the IAF planes that were now approaching their targets: Attack!

The IAF aimed to drive the REAF from the skies and win complete air supremacy. Three IAF Spitfires, the type having been chosen to confuse Egyptian gunners on the ground, and two Beaufighters achieved complete surprise as the Egyptians thought that the UN truce was still in effect. No REAF fighters got airborne in time to intercept the attackers, and no anti-aircraft guns opened fire until the IAF Beaufighters dropped several bombs, one exploding in front of a hanger. The Israeli Spitfires then joined in by strafing several parked aircraft, claiming to have damaged between two and four REAF Spitfires. Egyptian gunners then opened up with a vengeance, damaging one Beaufighter which had to make a forced landing at Tall Nūf on its way home.

The Israelis had enjoyed another, perhaps unanticipated advantage during this first attack on the al-ʿArīsh air base. It was a legacy of the tragic Ramat Dawid incident of nearly five months earlier. The REAF, it appears, was still very concerned about possible unintentional clashes with the British RAF. As Ṣāliḥ Maḥmūd Ṣāliḥ explained:

> I was then in charge of air operations at the al-ʿArīsh air base, under the late Air Commander Col. Muḥammad Ḥāfiẓ Muḥammad, when we spotted the approach of three Dakota (actually two Beaufighter) aircraft. The aircraft started to take an attacking position, attempting to attack the base. I ordered the anti-aircraft force to engage with the targets as they entered the active zone of the airfield, but the base commander stopped the order, fearing that those aircraft might be British. I ordered the late Lt. Pilot ʿAbd al-Ḥamīd Abū Zayd to follow the aircraft and identify them before engaging with them. The base commander stopped this order too, until we were surprised by bombs aimed at the hanger. As it happened, the Israelis' poor aim saved the hanger, and the damage was trivial, but our ground-air defenses and our pilot Abū

> Zayd could have shot down the three (sic) aircraft very easily. . . . As for this attack on the al-ʿArīsh air base, there was no effective damage, neither in lives nor equipment, for the Jewish bombs were small and ineffective, falling between the hanger and the camp.[18]

Meanwhile IAF B-17s sent against the Egyptian brigade HQ at al-Majdal missed their target altogether, their bombs falling on the nearby Arab village of Jurah. The biggest of the planned Israeli air raids was that against the Egyptian HQ in Gaza. This consisted of four C-46 "bombers" and three C-47 "bombers," comparable to the REAF's own modified C-46s and C-47s, escorted by four Avia C210 fighters. These aircraft again failed to locate their target, the Avia escort being headed off by a flight of Egyptian Spitfires led by Abū Zayd. Meanwhile the IAF C-47s had been delayed, missed their rendezvous with the Avia escort and the C-46s, and dumped their bombs in Gaza harbor. For their part the IAF C-46s also failed to identify the Egyptian Army HQ and so settled for bombing railway lines at Khān Yūnus and Rafaḥ.

The Israelis' loss of one Beaufighter during these unsuccessful operations meant that the following day's ground assault on ʿIrāq al-Manshīyah had to go in without air support. This, plus a spirited defense of the village led by Major Jamāl ʿAbd al-Nāṣir (Nasser) foiled the Israeli attack. Three tanks of the assault force were destroyed, all the others suffering damage, while some Israeli units lost in excess of 30 percent casualties.

At 7:00 A.M. the following morning a single Israeli B-17 again attempted to bomb al-Majdal. At 9:30 A.M. three B-17s, without their planned fighter escort, attacked al-ʿArīsh once more. This time the REAF had a patrol of two Spitfires in the air, but their attention was diverted by an IAF reconnaissance Avia C210. The Avia claimed to have shot one down, but the REAF suffered no casualties on that day. On the other hand the Israelis lost one Avia flown by a foreign Mahal pilot, Maurice Mann. He was injured in a crash-landing near Lydda after his engine started to fail and his cockpit filled with smoke. Mann was convinced that he had been the victim of mechanical failure. The REAF, however, stated that it had shot down two enemy Messerchmitts (Avia C210s) on the sixteenth, "one crashing on land, the other into the sea."[19]

A second Israeli Avia was indeed lost that day, and with it Israel's best-known fighter pilot. One of the few positive results of the day's activities for the Israelis was that Egyptian front-line troops were now pulling back from Isdūd toward al-Majdal. Two IAF Avia C210s were sent to attack them. The Israelis reported no air combat, or at least none has been admitted in official IAF sources, but one aircraft spun into the ground as its approached the Herzliyya air base, Modi Alon commander of No. 101 Squadron being killed.

The Egyptians looked upon the events of 15–16 October with considerable, if perhaps somewhat premature, satisfaction. This was clearly reflected in Cairo newspapers where the exploits of the Egyptian Air Force had, for several months, been taking pride of place in reports on the war. The REAF had certainly recovered very quickly from the unexpected Israeli air raids, and its handful of aircraft at al-ᶜArīsh made more than twenty combat sorties on 16 October alone.

At the very time that the Israelis were making their carefully prepared all-out attack on Egyptian positions in Palestine and at al-ᶜArīsh, Group Capt. Ṣāliḥ Maḥmūd Ṣāliḥ of the REAF gave a "top secret" report on the Egyptian Air Force's current operational strength and serviceability to the British air attaché in Cairo. It was clearly a very inflated and optimistic assessment designed to impress the air attaché, who duly reported it back to London without comment.[20] The most likely reason for the REAF releasing this misleading information was in the knowledge that it would probably be passed to the Americans, and thus almost inevitably to the Israelis.

In reality the REAF was suffering from a chronic shortage of aircraft, spare parts, and ordnance. Combat losses, accidents, and maintenance problems brought on by the lack of spares had reduced the air force's numerical strength and its ability to sustain flying operations. Nevertheless, the pilots and groundcrews at al-ᶜArīsh continued a high sortie rate with the aircraft available. Other REAF units were also sent to this front-line base to maintain the pace of operations. Meanwhile the IAF's operational strength had risen to nearly one hundred aircraft, including about ten bombers, twenty-five fighters, fifteen multi-engined transports, and nearly fifty assorted light aircraft. Arab fire and accidents had, however, claimed the lives of more than a dozen IAF aircrew and many aircraft. The actual number of aircraft has not been released to this day.

The IAF continued to strike at Egyptian positions at Gaza, al-Majdal, Faluja, and now Beersheba, each being the target of around twenty sorties between 15 and 22 October. Al-ᶜArīsh suffered no less than twenty-nine raids, while Bayt Jibrīn, Isdūd, and ᶜIrāq Suwaydan were also bombed. This sustained offensive inevitably forced the REAF's small fighter force onto the defensive, although ground attack missions were still flown in support of hard-pressed Egyptian ground forces. REAF aircrew were not only outnumbered but faced a more experienced enemy. Egyptian Spitfire pilots were now ordered to avoid air combat except in defense of the al-ᶜArīsh base and to concentrate on ground attack missions, particularly in support of Egyptian forces increasingly cut off in the Faluja and ᶜIrāq Suwaydan area. This zone would soon develop into an isolated piece of Egyptian-held territory known as the "Faluja Pocket." Here Israel's high-flying B-17 bombers, escorted by Mustangs and Spitfires, regularly pounded a series of Arab villages, causing panic among the Palestinian refugees crowded into this area, many of whom attempted to flee east to the Judean foothills.

Israel's drive against the flank of the Egyptian Army in southern Palestine was the most dangerous from a strategic point of view. Here, where the main Israeli thrust was trying to reach the sea and encircle the Egyptian Expeditionary Force, the war would be won or lost. First the Israeli Army had to break through the line of Egyptian positions that linked their main forces on the coast with the Transjordanian Arab Legion in the hills of Hebron, in what is now the southern part of the Israeli-occupied West Bank. The Egyptian line ran from al-Majdal near the Mediterranean coast eastward though ᶜIrāq Suwaydan, Faluja, and ᶜIrāq al-Manshīyah toward Hebron.[21]

The Israeli advance succeeded in breaking through Egypt's line and linked up with what remained of Israeli forces in the Negev. REAF Spitfires supported an unsuccessful attempt to retake the vital ᶜIrāq Suwaydan crossroads, but at 8:00 A.M. on 17 October the main Egyptian force on the Mediterranean coast began its withdrawal southward from al-Majdal toward Gaza. Five thousand Egyptian troops were left east of the Israeli breakthrough in what would later become the Faluja Pocket where they held out until the end of the war. On the coast the retreating Egyptian troops had to run the gauntlet of heavy enemy artillery fire as they moved south between Israeli-held hills and a strip of impassable sand bordering the sea. At one point the shelling grew so intense that REAF Spitfires had to lay a smoke screen along the road, enabling most of the army including the 2nd Infantry Brigade HQ to get through unscathed.

The groundcrews at al-ᶜArīsh were now working flat out, stopping only to dive for their shelters when Israeli aircraft roared in to bomb or strafe the field. Yet as fast as they worked, there were never enough planes ready for the demands made upon Wing Commander al-Malayjī's little force. Further east another bloody battle was raging in hills overlooking the strategic Israeli airfield of Ruḥāmah. For seven days the slopes were taken and retaken by each side as the IAF and REAF both threw in all the planes that were available. On one occasion a short lull in the fighting elsewhere allowed al-ʿArīsh to send no less than eight Spitfires to hammer an Israeli HQ on the ancient *tall* of al-Najīlah, destroying many vehicles. Despite several heavy attacks by Egyptian troops, in the end it was the Israelis that held Ruḥāmah.

October 19th was a day of particularly heavy action for the REAF. The UN Security Council had again reconvened to discuss the Palestine situation. If the Israelis wanted to take complete control of the Negev and to break the Egyptian Army before a new UN cease-fire was called, they would have to act fast. As the Egyptians continued to retreat southward out of this trap, Egypt used every means to keep the coast road open. Some troops were also evacuated in small ships and fishing vessels. An Israeli naval force consisting of the corvette *Wedgewood,* the escort vessel *Haganah,* and the smaller gun-runner *Nogaw* then tried to cut this escape route by attacking Egyptian craft off al-Majdal.

The clash that followed provided another example of how the Israelis convinced themselves, and most of their supporters abroad, that they were fighting a far more powerful foe—that the "David and Goliath" story was being replayed in Palestine. On this occasion the lightly armed Israeli ship *Nogaw* found herself facing an Egyptian vessel, the *Amīr Fārūq,* at 11:00 A.M. As she turned to retreat northward to rejoin the other Israeli ships, the *Nogaw* reported that she was being attacked by an Egyptian corvette. The Israeli ships thereupon summoned air support. In reality the little old *Amīr Fārūq* had served as an escort vessel in Egypt's Coastguard and Fisheries Administration since before the Second World War. She was still part of this Coastguard and Fisheries Administration but had been drafted in to help Egypt's newly formed navy in which she was listed as a sloop because she was armed with a solitary 6 powder gun and four machine guns. As such the *Amīr Fārūq* was indeed the most powerful ship in the Egyptian Navy!

The Egyptian *Amīr Fārūq* now met the other Israeli vessels and called for air support, the Israeli *Nogaw* reportedly being strafed by REAF Spitfires around 11:30 A.M. Earlier that morning Sq. Ldr. ʿAbd al-Ḥamīd Abū Zayd had led a formation of Spitfires to attack a concentration of Israeli troops threatening Gaza from the direction of Beʾerot Yiẓḥaq. At

11:20 A.M. he landed back at al-ᶜArīsh and promptly volunteered to fly to the assistance of the Egyptian ship. No spare aircraft were then available, although the solitary Hawker Fury and some Spitfires were being refueled and rearmed. At 11:50 two Spitfires took off and headed north, while Abū Zayd and his wingman fretted around the Fury and another Spitfire. The first two REAF Spitfires located the Israeli ships at approximately 12:10 P.M. and immediately attacked the *Haganah.* Although their bombs just missed, the concussion and splinters damaged the vessel, killed one sailor, and injured others. The *Haganah* fired back with machine guns, and its crew claimed to have shot down a Spitfire. In fact both returned safely to al-ᶜArīsh.

The *Amīr Fārūq* escaped in the confusion, but now other aircraft appeared on the scene, an IAF Beaufighter closely followed by Abū Zayd and his wingman in their Fury and Spitfire. The two Egyptians made one bombing run at the *Haganah* at 12:52 P.M., causing more damage and casualties. Then Abū Zayd turned to engage the Israeli Beaufighter. Maneuvering violently at masthead height, the two planes veered away from the ship. In desperation the IAF pilot dived low over the water and turned abruptly toward the coast. Abū Zayd followed, but his wingtip seems to have hit the water. The REAF Spitfire pilot did not see his leader crash. Nor did the crew of the Egyptian ship *Amīr Fārūq*. So the REAF operations diary of that date included only the following terse statement: "The Fury aircraft no. 701, flown by Squadron Commander Effendi ᶜAbd al-Hamīd Abū Zayd, has not returned back. The pilot had taken off to carry out raids to protect our Egyptian vessels. Now that the mission has been fulfilled, the aircraft and the pilot are announced missing in action as from today."[22]

It had been Abū Zayd's seventy-second operational mission since the war started on 15 May 1948, and he had fallen in combat with one of the IAF's most experienced pilots, Leonard Fitchett, who was himself shot down and killed the following day. The loss of Abū Zayd was, according to Air Vice Marshal Shafīq Ḥasīb, a serious blow to the pilots at al-ᶜArīsh.[23] He had shot down one Israeli aircraft confirmed during his brief operational career, while another was confirmed destroyed on the ground, and had several "probables" to his credit both in the air and on the ground. Another experienced REAF pilot was killed on 21 October, Flt. Lt. Mukhtār Maḥmūd Saᶜīd's Spitfire being shot down by an American volunteer near the frontier town of Rafaḥ.[24]

Egyptian personnel at al-ʿArīsh fought back despite these losses, and the REAF continued to attack the enemy, hitting Israeli positions at Dorot and Nirʿam in a determined attempt to relieve pressure on the vital coastal road as the Egyptian Army gradually withdrew. Supplies were now pouring into Israel by sea and by air. Nevertheless, the Egyptian government was still so sensitive to international opinion that it would allow no attacks on the main Israeli port of Haifa, much to the annoyance of the REAF. The situation was now growing daily more difficult for the Egyptian Army. In the Faluja Pocket the vital ʿIrāq Suwaydan police fort fell, and its defenders withdraw to Faluja and ʿIrāq al-Manshīyah. Here supplies of everything were running desperately low. This shrinking island of Arab territory was also crammed with tens of thousands of terrified Palestinian refugees—men, women, and children.

By now the UN's representatives were again trying to negotiate a cease-fire, and this time Egypt was happy to see the fighting stop. The Israelis, however, pressed forward aggressively, trying to take as much territory as possible before any cease-fire. Since fighting on the Transjordanian, Syrian, and Lebanese fronts had died down, the Israelis could concentrate almost all their efforts in the south. The Egyptian Army retreated steadily out of the trap of al-Majdal, but, after a tremendous effort, the Israelis cut their retreat by occupying the village of Bayt Hānum west of the coastal road on 21 October. Still the Egyptians would not give up. While REAF aircraft bombed and strafed the new Israeli positions, Egyptian Army engineers laid a new road of wooden logs and wire netting across the sand dunes between Bayt Hānum and the shore.[25]

At the UN Security Council meeting in Paris, the Israeli representative fought for time to enable the Israeli Army to complete its conquest of the south. British, Belgian, Chinese, and Syrian delegates pressed hard for a speedy cease-fire. This eventually came into effect at 3:00 P.M. on 22 October, with the British representative urging that Israeli forces be obliged to return to the positions they had held on 14 October. Despite the cease-fire, however, some Israeli units under Yigal Allon continued pushing further into the Judean hills toward Hebron. There they seized the Arab villages of ʿAjjūr and Bayt Jibrīn, finally completing the isolation of the Faluja Pocket.

By the time the new UN truce came into effect, the REAF was clearly in a very difficult situation. The problems of serviceability, lack of spares and insufficient trained groundcrew were acute. On the other hand, Egyptian pilots were gaining in experience. There were around twenty

Spitfire LF9s still operational, but only half of these were in a proper serviceable condition, plus a dozen time-expired Spitfire Mk. Vs. Operational support consisted of less than twenty-five transport and communications types, although the great majority of these were still serviceable. Other aircraft appear to have been held in store, but Egypt simply did not have enough pilots to fly them.

The Arab regular armies had, in reality, been on the defensive since the end of the first UN truce. But from now on they would find themselves not only outnumbered but outgunned both on the ground and in the air. The Israeli armed forces that had returned to the attack in June 1948 were no longer the desperate, ill-trained, and ill-equipped militias of a month before. They had tanks and many other armored vehicles, abundant if ill-assorted artillery, and above all a large, reasonably well-equipped, and certainly well-trained air force. The myth of an overwhelming Arab "Goliath" threatening a tiny Israeli "David" might persist for many decades but no longer had any truth in it, if it ever had. Meanwhile the unfortunate indigenous Palestinian militias of the Arab Liberation Army had become an irrelevance, at least in military terms.

The Arab armies would spend the rest of the war trying to save as much of the UN-designated Arab state of Palestine as they could. This they would do with greater success than has generally been recognized. The survival of any Arab areas in Palestine owed a great deal to the determination shown by outclassed, outgunned, and now outnumbered Arab air forces. In the south the Egyptians saved part of the Arab coastal region in the form of what is now known as the Gaza Strip. In the center the Transjordanian Arab Legion hung onto what has become the West Bank. Only in the north did the Lebanese and Syrians fail, despite hard fighting on the part of the Syrians if not the Lebanese, to save any of UN-designated Arab Galilee. All this region fell to the Israeli Army who, of course, regarded such a conquest as an act of "liberation."

## Operational Strength Claimed by the REAF in October 1948[26]

| Unit | Aircraft | Strength | Comments |
|---|---|---|---|
| No. 1 (Tactical reconnaissance) | Spitfire LF9 | 16 | Serviceability still high (up to 100) with about 8 aircraft in reserve for Nos. 1 or 2 Squadron |
| No. 2 (Fighter) | Spitfire LF9 | 16 | |

*Continued on next page*

*(Continued)*

| | | | |
|---|---|---|---|
| No. 6 (Fighter) | Spitfire LF9 | 16 | No. 6 Squadron Spitfires were Mk. V and were withdrawn for coastal patrol work in October |
| No. 3 (Communications) | Doves | 6 | |
| | Ansons | 9 | |
| No. 4 (Bomber) | C-46 | 16 | Converted C-46 "bombers," could carry up to 8,000 lbs. of bombs |
| No. 5 (Fighter) | Fiat and Macchi | 4 | Receiving Italian aircraft |
| No. 7 (Transport) | C-47 | 8 | |
| | C-46 | 2 | |
| No. 8 (Bomber) | C-47 | 10 or 12 | Converted C-47 "bombers" |
| No. 9 (Heavy bomber) | Stirlings | 6 | |
| No. 10 (Heavy bomber) | — | — | Awaiting Halifaxes recently ordered |

10

# A Losing Battle (1948–1949)

The fourth and final phase of the Palestine War was catastrophic from the Arab point of view. Israel focused almost all its military effort on the southern front and nearly succeeded in driving the Egyptian Expeditionary Force right out of Palestine. Only the fierce resistance of Egyptian troops besieged in the Faluja Pocket and the determination of the REAF not to give up the unequal struggle saved the situation. Even these heroic efforts might have been in vain if the Israeli Army and Air Force had not swept deep into Egyptian territory in the Sinai. Such an invasion of Egypt made military sense, as part of a wide flanking movement designed to surround the Egyptian forces still inside Palestine, but it was politically disastrous for Israel. Egypt asked for British assistance under the Anglo-Egyptian treaty, and Britain realized that it would have to support Egypt or see its prestige collapse throughout the Arab world. The British government also felt that its own military bases in the Suez Canal Zone were now threatened. As a result, the last few days of the Palestine War saw British and Israeli aircraft in combat over northeastern Sinai.

Britain now put its weight behind another UN cease-fire, and in January 1949 the Palestine War came to an end. The Arab armies and air forces were defeated in this first full-scale Arab-Israeli war, but they had managed to save a considerable part of the UN-designated Arab area of Palestine from Israeli conquest. These remaining zones were recognized by the international community as Arab rather than Israeli, and remain as such to this day, despite falling under Israeli occupation in 1967. Nevertheless, the proposed Arab state of Palestine did not come into existence. Instead the West Bank was annexed by Transjordan to form the Kingdom of Jordan, and the Gaza Strip was placed under Egyptian military rule, pending a solution to the Palestine problem.

## A Phony Truce

The third UN cease-fire in Palestine lasted several months, but by the time the final stage of the conflict broke out late in December 1948 there had been a fundamental shift in the balance of power between Arabs and Israelis. Arab military strength had been something of an illusion from the start, and now Israel's domination could no longer be denied. In the air the REAF found itself facing a ruthless enemy who enjoyed overwhelming advantages in equipment, numbers, and training.

The UN cease-fire held in the north of Palestine, but at Faluja, in a pocket of Egyptian-controlled territory, refugees and soldiers both planted crops in anticipation of a long siege. Medical supplies were desperately needed, and the only doctor was operating with razor blades. The situation in the medical wards was also precarious since there was an almost total lack of drugs. On 27 October an Israeli commando battalion captured the Bayt Jibrīn police fort, and the Faluja Pocket was now totally cut off. Meanwhile back in Cairo the Egyptian War Ministry, with an admirable show of confidence, announced that it had ordered nineteen Gloster Meteor F4 jet fighters to be delivered when the arms embargo was lifted.

Morale was still surprisingly high in the Egyptian forces, despite the impossible situation in which both army and air force now found themselves. Professional pride and native Egyptian stubbornness kept officers and men at their posts. This attitude was best summed up by Col. Sayyid Ṭāhā, commander of the Faluja garrison, when he met Yigal Allon, CO of the Israeli southern front at Gat. After Allon had summed up the military situation, Ṭāhā replied: "I cannot delude myself that by continuing my stand I will be able to change the balance of forces and save the situation on our front. However, there is one thing that I can save and that is the honor of the Egyptian Army."[1]

The REAF did a great deal to help the Faluja garrison hold out right to the end of the war. Soldiers and civilians were on minimum rations. At one time food stocks declined to a point where there was only enough to feed a thousand men for ten days, but for three weeks the REAF's Spitfires defied both the truce and intensive concentrations of Israeli anti-aircraft guns around the pocket to fly in urgent supplies of food, seeds for planting, medical supplies, ammunition, and cigarettes. The common soldiers' priorities were clear in the pages of Sayyid Ṭāhā's diary: "October 30. At 10:00 P.M. six Egyptian planes appeared and dropped their packages, accompanied by the triumphant cries of residents and soldiers. The visit of the planes did more than anything else to lift our morale. The Jews were taken by surprise. The dispatch included medicines and drugs in addition to ammunition and cigarettes. The soldiers were more interested in the cigarettes than the ammunition."[2]

Four of Egypt's new Macchi MC205V fighters were now stationed at al-ʿArīsh and newly cleared dispersal airstrips of beaten earth in eastern Sinai , but it was the ever reliable Spitfire LF9s that undertook these hazardous missions. Supplies were packed into empty drop tanks. The planes then came in at 7 meters maintaining a speed of 295 kph to release these tanks over the fields outside Faluja village. Although dropped without parachutes, the carefully packed drop-tanks seldom burst and their contents usually survived the delivery. Many of the Spitfires were damaged by ferocious Israeli anti-aircraft fire while making these drops, but none were lost. REAF teams were now searching the Western Desert for abandoned Second World War drop tanks, and when this supply ran out, REAF engineers at Almāẓah and al-ʿArīsh made their own.

As Israeli ground fire grew increasingly dangerous, daylight missions by Spitfires were halted in favor of a large-scale supply drop by a C-47 flying in during that short period of Middle Eastern twilight just after sunset. But the C-47 had to come in so low and at a moment of such poor visibility that the idea was abandoned as too dangerous after only one operation.[3] So Spitfires took up the task again, often protected by newly acquired Macchis. The supply effort was also supplemented by a secret camel caravan organization set up in liaison with the Arab Legion in the Judean hills west of Faluja.

The UN truce turned out to be little more than a fiction over the Mediterranean Sea as well. On 4 November ʿUmar Shakīb, CO of No. 3 Communications (Comm.) Squadron, met a pair of IAF fighters flown by Mahal volunteer pilots Boris Senior and Rudolf Augerten who were on a reconnaissance patrol near al-ʿArīsh. ʿUmar Shakīb continued the story:

> On the morning of 4 November 1948 I left the main air base at Almāẓah in an unarmed Dakota (C-47), serial number 807, with fifteen officers from the general staff, to fly directly to al-ʿArīsh air base. My aircraft hadn't recrossed the eastern Sinai Mediterranean coast when I was attacked by two Israeli fighter aircraft. They shot at my aircraft with their guns. I was hit in my face and my hand, and also a fragment lodged in the wall of my heart. Although I had lost a lot of blood, I aimed my aircraft directly toward the al-ʿArīsh base so as to land—such is a pilot's determination to survive! And we were worried that the base's anti-aircraft defenses would fire at us and hit us. But I was able to keep control of myself and fly almost at ground level to the base where I made

> a wheels up landing. I couldn't take my hands form the controls until we stopped skidding, but once all the passengers had clambered out, I lost consciousness. They took me to the base hospital for two emergency operations, after which I was taken on a train for the wounded to the injuries hospital at al-Jisā, which was then being used specifically for wounded soldiers. . . . The days passed, and my condition improved. Then Dr. Ḥasan Ibrāhīm, son of the late Dr. ʿAlī Pāsha Ibrāhīm, arrived and carried out another operation on me to remove the fragment that was lodged in the wall of my heart.[4]

To the men of the REAF it sometimes seemed as if the Israelis interpreted a UN cease-fire to mean that the Arabs should "cease-fire" and not even take the air, while the Israelis should be allowed to fly, fight, and do whatever they pleased. Some members of the UN cease-fire observation teams also got the same impression.

Wing Comdr. ʿUmar Shakīb's No. 3 Comm. Squadron never received the credit it deserved during the 1948 Palestine War. It had continued to ferry troops and supplies from the Nile Delta to the main Egyptian base at al-ʿArīsh on the far side of the inhospitable Sinai Desert, despite its aging aircraft and chronic lack of spares. Egyptian C-47s and C-46s sometimes are also said to have taken supplies up to Egyptian troops inside Palestine itself. Since the pilots of No. 3 Squadron were among the most experienced in the REAF, and were the only ones qualified to fly multi-engined aircraft, they also had to undertake night bombing missions in those C-47s and C-46s that had been converted to "roll-out-the-door" bombers. The pilots later mentioned by surviving colleagues were ʿAbd al-Laṭīf Baghdādī, Shafīq Ḥasīb, Muḥammad ʿAdlī Kafāfī, ʿAbd al-Bakr Mursī, ʿUmar Shakīb, Ḥalīm Ṭāhir Zakī, and several known only by their first names—Abū Bakr, ʿArabī, Najāt, Saʿīd, and Zaqīlah. These may also have included some civilian pilots from the Egyptian airline, Misrair. On 20 November the IAF shot down a high-flying British Mosquito reconnaissance aircraft of No. 13 Squadron over the Mediterranean.

While Israeli Air Force strength continued to grow, the REAF's stock of combat-weary ex-RAF aircraft was deteriorating rapidly. Spares were almost completely used up, and damaged aircraft had to be cannibalized to keep others operational. The new Italian Macchis and Fiats were arriving very slowly, but on 22 November the first Fiat G55 was ready so that it could be tested against one of the REAF's Macchi MC205s during the official hand-over ceremony.[5] Problems had developed with the air compressors on some of the REAF's Macchi MC205Vs, but fortunately these could be replaced by compatible parts from Spitfires, this modification subsequently becoming standard on Egyptian Macchis. Meanwhile

Macchi technicians and delivery pilots flew to the Egyptian forward air base at al-ʿArīsh several times, perhaps prompting an unfounded rumor that the Arab air forces were recruiting "ex-Luftwaffe aces." But this was not without its dangers, an REAF Dakota with Italian engineers on board being hurriedly recalled to Almāzah on 7 December because Israeli fighters were reported over northeastern Sinai. The Fiat team seems to have been far slower in getting its aircraft operational, perhaps reflecting the Fiat company's greater concern for the international arms embargo. By December only their two-seater Fiat G55B had yet flown. The only other reinforcements that the REAF received was a flight of three Hawker Furies, plus their pilots, seconded to Egypt by the Iraqi Air Force. Two were held back to defend Cairo, while one was sent on rotation to al-ʿArīsh to replace the Fury prototype in which Squadron Leader Abū Zayd had died.[6]

The REAF's small bomber force also suffered a serious loss during November 1948. On 11 November Wing Commander ʿAdlī Kafāfā returned to Almāẓah from a two-day tour of duty flying C-47s at al-ʿArīsh. He found some Stirling bomber pilots practicing emergency procedures. There had already been a series of mechanical problems with these aircraft, and they were profoundly mistrusted by their crews. Hoping to restore the confidence of his men, Kafāfī took up one of the Stirlings with Sq. Ldr. Muṣṭafá Ṣabrī ʿAbd al-Ḥamīd, but they had not completed a circuit before the aircraft simply blew up in the air, both men being killed. The reasons for this disaster were never discovered, although rumors of enemy sabotage naturally circulated.[7]

## Israel Invades Egypt

The Israeli Army was by now poised in strength near the Egyptian frontier. Everyone knew that it would soon strike another blow, and this came on 22 December with "Operation Ḥorev." The Israelis wanted to obliterate the REAF and push the Egyptian Expeditionary Force right out of southern Palestine. On the ground they achieved almost total success, driving a ferocious left hook well into Egyptian territory. Israeli forces seized the frontier village of al-Awjah, the REAF's dispersal strip at Biʾr Lahfān, and finally an outlying part of the main al-ʿArīsh air base. As a result the REAF had to evacuate the entire base. The Israelis moved so fast that a grounded civil De Havilland 89A Dragon Rapide fell into their hands along with an unserviceable Spitfire LF9 which the Israelis then towed away. This aircraft was, however, severely damaged when its tail fell off the Israeli towing truck. One of the main reasons for British RAF reconnaissance aircraft entering the war zone a few days later was, apparently, to locate this captured Spitfire.

In the air the situation was rather different. Despite their now overwhelming numerical and qualitative advantage, the IAF's attempt to drive the REAF from the skies failed. The IAF's strength stood at more than one hundred aircraft, including fifty bombers and fighters, compared with Egypt's thirty or so assorted aircraft, only half of which were fighters. The Israeli plan had been to wipe out the REAF on the ground at al-ʿArīsh. Wing Commander al-Malayjī had, however, dispersed his valuable operational aircraft in expectation of just such an attack and in readiness for the moment when the REAF could hit back. Heavy Israeli air attacks only destroyed one Macchi MC205V and a C-47 at al-ʿArīsh. The Macchi's pilot, Shalabī al-Ḥinnāwī, was, in fact, caught on his final landing approach with wheels down by an Israeli Spitfire. Despite terrible wounds, al-Ḥinnāwī made a crash-landing next to the runway.[8] But with so few combat aircraft available, there was little the REAF could do to alter the outcome of the battle. Nevertheless, even the Egyptians' loss of their main forward base at al-ʿArīsh did not stop the REAF striking back rapidly, though with only limited success.

Most Egyptian fighter pilots still preferred to fly familiar Spitfires rather than their newer Italian equipment. The Macchis were maneuverable and fast, but REAF pilots had little time to familiarize themselves with these new aircraft, and they also criticized the Italian aircrafts' unreliable Daimler-Benz DB 605 engines. Nevertheless, Egyptian pilots fought back courageously, bombing the advancing Israeli troops whenever possible and even attacking enemy airfields, including the main IAF fighter base at ʿĀqir on 23 December. Egyptian ground fire also claimed an Israeli aircraft over the Faluja Pocket on 22 December.

Meanwhile the REAF took on the IAF's Spitfires, Mustangs, and Avias whenever possible, although Egyptian air combat claims have been denied by the Israelis. On 28 December, for example, the REAF claimed the downing of two Israeli aircraft, with two others damaged, for the loss of one Egyptian machine.[9] An IAF reconnaissance Piper Cub, flown by Zvi Ziebal, was in fact shot down by a flight of REAF Spitfires over its own base at Beersheba on the twenty-eighth, while an Israeli Avia fighter was seriously damaged in a large air battle over the Faluja Pocket. The

Egyptian loss was a Macchi flown by Pilot Off. ʿAbd al-Fatāḥ Saʿīd, who was killed by the IAF Mahal pilot J. Doyle flying a Spitfire during an IAF air strike against the dispersal strip at Abū Awayjilah on 28 December. The next day the REAF again attacked advancing Israeli ground forces and claimed to have destroyed three Israeli aircraft without loss. In fact no planes appear to have been shot down by either side on 29 December, though there was considerable air action. On 30 December the REAF claimed two more victories without loss, these being denied by the Israelis, but during a large IAF attack on the Birʾ Hama airstrip on 31 December Sq. Ldr. Muṣṭafá Kāmil ʿAbd al-Wahhāb and Flt. Lt. Khalaf al-ʿArūsah, both said to be flying Macchis, were shot down and killed as they tried to take off.[10] A third Macchi claimed by the Israelis was possibly damaged.

From then until the end of the war the REAF issued daily communiqués which basically repeated that Egyptian aircraft had attacked the advancing Israeli forces, had carried out "reconnaissance in depth," and had suffered no losses.[11] An Israeli P-51D Mustang claimed to have shot down a REAF aircraft over Sinai on 5 January, but the Egyptian pilot may have bailed out safely as the REAF suffered no casualties that day. Meanwhile the only loss officially admitted by the Israelis during these last days of fighting was the IAF Piper Cub shot down on 28 December.

Everything was now being thrown into the unequal fight as the REAF struggled to save the army from encirclement. Improving standards of combat flying by the REAF may also have contributed to the rumor in the IAF that ex-Luftwaffe "aces" had enlisted in the Egyptian ranks. It was, of course, completely untrue. The Stirling heavy bombers even flew occasional daylight missions, including an attack on Mishmar HaʿEmeq, next to the major IAF base at Megiddo in northern Israel, on Christmas eve. Another four-engined Egyptian aircraft attacked one advancing Israeli column, making no less than six passes and escaping unscathed. It was almost certainly a Stirling, though by this time Egypt's first Handley-Page Halifax C8 may have been operational.

At the very end of 1948 the Israelis made yet another unsuccessful attempt to crush the Egyptian battalions holding out at Faluja. The REAF played its part in the defense of the pocket, and a dogfight involving many aircraft developed over Faluja village. This time the outcome was more balanced and may even have been to the Egyptian pilots' advantage. The Israelis claimed to have shot down one REAF Spitfire, although the only Egyptian aircraft reported lost on 28 December was a Macchi over Abū Awayjilah in the Sinai peninsula. An IAF Avia C210 also fell away from Faluja trailing smoke. This may have been the aircraft of No. 101 Fighter Squadron which Israeli sources unofficially admitted losing during the final phase of the Palestine War, without supplying a precise date.

The Arabs still took the fight to the Israelis, and on 2 January 1949 the outskirts of West (New) Jerusalem were bombed by a single plane. Four days later Egypt agreed to begin negotiations for an armistice, though clashes in the air continued right up to the final hours. On 7 January the fighting ceased. That same day a flight of REAF Macchis and possibly also Spitfires intercepted Israeli P-51s which were escorting an IAF Harvard which was in turn was dive-bombing Egyptian positions at Dayr al-Ballāḥ. A fierce dogfight followed in which aircraft hits were claimed by both sides.

The IAF's dominance could no longer be challenged by the last week of the war, apparently not even by the RAF. Political leaders in London considered that Israel's advance into Egypt posed a threat to the British bases and airfields in the Suez Canal Zone, which was now within IAF range. So additional RAF fighters, bombers, and reconnaissance aircraft were brought in, while the RAF's No. 205 Group set up a Joint Fighter and Anti-Aircraft Operations Room to coordinate the defense of the Canal Zone.[12]

The Egyptian minister of war, Ḥaydar Pasha, was fully aware of the serious situation in Palestine and summoned the British air attaché on 31 December to ask for British help.[13] The REAF had just been forced to evacuate al-ʿArīsh and, apparently, some dispersal airstrips in northeastern Sinai. Egyptian aircraft now had to operate at extreme range from bases west of the Canal Zone. To the minister's obvious relief, the air attaché, Air Commodore Hayes, told him that Britain would provide refueling facilities at RAF airfields in the Canal Zone and would now allow Egyptian pilots to land if their aircraft were at risk from serious damage or loss. Such facilities were desperately needed, as Hayes made clear in his subsequent report: "It is recorded that several Royal Egyptian Air Force aircraft landed on taxi strips of disused airfields having insufficient fuel to land at an airfield in use, and others having landed normally could not taxi away as they were dry of fuel."[14]

These included Macchis, Fiats, and Spitfires. The problems faced by existing Egyptian aircraft operating from bases west of the Canal Zone had been apparent as early as 12 June. Then three REAF Spitfires had been obliged to land at the RAF base of Fā'id, short of fuel following a patrol to check Israeli observation of the first UN truce. Not that the British gave favors for nothing. They had earlier permitted Egyptian personnel to operate, on loan and under RAF supervision, three Type 63 early warning radar sets in the Canal Zone both to supervise RAF training activity and to check for Israeli attempts to hide among British aircraft and thus penetrate Egyptian airspace. In return Egypt had relaxed some of the flying restrictions imposed on British night flying near the Canal Zone.

The Egyptians had, in fact, only asked for the loan of two such radars because Israeli aircraft were already making long-range reconnaissance flights east and west of the Suez Canal. This time it was the British Foreign Office, probably concerned about the political as well as military impact of such flights, that urged the RAF's Middle East Command to agree to the Egyptians' request. At the same time the REAF urgently asked for long-range fuel tanks so that their Spitfires could reach the battle zone from bases west of the Canal.[15] On 5 January Britain also allowed the REAF to take over a small disused RAF airfield at al-Ballāḥ in the northern part of the Canal Zone which now became the REAF's Operational HQ.[16] The REAF's two fighter squadrons, No. 2 equipped with Macchi MC205s and No. 5 with Fiat G55s, were then based at al-Ballāḥ. Only the Macchis were yet operational, although two Fiat fighters reached operational status just before the fighting ended. According to some reports, No. 1 Tactical Recce Squadron was also stationed at al-Ballāḥ for a short while.[17]

London was so concerned about the Israeli incursion into Sinai that the chief of air staff ordered the RAF commander in chief, Air Marshal Sir William Dickson, to monitor the situation with a series of photo-reconnaissance flights. This information would then be passed on to the UN truce supervision teams.[18] The potential for confusion was clear. There were, for example, two joint RAF-REAF reconnaissance missions over northeastern Sinai. In each case the RAF Spitfires took off at dawn before making their rendezvous with REAF Spitfires. Together these aircraft established that the Israelis had reached Abū Awayjilah 17 miles inside Egyptian territory.[19]

The UN-negotiated cease-fire was due to come into effect at 2:00 P.M. on 7 January. That day the RAF ordered another reconnaissance mission by four Spitfires of No. 208 Squadron, this time without REAF participation. Egyptian Spitfires of No. 1 Squadron were, however, on a ground attack mission against an Israeli column in the same area around the same time. Not only that, but five RAF Tempests from No. 213 Squadron were also flying over northeastern Sinai. The RAF Tempests saw the REAF Spitfires make an apparently very successful strike against Israeli Army vehicles, leaving great columns of black smoke in their wake, but did not of course take part. IAF fighters summoned to protect the Israeli column arrived just after the Egyptian Spitfires and British Tempests had both departed, and found themselves above three RAF reconnaissance Spitfires from No. 208 Squadron. These had arrived only minutes before and were now circling around the parachute of a fourth 208 pilot whose aircraft had been brought down by Israeli ground fire. As they watched their comrade's safe descent they were jumped by the Israeli Spitfires. At this time British and Israeli Spitfires not only had the same camouflage, but some even had the same red airscrew spinners. Clearly surprised and confused, all three British aircraft were shot down within seconds. Two pilots bailed out safely, while the third was killed.

That afternoon, with the cease-fire either due or technically already in effect, seven British Tempests from No. 213 Squadron and eight from No. 6 provided cover for six Spitfires from No. 208 who were sent to look for their missing comrades Again the group was bounced by Israeli Spitfires, and again confusion was caused by the similarity between IAF and No. 208 Squadron aircraft. One Tempest of No. 213 Squadron was shot down. The guns of most of the British fighters were not cocked, presumably because of the cease-fire.

The reaction of Egyptian pilots to these incidents was naturally one of sympathy. But there was also amazement that the British simply stopped the reconnaissance flights and did not retaliate against the Israelis.[20] In reality, of course, there was a great deal going on at a high political level with the British threatening to intervene militarily by evoking the 1936 Anglo-Egyptian treaty if the Israelis did not withdraw. Israel clearly took this threat seriously and eventually evacuated northeastern Sinai. The Egyptians were also able to retain control of the small section of Palestine still known as the Gaza Strip.

## The REAF's Performance over Palestine: An Analysis

The role and achievements of the Egyptian Air Force during this first full-scale Arab-Israeli war were naturally analyzed in detail in both Egypt and Israel. The only British summary of the REAF's performance to have been released was that drawn up by Air Commod. G. W. Hayes, the air attaché in Cairo throughout the Palestine War. This effective and objective document is a mixture of damning criticism and considerable praise.[21] It was also drawn up by a man who not only knew what he was talking about but was personally acquainted with the men involved. Above all it was written without the deep anti-Arab prejudice that has debased so many subsequent accounts of the air war over Palestine in 1948 (see Appendix 2 for extensive extracts from Air Commodore Hayes' report).

The REAF, unlike the Egyptian Army, was eager to continue the fight and clearly did not regard itself as a beaten force. Yet there was no denying Egypt's defeat. Israel's true strength had been brutally demonstrated to all its Arab neighbors, and no one saw this more clearly than the officers and men of the REAF. The determination they had shown over the past eight months had impressed many observers, including King Fārūq who told the British embassy in Cairo that Egypt should be proud of its air force. The British air attaché was more selective in his praise, noting the REAF's high morale, initiative, and keen desire to enter battle. By local standards, he said, the air force was efficient and had given valuable support to Egyptian ground forces as well as carrying out independent action. He made particular note of supply dropping missions by Egyptian Spitfires to the beleaguered garrison of Faluja and of night operations by Stirlings and other bombers, despite the REAF's very limited experience of night flying. According to Air Commodore Hayes, the REAF's operations during the Palestine War had been on a "shoe-string" basis and the Egyptians had done well with the limited equipment available, both as regards salvage, repairs, and maintenance, and in active operations.[22]

The British air attaché's report on the REAF also included a report by Sq. Ldr. J. R. Baldwin of the RAF, who had been in charge of several training courses given to selected Egyptian aircrew immediately after the Palestine War ended. Baldwin's report was written only five months after the cease-fire, and although he came to like his Egyptian pupils, his report went a long way toward explaining the REAF's relative lack of success in air combat over Palestine.[23] For example, tactics and formations were of pre-Battle of Britain (1940) RAF type. The importance of understanding angles and velocities for air-to-air gunnery was a virtually unknown quantity to REAF pilots. Although they had obviously shot at ground targets during the Palestine War, most Egyptian airmen had never been through a properly organized aerial gunnery program. As a result, most missed their targets in their first shoots during the RAF courses. Egyptian pilots also had bad radio discipline, a problem that inhibited efficient air combat, while standards of instrument flying were poor, probably because of the normally excellent weather found over Egypt. On the other hand, Baldwin considered the quality of personnel to be good, particularly in the lower ranks, but criticized a lack of good leadership among flight lieutenants and above. Nevertheless, he noted "very considerable improvement" even after one brief operational training course.[24]

The British air attaché also made a detailed estimate of the REAF's strength as of 1 February 1949. At this point combat attrition had not yet been replaced, and, according to this assessment, later confidentially confirmed in outline by REAF sources, Egypt had only eighteen operational fighters out of a total of forty operational aircraft.[25]

At the start of the war the REAF's role had been largely restricted to supporting the Egyptian Army's Expeditionary Force in Palestine. A limited "strategic" bombing campaign against military and communications targets around Tel Aviv was probably expected to have more of a moral than a military impact. During the first phase of the war the air force carried out these tasks effectively, though the Egyptian Army was rarely in a position to take full advantage of the REAF's efforts. The sudden and unexpected appearance of a modern Israeli fighter force operated by highly experienced, largely foreign personnel then forced a rapid reassessment of the REAF's role. For the rest of the Palestine War the REAF continued its close support of the Egyptian Army, though at an increasing cost in aircraft. REAF fighters now had to protect such ground attack missions, the reduced operations of Egypt's unarmed multi-engined bomber force, and the heroic efforts to deliver supplies to the beleaguered Faluja Pocket. Above all, REAF fighters had to defend their own al-ᶜArīsh base and, later in the war, its satellite desert airstrips in northeastern Sinai. The creation of a "heavy bomber" force did not achieve the desired results, although many remarkable raids were made, mostly at night.

The final phase of the war saw the REAF lose its forward landing grounds in Sinai. As a result it had to operate from bases west of the Suez Canal, until, in the very final days, it obtained an airfield in the British-occupied Canal Zone. Nevertheless, the REAF's role in close support of the Egyptian Army remained basically the same. In fact this support became increasingly vital as the Egyptian Army fought to extract itself from a vicious Israeli encircling movement. REAF casualties mounted during the last stage of the Palestine War, but this had little if any impact on REAF morale. Israeli success in air combat had also been exaggerated. For example, only three Egyptian Macchis had been written off as a result of air combat or accident, and no Fiats had been lost. On the other hand, six MC205Vs were under repair at the end of the war, five of them in a condition suggesting possible battle damage.[26] In fact the REAF ended the war as a much more experienced and in some ways more self-confident force than it had entered the war. It may also have been an embittered force which felt betrayed both by its own government and by an outside world that, for reasons the Arabs failed to understand, offered more sympathy to the Israelis than to their Arab victims.

## Operational Strength of the REAF, 1 February 1949[27]

| Unit | Aircraft | Numbers | Location | Comments |
|---|---|---|---|---|
| No. 1 (Tactical reconnaissance) | Spitfire IX | (6 + 3 unserviceable) | al-ᶜArīsh | After Israeli withdrawal |
| No. 2 (Fighter) | Macchi MC205 | (6 + 0 unserviceable) | al-Ballāḥ | Not fully operational |
| No. 3 (Communications) | C-47 | (3 + 0 unserviceable) | Almāẓah | |
| | Anson | (4 + 0 unserviceable) | | |
| | DH Dove | (2 + 2 unserviceable) | | |
| No. 4 (General reconnaissance) | Anson | (4 + 0 unserviceable) | Dākhilah | |
| | DH Dove | (2 + 0 unserviceable) | | |
| No. 5 (Fighter) | Fiat G.55 | (6 + 0 unserviceable) | al-Ballāḥ | |
| No. 6 (Fighter) | | | | Nameplate, Spitfire Vs withdrawn |
| No. 7 (Transport) | C-46 | (4 + 0 unserviceable) | Almāẓah | |
| | C-47 | (2 + 0 unserviceable) | | |
| No. 8 (Bomber) (unnumbered bomber)[28] | Stirling | (0 + 6 unserviceable) | Cairo West | |
| | Halifax C8 | (0 + 1 unserviceable) | Heliopolis and Almāẓah | |
| | C-47 | (4 + 0 unserviceable) | | |
| | C-46 | (3 + 3 unserviceable) | | |
| | Bonanza | (2 + 0 unserviceable) | | |

11

# Reactionaries and Revolutionaries (1949–1953)

The Palestine War had been a disaster for the Arabs, most obviously for the indigenous Arab population of Palestine. Huge numbers had been forced to flee as refugees, and the proposed Arab state of Palestine did not come into existence. The Syrian government would soon be overthrown in a military coup. King ʿAbd Allāh of Jordan (incorporating both Transjordan and the annexed West Bank of Palestine) would be assassinated. King Fārūq of Egypt would be overthrown by Colonel Nasser's Free Officers movement in 1952, and a few years later the Iraqi monarchy would be toppled in a bloody coup.

Much of the Egyptian officer corps, particularly that of the army, felt humiliated by the defeat in Palestine. Many officers also believed that they had been badly let down by their own government. The armed forces had been thrown into a war ill-trained and ill-equipped as part of an ill-conceived political gamble. A large part of the Egyptian officer corps felt that King Fārūq no longer deserved their loyalty, and as a result disaffected groups such as Nasser's Free Officers movement grew stronger. Britain's unwillingness to help the REAF modernize also led to bitterness in the air force. At the same time, "anti-colonialist" and "anti-imperialist" ideas made many in the Egyptian officer corps see their country's future in terms of the third world rather than as part of a Mediterranean dominated by Europe. However, the palace clique that controlled Egypt made little effort to answer criticism, instead resorting to bribery, corruption, and occasional oppression in an attempt to stifle dissent.

## Expansion, Recruitment, and Training

The heavy fighting of the final two weeks of the Palestine War had reduced the REAF's front-line strength. The Israelis eventually withdrew from Sinai as part of a cease-fire agreement; nevertheless, only a token unit of Egyptian fighters and transports was sent to al-ʿArīsh to observe Israeli actions. Other squadrons were concentrated at Ḥulwān, Almāẓah, Dākhilah, and al-Ballāḥ.

Further deliveries of Italian aircraft gradually enlarged the REAF fighter force. To follow up the first much-delayed batch of twenty-four Macchi MC205Vs, Egypt ordered a second batch of eighteen in February 1949 plus a third batch of twenty in May of that year.[1] British fears that Egypt might turn to Italy as its major aircraft supplier proved unfounded; some of these aircraft were later resold to Italy to serve as advanced trainers. Italy also overhauled some REAF war-weary C-47s, while Britain and the United States continued to enforce the existing arms embargo.[2]

General al-Shaʿrāwī, an army officer, was still in command of the REAF, but air force officers hoped to have one of their own take control in the near future. Professional relations between the REAF and the RAF remained good despite increasing tension between their governments. The RAF organized operational training courses for the pilots of Nos. 1 and 5 Squadrons immediately after the Palestine War ended, the Egyptians' use of al-Ballāḥ airfield in the Canal Zone encouraging close collaboration. At first, Wing Commander Tawfīq, the REAF station commander, had wanted only ground attack training, but under British prompting the Egyptians eventually went through a whole course in air fighting, tactics and formations, navigation, and weapons training using both Spitfire LF9s and Fiat G55s.[3] At this point the British government preferred military cooperation with Egypt rather than Iraq or Syria as the REAF had shown itself better able to make proper use of its equipment in the recent Palestine War.[4] The REAF had already inquired about the possibility of British advisors returning to Egypt, though only in a civilian capacity because the Egyptians' fear of losing face was still a problem. In private, however, senior and junior officers admitted the REAF's need for guidance. As the British air attaché in Cairo noted: "All the training that was arranged was done on the Service to Service "old Boy" basis as the Egyptians did not seem prepared to publicise the fact that the Royal Air Force was assisting the Royal Egyptian Air Force."[5] London was similarly reticent in the light of growing domestic and American sympathy for Israel. After winter rains flooded the small al-Ballāḥ airfield, the entire training program moved to the REAF air base at Ḥulwān, while the REAF pressed for use of less waterlogged Canal Zone airfields. On this, however, the British refused to budge.

In 1949 King Fārūq tried to boost his popularity by proclaiming the need for a pan-Arab air force. Two to three thousand planes would support a pan-Arab army of a million men modeled around a training division supervised by ex-Wehrmacht German officers.[6] Few ideas could have been more calculated to alienate British and American public opinion and to increase western sympathy for Israel. On a more realistic level the REAF embarked on a publicity program to boost morale and improve the air force's standing with the Egyptian public. A collection of historical aircraft was assembled from various airfields, and every major Egyptian Air Force type was said to have been represented—until most were destroyed in the Anglo-French attack of 1956.[7]

The REAF's confidence was heightened by the delivery of several planes that had been held up by the UN's arms embargo, while new orders were hurriedly sent to Britain. Egypt, despite its ambiguous relationship with Britain, still preferred British weaponry. In the summer of 1949 the Egyptian government proposed an expansion plan calling for three to six fighter squadrons, two light bomber squadrons, one medium bomber squadron, and two transport squadrons. Britain, however, urged Egypt to concentrate on a defensive role by building up its fighter units.[8] After a great deal of haggling, agreement was finally reached on an Egyptian "shopping list" that included Avro Lancaster B1 and Handley Page Halifax A9 four-engined bombers, Gloster Meteor F4, and De Havilland Vampire FB52 jet fighters. Hawker Sea Fury FB11 and Supermarine Spitfire F22 piston-engined fighters were soon added to this list.[9] Egypt's version of the Gloster Meteor F4 fighter was essentially the same as that serving in the RAF, though with provision for desert survival equipment in the rear fuselage.

The first of a small number of Handley Page Halifax C8 transports had arrived in Egypt in January 1949, having been purchased for meteorological use, air-sea rescue, and bomber training. They were soon converted to bombers with an improvised 0.50 inch calibre defensive armament.[10] The REAF was plagued by its inexperience in operating heavy aircraft, and this time the RAF did not help, as such "offensive" bombers were regarded as politically more sensitive than "defensive" fighters. So the Egyptians had to learn by experience, always a hazardous approach.[11]

Egypt received its first jet in 1949. The Israelis, at least according to official reports, did not get theirs until February 1953.[12] The REAF's first jets—a Meteor F4 fighter and Meteor T7 two-seater trainer—flew to Cairo on 27 October 1949. The political pressure led to an associated training program being cancelled three weeks later. Meteors continued to arrive singly and in pairs throughout 1950, the British government sometimes embargoing deliveries, then releasing them.[13] This undercut REAF training and expansion schedules and undermined Britain's reputation for reliability. It encouraged Egypt to look elsewhere for military supplies, but this growing dissatisfaction would not lead to a major shift in policy until after the Revolution of 1952. Though Egypt was receiving jet fighters, it also bought piston-engined fighters, including nineteen reconditioned Spitfire F22s and one T9 trainer, delivered in 1950.

In 1949 Egypt had ordered twenty De Havilland Vampire FB52 jet fighter-bombers, the first of an agreed but never attained total of sixty-six FB52s, and in December 1950 the first example arrived. The Vampire would earn a special place in the affections of REAF pilots, generally being regarded as a delight to fly.[14] By operating these relatively simple single-engined aircraft the Egyptians were able to build a cadre of jet aircrew and, even more important, maintenance personnel. Sixty-two Vampires were, in fact, eventually delivered to Egypt. Some were built in Britain, but the rest were constructed under license by Macchi in Italy, being sold to Egypt via Syria in circumvention of the British arms embargo. Four Egyptian fighter-bomber squadrons operated these aircraft, and shipments ended only with the Suez Crisis of 1956. In October 1949 Egypt had also ordered twelve Vampire NF10 two-seater night-fighters, but this order was first held up and then cancelled when a new embargo came into effect. Eventually Britain did deliver a dozen Meteor NF13s to fill this yawning gap in Egypt's defenses.

Egypt also tried to develop airborne forces, and in the summer of 1952 ordered three thousand British-made Irving Type X (paratroop) parachutes. But several more years would pass before the Egyptian Army had its first parachute battalion. An Air Force College was, however, successfully established at Bilbays.[15] For its part the Egyptian Army was also attempting to upgrade its anti-aircraft capability, wishing to acquire three types of radar stations, but difficulties with the British arms embargo meant that Britain's reliability as a source of equipment was under strain in the army as it was in the REAF.[16] Only one radar station of uncertain value had been set up at al-ʿArīsh by the end of 1951,[17] and by

1952 Egypt was looking elsewhere even for light anti-aircraft guns, one order being placed in Switzerland.[18]

By the start of 1950 the REAF's strength had increased considerably to 281 aircraft, 89 being operational types.[19] Front-line fighter units still flew Fiats or Macchis as the Meteors were not yet operational, while the remaining Spitfire LF9s had been transferred to the advanced training role. The REAF's expansion plans meant that recruiting and training more aircrew was now a priority. Many men who would make names for themselves fighting the Israelis in the 1960s and 1970s joined the air force at this time.

According to confidential British reports, current REAF personnel included excellent material with high morale and good general discipline, but the air force still suffered from poor leadership, low operational standards, and bad radio-telegraphy discipline among the pilots.[20] Other aspects of training were simply unavailable to the Egyptians. Britain stopped Egypt from obtaining night-fighters for many years, and the REAF was not permitted proper night flying facilities. The REAF tried to get around these obstructions by giving its pilots and navigators night-fighter training on unsuitable Ansons, Doves, and Beechcraft. The REAF also bought a number of Czechoslovakian Mraz M-1D Sokols in an effort to circumvent another threatened British arms embargo, as well as eight Supermarine Sea Otters.

Relations between the RAF and the REAF were, at this time, complicated. In some ways they still reflected the old imperial relationship, with a patronizing but supportive attitude on the part of the British and a respectful reticence on the part of the Egyptians. At the same time the Egyptians were growing both in confidence and in their resentment at British interference. It was to be a typical relationship between the military personnel of a developed nation and those of an underdeveloped third world country, a pattern that would be repeated in many parts of the world over the next half century.

## Getting Rid of the British

The existence of foreign forces and air bases in the Suez Canal Zone was unacceptable to an Egyptian government that was eager to establish its position as leader of the Arab countries. Egypt therefore focused its attention on removing this final British military presence. To western countries, however, the Suez Canal remained a vital strategic waterway.

As the cold war developed, Britain felt little confidence in Egypt's ability to defend the canal against a Soviet threat. At first this merely led to combined air defense exercises, but even these were politically sensitive in Egypt and were done with the minimum of publicity.[21] During 1949, technical discussions had also worked out a joint air defense arrangement for the Canal Zone. This called for joint British and Egyptian maintenance of air bases which could, if needed, accommodate twenty RAF and allied squadrons. To Britain this was a cooperative Anglo-Egyptian defense system, but the Egyptian government, while agreeing to the maintenance of the air bases, insisted that the five RAF squadrons then based in the Suez Canal Zone be withdrawn to other countries in the region. If war threatened, the treaty would permit these RAF units to return. The Egyptians also wanted the British to equip five REAF jet fighter squadrons and train them to defend the canal. British support for this concept rapidly evaporated, however, when an Egyptian general election voted the more nationalistic Wafd Party into power in Cairo late in 1949.

All these political maneuverings caused the supply of new British aircraft to the REAF to be both intermittent and unpredictable. Meanwhile King Fārūq own royal flight was growing into what sometimes seemed to be a "private air force."[22] There was even a special REAF unit known as the Fārūq Squadron with Vultee BT-13 Valiants which were used for anti-malarial DDT spraying.

In May 1950 the embargo on arms sales to Egypt ended, despite the failure of Middle East peace talks. This followed a Tripartite Declaration issued by Britain, France, and the United States in which they guaranteed the existing Arab-Israeli armistice lines against future violation. Events were now proceeding on two quite separate levels. The British government negotiated, imposed, and then lifted arms embargoes while trying to force Egypt to act in accordance with Britain's wishes. At the same time the REAF and RAF continued to cooperate in a spirit of respect and even friendliness despite the political problems. Egypt had been investigating the possibility of a senior retired RAF officer advising the REAF on its future plans. On 9 January 1950 Air Marshal Sir William Dickson, who was soon to retire as commander in chief of RAF Middle East Forces, was asked by the Egyptian minister of war whether he would like this job. Dickson laughed and pointed to his bald old head, whereupon the minister also burst out laughing, thumped the table, and cried out, "You'll do!"[23] This fascinating proposal came to nothing, but four days later Air Marshal Dickson visited the REAF officers' mess at Heliopolis and handed over to Air Vice Marshal ʿAbd al-Munʿim Mīqāti a silver model of an eagle in flight. This was to be the prize in an Egyptian intersquadron air gunnery and bombing competition. In reply Mīqāti announced, somewhat fatefully perhaps, that "there will never be a breakdown in the true friendship between the two air forces."[24]

Official British documents show that some people in the Foreign Office knew that constantly interrupting supplies was undermining Britain's influence in the Egyptian Air Force. But the degree to which the Egyptian armed forces were growing impatient, both with their own government and with Britain, does not seem to have been appreciated. How far such frustration contributed to the success of Colonel Nasser's Revolution of 1952 remains unknown, but may have been considerable. British aviation publications showed no sign that they sensed trouble, nor did British aircraft manufacturing companies. Instead these concentrated on making as much publicity as they could out of their delivery flights to Egypt. In fact, selling war planes to the REAF appeared to be regarded as an extended air race, with Hawker and De Havilland machines trying to knock a few minutes off the latest London-to-Cairo record flight time.

Sometimes British insensitivity reached extraordinary proportions. For example, in June 1951 Egyptian and British forces held full-scale joint maneuvers to test the Suez Canal's defenses against a hypothetical invader coming from the north. The maneuvers were given the rather inappropriate name of "Exercise Contentment," and RAF aircraft based in Cyprus played the part of aggressors in what could be seen as a remarkable preview of the Suez War of 1956. The REAF's No. 1 Squadron flying Furies flew alongside "defending" RAF squadrons based in the Canal Zone, while the REAF's new No. 20 Squadron's Meteors defended the western flank from their base near Cairo.

The British reported that the Egyptians played their part admirably in Exercise Contentment, but once again political tensions soon overshadowed professional cooperation. In October 1951 the Egyptian prime minister announced that his country was abrogating the Anglo-Egyptian treaty of 1936. Four days later Britain, France, the United States, and Turkey offered Egypt a partnership in their new Middle Eastern Defence Organization, but as Egypt was still trying to get foreign troops out of the country, this ill-timed proposal was turned down. The country had no wish to see the British presence in the Suez Canal Zone replaced by four non-Arab armies.

During 1951 anti-British resentment in Egypt flared into violence. In the British-occupied Canal Zone, British soldiers were murdered; Egyptian police clashed with Egyptian civilians and with British troops, and the cycle of terror deepened. King Fārūq recognized the mood of his people and in December 1951 ordered all REAF officers to return home from training courses in Britain, almost all apparently doing so.

The crisis worsened. Britain sent more troops and aircraft to the Canal Zone, a squadron of Lincoln bombers and two Vampire squadrons arriving early in 1952. Contingency plans were even drawn up for an attack on the REAF to cover a British ground advance against Cairo, plans that would later be looked at again, during the crisis of 1956. Meanwhile REAF combat aircraft always flew fully armed, with other armed aircraft held on standby.

## The REAF on the Eve of Revolution

The majority of REAF officers took a "neutralist" position as far as the worsening cold war between the West and the Soviet bloc was concerned. Feeling naturally ran higher on Egypt's relationship with Britain, and while personal contacts between the REAF and RAF remained friendly, the Egyptians clearly wanted the British out of their country. They also hoped that their own air force would grow much stronger. Yet this did not look likely in the near future. The British government still used various pretexts to delay delivery of an extraordinary list of equipment that Egypt wanted to purchase. By May 1952 this list included no less than 161 aircraft, enormous numbers of spare engines, static Link Trainers, parachutes, tires, maintenance equipment, and assorted other items.[25] The Egyptian Army and Navy were suffering the same problems. Not surprisingly, frustration and resentment against Britain, even against the West as a whole, grew steadily worse throughout Egypt's armed forces.

On paper the REAF's strength now looked quite formidable,[26] but in reality Egypt lacked sufficient trained pilots to make use of its front-line strength. The British-trained Fury and Meteor squadrons were regarded as the best, No. 20 with its Meteors being the elite, while the locally trained Vampire pilots were not yet up to the same standard. Most conversion training was done in the squadrons themselves. There was neither a fighter control nor an early warning system, except for a barely serviceable radar unit near the Israeli frontier at al-ʿArīsh. Pilots averaged only ten to twenty flying hours per month. The bomber squadrons had no live bombing nor air firing experience and were in no sense really operational. Since the REAF had no proper maintenance facilities, operational aircraft were either "put on the shelf" or continued flying when they were due for major overhauls. Leadership was, in the opinion of the British air attaché, very weak at the top. The attaché regarded Mīqātī as very knowledgeable but not very strong, allowing himself to be overruled by Group Captain Jazzārīn who, in the air attaché's opinion, now acted as the minister of war's mouthpiece. Conditions on REAF airfields were now less crowded since many units had been dispersed away from Almāẓah. The Egyptians were also upgrading three landing grounds in Sinai by laying concrete runways. In general, however, British observers considered that the REAF's operational standards were declining rather than improving.[27] Meanwhile the new "Egyptian Aircraft Factory" at Heliopolis had also just assembled its first Bücker Bestmann primary trainer, a type to be known as the *Jumhūrīyah* following the Revolution of 1952.

Some Egyptian officers concentrated solely on their professional duties while others took a closer interest in politics. This would eventually lead to a division within the Egyptian Army, Air Force, and Navy between "political" and "professional" officers, a division that led to disaster in 1967. In 1952, however, there was no evidence that the politically motivated Egyptian officers were neglecting their military duties. King Fārūq, ever eager to court popularity by trying to push the British out of Egypt, failed to recognize the serious discontent among his officer corps. Even some senior men who had a stake in the status quo were growing very restless with the political situation.

Unknown to the king, and to most of these senior officers, a number of middle-ranking Egyptian officers were planning a revolt. Nasser's Society of Free Officers had been born many years earlier in the stern fortress of Manqabād overlooking the Nile in Upper Egypt. Many army and air force officers had now joined its ranks, though the navy seems to have remained almost entirely loyal to King Fārūq until the end. A crackdown by the king's security police in 1949 had necessitated a reorganization of Colonel Nasser's Society of Free Officers, and a secret central executive committee of ten was set up. Six members, including Nasser, came from the Egyptian Army, and four from the REAF: Wing Comdr. Jamāl al-Dīn Muṣṭafá Salīm whose army brother was also on the committee, Wing. Comdr. ʿAbd al-Laṭīf Baghdādī, Sq. Ldr. Ḥasan Ibrāhīm, and Munʿim ʿAbd al-Raʾūf who had tried to fly Gen. ʿAzīz al-Miṣrī out of Egypt during the Second World War. The membership of this committee changed little over the following three years and was to form the basis of the later Council of the Revolution. Other REAF officers played a leading role in the movement, and they included Wing. Comdr. ʿAlī Ṣabrī, chief of Air Force Intelligence, who had particularly good connections with the U.S. air attaché in Cairo. This would be a factor of some importance when the Revolution came in 1952.[28]

## The Egyptian Revolution of 1952

The Free Officers movement made the final decision to overthrow King Fārūq at the end of June 1952. Only one REAF officer was present at this fateful meeting, Ḥasan Ibrāhīm, who had recently been promoted to wing commander. The other REAF members were with their squadrons. Zero hour was to be midnight, 21 July 1952, but preparations took longer than expected, and the uprising was delayed for twenty-four hours. The plan called for Wing Commander Baghdādī to seize control of Almāẓah, the REAF's main base. Then he and Wing Comdr. Ḥasan Ibrāhīm would brief the pilots stationed there on the coup. None doubted that the REAF would support their revolution. Meanwhile Wing Comdr. Muṣṭafá Salīm was to fly his brother, Ṣāliḥ Salīm, and two other army officers, ʿAbd al-Ḥakīm ʿAmr and Anwar Sadat, to the Sinai. There Muṣṭafá Salīm was to take control of the al-ʿArīsh air base and win over the squadrons based in that area.[29]

There were many reasons for the military coup of 1952 which overthrew King Fārūq. Not only had the Egyptian armed forces been humiliated by their recent defeat in the Palestine War, but they, and the country at large, felt that the monarchy was still too subservient to the British who occupied the Suez Canal Zone. The country's ruling class was regarded as feudal, oppressive, and corrupt. Some of these criticisms were certainly valid, and the 1952 coup did usher in a completely different era for Egypt. The Revolution did not solve all the country's problems and in some respects added new ones. Yet the staunchly Arab nationalist government of President Nasser which soon followed undoubtedly revived Egypt's pride and the self-confidence of its armed forces.

The Free Officers' coup took place in the early hours of 23 July 1952. It went exactly according to plan and was nearly bloodless. Most REAF officers and men enthusiastically welcomed the Revolution, and from dawn onward Egyptian Meteors, Vampires, Lancasters, and Halifaxes flew above Cairo, Alexandria, and the main Nile Delta towns in a show of support. General Najīb, one of the revolutionary movement's leaders, considered that such a show of force would be a powerful psychological weapon, and the aggressive low-level flying of the fighters clearly discouraged any resistance. They certainly had their effect on ʿAlī Māhir, a leading politician whom Anwar Sadat persuaded to cooperate with the coup leaders to form a new government.[30]

REAF transport squadrons were also busy ferrying various members of the Free Officers' central committee, now known as the Council of Ten, from place to place during the complicated negotiations that succeeded in cementing the 1952 coup. As such they probably did a great deal to avoid unnecessary bloodshed.[31] Other REAF aircraft regularly overflew the Suez Canal Zone, observing the British air and ground forces stationed there. The coup leaders were clearly worried about possible British intervention, but this never took place. Nevertheless, Squadron Leader ʿAlī Ṣabrī was busily making contact with the most important foreign embassies in Cairo. As chief of Air Force Intelligence he knew Lt. Col. David L. Evans, assistant air attaché at the U.S. embassy. The Free Officers were well aware of the importance of the American reaction to their coup, and ʿAlī Ṣabrī was given the vital task of reassuring the American ambassador that Egypt's new rulers were not anti-western. ʿAlī Ṣabrī also made sure that the British embassy knew that the Free Officers movement had drawn up plans for guerrilla resistance should Britain interfere.[32]

## The REAF's Old Guard Retires

Those not immediately involved in the coup of 23 July were taken by surprise, though not entirely so. Air Vice Marshal Mīqātī was the most senior REAF officer, although overall command of the air force had still been given to an army man. He had been waiting for a staff car to arrive and take him to military headquarters. When it failed to arrive Mīqātī telephoned HQ and only then learned of the Revolution, which also explained the unusual amount of air activity that morning. But it did not clarify the air vice marshal's own uncertain position. Some of his fellow senior officers had already been arrested. No one came for Mīqātī, so he put on his best uniform and went to see General Najīb and the other coup leaders. Was he to be arrested, or was he to continue work? Neither, was Najīb's reply. Mīqātī was known to have a very low opinion of King Fārūq and his palace clique. The air vice marshal's long years of work in support of the REAF were greatly appreciated, so he would certainly not be arrested. On the other hand, Mīqātī was clearly a member of the "old guard." Instead Air Vice Marshal ʿAbd al-Munim Mīqātī was forced to retire.[33]

Other REAF officers of the "old guard" had been more closely associated with King Fārūq's palace clique and were not treated so leniently. Early on the morning of the coup, Ḥasan ʿĀkif, the king's chief pilot, arrived at Almāẓah to prepare the royal C-47 for a scheduled flight.[34] The base was already in revolutionary hands, and ʿĀkif was refused admission. Ḥasan ʿĀkif remained loyal to his king to the end, traveling overland to Fārūq's palace outside Alexandria. There he guarded the king as the latter tried to raise support in the Egyptian Navy. This failed, and Ḥasan ʿĀkif followed Fārūq into exile, being one of the eight men from the royal entourage that the new revolutionary leaders insisted had to go.

Once the Free Officers group had firm control, there was a debate about what to do with King Fārūq, who had to leave. Wing Commander Jamāl Salīm argued in favor of putting him on trial, but most of the coup leaders accepted Colonel Nasser's advice that Fārūq be sent into exile while his infant son, Aḥmad Fuʾād, be proclaimed the new king of Egypt. In fact, it was Jamāl Salīm and General Najīb who bade Fārūq farewell when he sailed from Alexandria in the royal yacht *Mahrūsa.* This ship subsequently returned and still sails as the Egyptian presidential yacht, while at the same time being one of the oldest and most historic vessels afloat.

Jamāl Salīm later became Egypt's minister of communications and deputy prime minister before leaving the government in June 1956. Wing Comdr. ʿAbd al-Laṭīf Baghdādī became minister of rural and municipal affairs before taking over as minister of war. Wing Commander Ḥasan Ibrāhīm became minister of national production before leaving the government in June 1956.[35]

These enthusiastic and technically trained air force officers brought a new outlook to the stagnant Egyptian political process. But other air force officers found it harder to accept Egypt's new direction. ʿAbd al-Munʿim ʿAbd al-Raʾūf, though an early member of the Free Officers movement, had also been closely associated with the Muslim Brotherhood, an Islamic "literalist" and revolutionary organization. This group strongly disapproved of the Free Officers' program of socialist-inspired economic and political reforms. Several members tried to assassinate President Nasser in 1954; ʿAbd al-Raʾūf was implicated and fled the country.

Air Commodore Ḥasan Maḥmūd was in a London hospital having his appendix removed when news of the July coup reached him. His next surprise was a telephone call to his hospital room in which Egypt's new prime minister asked him to take charge of the Egyptian Air Force. After serving as a group captain during the Palestine War, Ḥasan Maḥmūd had been the Egyptian air attaché in London for several years. His British contacts were wide-ranging and friendly, and this was expected to be a great help in his new job. Unfortunately the new Egyptian government's hopes of a new and improved relationship with Britain were unfounded. The British, it seemed, were no more sympathetic to the new Egypt than they had been to the old. As soon as Ḥasan Maḥmūd arrived back in Cairo he agreed to take command of the air force, but only on condition that no officer should be promoted ahead of his peers merely for being on the right side in the recent coup. Nasser and the other Free Officers agreed.[36]

Things went well until July 1953 when the monarchy was finally abolished and Egypt was declared a Republic. General Najīb took over as the country's first president, although Colonel Nasser retained real power as head of the governing Revolutionary Council. Unfortunately Maj. ʿAbd al-Ḥakīm ʿAmr then succeeded Najīb as commander in chief of the Egyptian armed forces, promptly being promoted from major to major general. This caused considerable resentment among many other officers who saw it as a "political" promotion. Air Commodore Maḥmūd was one of those who felt that political loyalty was being confused with professional competence and so resigned as head of the air force. He appears to have been briefly succeeded by ʿAbd al-Laṭīf Baghdādī, a member of the original Revolutionary Council, at least until Maḥmūd Ṣidqi Maḥmūd took command late in 1956.

There were also many others in the Egyptian Air Force who felt unfairly treated by the new government. Many senior officers were forced to retire so that junior men, more loyal to the new regime, could be rewarded with a promotion. Such men were often innocent victims of a perhaps inevitable purge, yet their sacrifice sometimes left bitterness in families that had provided Egypt with loyal officers over many generations.[37] A series of investigations and prosecutions known as the Defective Arms Trials also started soon after the 1952 Revolution. These attempted to uncover supposed profiteering by various military officers and government officials during the 1948–49 Palestine War. The REAF's unsatisfactory Short Stirling bombers featured prominently, and among those accused was Prince ʿAbbās Ḥalīm, a cousin of King Fārūq, who had been closely involved in various unofficial procurement programs for the REAF. The prince was cleared of all charges along with the thirteen others accused, but these events also contributed their share of bitterness.

The years from the end of the Palestine War to the declaration of an Egyptian Republic in 1953 saw huge political changes in the country. Economic and social upheavals would come later. For the Egyptian Air Force it was a time of high hopes, almost all of which were dashed on the rocks of political corruption and incompetence at home and of blinkered British intransigence abroad. The REAF had grown considerably in size but may actually have declined in competence. Egypt's new revolutionary government hoped for a new relationship with the outside world, even for an honorable peace with Israel, aims that Colonel Nasser shared with the young King Hussein of Jordan. This view was certainly shared by the Egyptian Air Force which as yet showed no inclination to change from western to Soviet equipment. But once again the short-sighted attitude of Britain , and of many other western governments, coupled with an Israeli determination to sabotage any warming of relations between Egypt and the West, meant that the air force's hopes were unfounded.

## Operational Strength of the REAF in January 1952[38]

| Unit | Location | Aircraft | Strength |
|---|---|---|---|
| No. 1 (Fighter reconnaissance) | al-ʿArʾīsh | Hawker Fury | (9 + 3 unserviceable) |
| No. 2 (Fighter bomber) | Ḥulwān | Spitfire F22 | (10 + 9 unserviceable) |

*Continued on next page*

*(Continued)*

| | | | |
|---|---|---|---|
| No. 3 (Transport) | Almāẓah | C-47 | (10 + 2 unserviceable) |
| No. 4 (Navigational training) | Dākhilah | C-47 | (1) |
| | | Beechcraft | (4 + 2 unserviceable) |
| | | Sea Otter | (0 + 5 unserviceable) |
| No. 5/6 (Advanced training) | Bilbays | Fiat G55 | (4 + 15 unserviceable) |
| | | Macchis MC205 | (4 + 8 unserviceable) |
| No. 7 (Transport) | Ḥulwān | C-46 | (10 + 6 unserviceable) |
| No. 8 (Bomber) acting as a conversion unit for No. 9's Lancasters | Cairo West | Halifax | (4 + 4 unserviceable) |
| No. 9 (Bomber) | Almāẓah | Lancaster | (5 + 4 unserviceable) |
| No. 10 (Headquarters communications) | Heliopolis | Bonanza | (3 + 0 unserviceable) |
| | | Dove | (5 + 0 unserviceable) |
| | | Anson | (2 + 1 unserviceable) |
| No. 20 (Fighter training) | Almāẓah | Meteor F4 | (9 + 1 unserviceable) |
| | | Meteor T7 | (1 + 0 unserviceable) |
| No. 30 (Fighter) | Almāẓah | Vampire | (20 + 5 unserviceable) |
| Fārūq Health Squadron | Heliopolis | Vultee Valiant | (5 + 1 unserviceable) |
| Royal flight | Almāẓah | C-47 | (2 + 0 unserviceable) |
| | | C-46 | (1 + 0 unserviceable) |
| | | Beechcraft | (2 + 0 unserviceable) |
| | | Grumman Mallard | (2 + 0 unserviceable) |
| | | Westland S51 | (2 + 0 unserviceable) |
| | | Feisler Storch | (2 + 0 unserviceable) |
| Flying training college | Bilbays | DHC Chipmunk | (9 + 10 unserviceable) |
| | | Mraz Sokol | (10 + 20 unserviceable) |
| | | Miles Magister | (2 + 8 unserviceable) |
| | | Harvard | (12 + 12 unserviceable) |

## 12

# "Czech" Arms (1953–1956)

Following the 1952 Revolution, the new Egyptian government hoped to improve its relations with Britain and the West in general, perhaps even reaching a just peace with Israel. But Egypt's new leadership also intended that their country become a leader of the Arab world. In the end none of these goals was achieved. The continuing state of war between Israel and the Arab states continued to overshadowed Egypt's foreign relations with the West. As a result the country turned to the Soviet bloc for military and political support. Within four years of the Revolution, Egypt, now led by President Nasser, was widely perceived as an increasing threat to western interests.

## 1953–54: The Failed Honeymoon with Britain

The REAF (or EAF as it soon became) remained essentially pro-RAF, if not pro-British, and hoped for better relations. Despite its support for the Revolution, it was a conservative force and looked to the RAF as a model of the technical competence it hoped to achieve for itself. Nevertheless, the British air attaché in Cairo recognized that London would have to do something quickly about the Egyptians' supply problems, or, as he said, "the rot would set in" with "bitterness towards us."[1]

There had, in fact, been a marked improvement in relations after RAF units stood down from the high state of readiness to which they had been brought during the Revolution. The British government stated that it wanted to build a new relationship with the Egyptian government, and this was given substance by the sale of twelve Meteor F8s. Yet in reality British policies toward Egypt remained wary, as illustrated by a message sent by Prime Minister Winston Churchill to his secretary of state at the Foreign Office in London on 27 January 1953: "Please do not let these jets go to Egypt until you and I have talked it all over together, I hope at noon on Thursday. Necessary technical hitches should be made to occur."[2] Four of these aircraft eventually reached Egypt in February 1953, though with ballast instead of cannons. Britain then again imposed an embargo on further deliveries of new weapons to Egypt. Five of the remaining Meteors from this order went to Brazil and three to Israel. Meanwhile Egyptian and British forces continued to cooperate. There were no more full-scale joint maneuvers, but late in March 1953 Egyptian anti-aircraft units joined the RAF regiment in a three-day air defense exercise while Egyptian Air Force officers took part as observers.

Negotiations on the future of British military bases in the Suez Canal Zone continued. Prospects of a British withdrawal looked good. The United States hoped to bring Egypt into a pro-western regional alliance, and discussions about this possibility were already under way. British military planners and diplomats realized that it was counterproductive to hang onto Canal Zone bases that only fueled Egyptian hostility. Two plans were consequently drawn up by Foreign Minister Anthony Eden. The British government hoped that Egypt would accept "Plan A" which called for 7,000 uniformed British troops to be stationed in the Canal Zone and for a joint Anglo-Egyptian air defense system. "Plan B," which most planners considered more realistic, would place the Canal Zone bases under Egyptian control though with a small number of British or allied technical staff to supervise their maintenance. "Plan B" also envisaged the EAF being solely responsible for the air defense of the Suez Canal. Again EAF capabilities and operations were placed at the heart of Egyptian foreign policy decision making. Negotiations between Britain and Egypt began in Cairo on 28 April 1953 with Wing Comdr. ʿAbd al-Laṭīf Baghdādī, a member of the Revolutionary Council, leading the Egyptian team. A week later these talks collapsed when Egypt refused to allow any uniformed British troops to remain on Egyptian territory.

Meanwhile Egyptian leaders put great effort into improving relations with other Arab states. Here the EAF was used as an instrument of foreign policy. For example, Saudi Arabian Air Force pilots were invited to train in Egypt in 1953, and several Vampires were also handed over to the Saudis. Continuing arguments over the Canal Zone led to a British arms embargo being reimposed in 1953, although Egypt was able to obtain thirty Italian Macchi-built Vampire FB52 fighter-bombers via Syria.[3] In the summer of 1953 there were further clashes between British troops and Egyptians, and in September Israeli forces unexpectedly occupied the al-Awjah demilitarized zone on the Negev-Sinai border, further increasing tension.

## 1954–55: Short-lived Cooperation in Training

Talks between Egypt and Britain then restarted, and an agreement was finally signed in October 1954. The Anglo-Egyptian treaty of 1936 was ended by mutual consent, and Britain agreed to withdraw all troops from Egypt within twenty months. Thereafter Britain would be allowed to station 2,200 civilian technicians in Egypt to maintain several bases near the Suez Canal for a further seven years.

Britain now resumed arms sales to Egypt. In early 1955 the EAF received eight Meteor F8s and six Meteor NF13 night-fighters. These deliveries finally enabled the EAF to transfer its piston-engined fighters to the advanced training role. Several RAF squadrons were still stationed in the Canal Zone, but now that a timetable for British withdrawal had been agreed upon, confrontation was replaced by the cooperation seen in earlier years. No. 208 Squadron RAF, with its advanced Meteor FR9s, worked closely with the EAF's jet conversion unit of Vampire FB52s and T55s which, under the terms of the new agreement, had moved into the ex-RAF base at Fāʾid.[4]

The Egyptians had only just converted to Vampires and were very much in need of air-to-air gunnery practice. Wing Comdr. M. G. Bradley (ret.) commented: "After one or two bad frights from ricochets and other 'zero degree angle-off' incidents, we were asked to send our squadron Pilot Attack Officer down to Fāʾid to check their gunsight harmonization and give them any other possible help. Legend has it that the first Vampire our chap climbed into and switched on the sight, it wound itself up its slide and fell off the top into his lap!"[5] The British pilot in question was Flt. Lt. Chris Bushe, who, while admitting that his recollections were "coloured by the circumstances of the time," went on to write:

> As we were specifically banned from making any passes at EAF aircraft, should we even see one, we had little idea about them or how they flew. After two days of lectures at EAF Deversoir, attended by pilots from other units, we got down to the practical bit and harmonisation was high on the priority list. That was when the gunsight ended in my lap. . . . Their flying generally was of a standard accepted as being "average" with us. The whole thing was

> rather slap-dash with checklists not carried out on occasions. . . . I must admit that they were very keen about the cine-gunnery that we did, and enthusiastic that 208 should mix it with them if ever we met in the air. This point was made by other squadrons that attended. In my innocence I took them at their word and some days later bounced a section of NF13s—just one pass. No more than 15 minutes later after landing, the CO and I were summoned to the Station Commander's office for me to be given a dressing down for that one pass.[6]

Once again it was clear that EAF aircrew were keen to keep up contacts, both professional and social, with their RAF counterparts, but that their political leaders were far less enthusiastic.

The EAF was undoubtedly having trouble training a new generation of experienced aircrew and support personnel. There were plenty of volunteers and training aircraft, though these included a wide variety of types. The chief difficulty lay in a lack of support equipment, outdated methods, and inexperienced instructors. The EAF advertised for twenty ex-RAF instructors to strengthen its flight training program. Only six British instructors accepted the Egyptian offer, and when they arrived at the Bilbays Air College in October 1955 they were surprised by what they found. The condition of training aircraft and base facilities was poor, and even flying clothes were in short supply. The EAF was making a determined effort to improve this situation, and many changes were already under way to upgrade the base and its training equipment.

Egyptian pilot and navigator candidates came to Bilbays after spending two years at the Military Academy during which time they had experienced at most three short flights in a Chipmunk. These fledgling pilots were then plunged into an intensive training program flying the thirty Chipmunks and twenty-seven new Jumhūrīyahs stationed at Bilbays. Cadets were taught by EAF pilots, most of whom had only a few hundred hours of flying experience themselves. Future navigators attended hundreds of hours of classes and then trained on C-47s or other transports.

It was a demanding program, and if a cadet failed to master the art of flying after twenty hours in the air he was considered to be of below average ability. EAF cadets were, in fact, rushed through their training. Those who made the grade flying Harvards were then sent straight on to the jet conversion course located at Fāʾid, an air base near the shore of the Great Bitter Lake. Here student pilots flew piston-engined Fiat G55 and Spitfire fighters and then Vampire T55 or Meteor T7 jets. Few student pilots were assigned to bomber or transport units as these spots were filled by older, more experienced pilots.

Although many changes were made to improve the quality of Egyptian pilot training, the EAF refrained from dramatically altering its program because this would have slowed down the pace of training. The limited pool of technically trained Egyptians, especially those available to the air force, continued to inhibit expansion. At the moment, numbers were considered to be more important than quality. Egypt needed pilots quickly to fly the large number of new planes that the government was about to buy from a completely new source of supply. By the end of 1955, just before this new equipment arrived, the EAF had a manpower strength of about four hundred aircrew plus three thousand ground support personnel. It possessed almost one hundred combat aircraft, over 75 percent of which were operational, with various obsolete aircraft held in store. In addition, there were a little over 200 training and support planes, this equipment being almost entirely of western manufacture.

At this stage the EAF planned to have four fighter-bomber squadrons equipped with Vampires, while one squadron still flew Meteor day-fighters. A single EAF flight was having difficulty reaching operational status with Meteor NF13 night-fighters. When these aircraft arrived in the middle of 1955 the unit had few trained crews and very little of the complex support equipment needed to maintain a night-fighter's radar system. A single heavy bomber squadron, No. 9, worked hard to keep three Lancasters and three Halifaxes flying. The three transport squadrons were in better shape, operating a fleet of about forty C-47s, C-46s, Ansons, and Doves.[7] What the EAF lacked was adequate technical support and a coherent operational planning structure.

## Egypt Desperate to Diversify

In 1954 the leaders of Egypt and Israel opened a secret dialogue concerning a possible peace agreement. This ended when David Ben-Gurion returned to power in Israel and such contacts came to an abrupt end.[8] The new Israeli government took over in the wake of the "Lavon Affair," a campaign of terrorist bombing against western interests in Egypt carried out by the Israeli secret service to undermine any improvement in Egypt's relations with Britain and America. As a result, Egypt continued to bar Israeli shipping from the Gulf of Aqaba and maintained its embargo against foreign companies that traded with Israel.

The Israeli policy of "active defense" against Egypt led to preemptive strikes against Palestinian bases within the Egyptian-ruled Gaza Strip. The Egyptians had been blocking almost all infiltration into Israel by Palestinians, but Israeli commandos still attacked Egyptian Army positions on the Gaza to al-Brayr road on 28 February 1955. Thirty-seven soldiers and two civilians were killed, thirty soldiers and one civilian being wounded. On 21 July 1955, four Israeli planes overflew the Gaza Strip and were met with anti-aircraft fire.

The Egyptian government retaliated by allowing Palestinian guerrillas based around Gaza to attack settlements in southern Israel, and the rise in tension inevitably led to a renewed arms race between Israel and Egypt. Egypt's leaders felt that they had to counter Israel's growing power and increasingly aggressive actions with an all-out effort to strengthen the Egyptian armed forces, above all the air force. Even so, Egypt lost this arms race, and before long Shimon Peres, then director general of the Israeli defense Ministry, would declared that "we have provided Israel with superiority over Arab states" and that the Israeli Air Force's new modern aircraft "gave us a new air superiority."[9]

In a message to London, the British air attaché in Cairo warned that anger was already spreading through the EAF as a result of the slow and intermittent delivery of British-built combat aircraft.[10] An increasing number of Egyptian officers also believed that such British delaying tactics reflected support for Israel. Meanwhile the Egyptian government continued to look for other sources of supply. Egypt had asked the Americans for F-86 Sabre jet fighters in 1953 but were refused. Israel also asked for these top-line fighters but was similarly refused. In 1955 Egyptian military buyers negotiated with Sweden to purchase Saab B18B piston-engined bombers, Norway for ex-Norwegian Air Force Vampire jets, and Britain to buy various miscellaneous aircraft. Another Egyptian mission visited Spain, but none of these discussions were to pay off. Britain even refused to provide Egypt with drop tanks for EAF Meteors and Vampires on the grounds that this would give them offensive capability.[11]

In 1955 Egypt did receive, from Canada, a shipment of Harvard trainers. But this sale again caused controversy. The problem was finally laid to rest in 1956 when the Canadian prime minister announced that "The

Egyptian order for Harvards had first been discussed with Britain and the USA, and neither country had raised any objections."[12]

## Turning to the Soviets

Despite these efforts to diversify its sources of military supplies, Egypt felt increasingly bitter toward unreliable foreign suppliers. Anger within the EAF officer corps was reaching a serious level, and President Nasser's government may have felt threatened by this rising chorus of discontent. The Egyptian armed forces clearly felt threatened by Israel's new aggressive policies and humiliated by their apparent inability to defend the nation's borders.

Soviet support for Israel in 1948 had not given it political leverage, and the Soviet Union was aware of these feelings inside Egypt. The Soviet Union therefore started to court Arab states that it had previously been describing as reactionary, feudal puppets of the British Empire. The new revolutionary and enthusiastically anti-imperialist government in Cairo was an obvious target for this Soviet political offensive. Western policies toward Egypt also made the Soviet task easier. There was plenty of resentment for the Soviets to seize upon, though as yet there was little anti-western feeling among the educated officer class of Egypt.

Beginning in June 1955 President Nasser met several times with the Soviet ambassador in Cairo to discuss possible arms sales. When the British government discovered these secret discussions they tied any future British-Egyptian arms deals to Egypt's participation in an alliance against the Soviet Union, the Baghdad Pact. The United States also refused Egyptian requests for arms unless it signed the Baghdad Pact and allowed an American military mission into the country. This would have sidetracked goals for Arab leadership. Most Egyptians, while they felt a far greater cultural sympathy for the West than for the Soviet bloc, earnestly wished to remain neutral in the cold war. The idea of a "third world" of neutral countries standing aloof from the East-West confrontation was already taking root, and Nasser started looking beyond the Arab world for political allies.

When information concerning Cairo's delicate negotiations on the Baghdad Pact appeared in the world's press, President Nasser believed that such information had been leaked by the British government in order to embarrass him. Furious, he announced an arms deal with

Czechoslovakia in exchange for rice and cotton. The Soviet Union was, in reality, the power behind this deal, and Czechoslovakia was basically just used as a front.

The announcement of what the western press called the "Czech Arms Deal" caused a diplomatic flurry in London but did not influence the gradual evacuation of British troops from the Canal Zone as this process had already been agreed upon. In Israel it led to stronger military co-operation with France and aggressive military action along the frontier. Israeli jets regularly crossed the border, while Palestinian guerrillas in Gaza traded raids with Israeli commandos. On 31 July 1955 Egypt retaliated. Four EAF Vampires crossed into Israel but were intercepted by a pair of IAF Meteors, and two Vampires were shot down.

In a speech to Parliament on 12 December 1955 Britain's new prime minister, Anthony Eden, declared that "Israel is not in my belief at a military disadvantage today in relation to any Arab state, or indeed to any combination of Arab states who are on her frontier."[13] Among the Arab states, however, there was a growing feeling that it was they who were at a military disadvantage and that the military gap was growing. Certainly Israel's new French-built Ouragan fighter-bombers outperformed the EAF's Meteors and Vampires, but the Egyptians hoped that newer Russian MiGs would redress the balance. Concern about the balance of power in the air had, in fact, played a significant part in a major realignment in Egypt's foreign policy.

## The MiGs Arrive

The first crated Soviet-built aircraft arrived in Alexandria on 1 October 1955 aboard the freighter *Stalingrad*. Subsequently Ilyushin Il-14 transports seem to have flown in military supplies from Bulgaria. These deliveries included eighty-six MiG-15 jet fighters and MiG-15UTI conversion trainers, thirty-nine Ilyushin Il-28 jet bombers, twenty Il-14 transports, and twenty-five Yak-11 primary trainers. In addition, the Egyptian Army and Navy received large amounts of Soviet-made weaponry. The EAF's first two MiG-15 squadrons were established at Almāẓah in December 1955. Soviet bloc instructors, Russians and Czechoslovakians, had already given Egyptian pilots hurried conversion courses using MiG-15UTI trainers at Kibrit air base. On 15 January 1956 a flight of MiG-15s flew over Cairo to display the EAF's new jets.

Despite turning to the Soviet Union for arms, the Egyptian government still hoped to remain genuinely neutral. Early in 1956 another request to purchase British military equipment—Bristol Sycamore helicopters, Avro Lincoln piston-engined bombers, and the latest Hawker Hunter jet fighters—was made to London. The British refused. Partly as a result of this as well as other British and American political actions, Egypt fell out of the pro-western orbit altogether. Once again the interests of the EAF had a considerable impact on broader political events.

Throughout the spring of 1956 there were repeated clashes along the Sinai and Gaza Strip borders. Egyptian and Israeli jets often violated each other's frontiers while collecting reconnaissance information. On 12 April 1956 a pair of EAF Vampires was ambushed by two of Israel's new Ouragan fighters. One Vampire was hit, though the pilot managed to make an emergency landing and was captured.

Meanwhile the EAF's main priority was to train and expand its ranks and to bring the new Soviet jets into service. According to an annual report drawn up by the British air attaché in Cairo, the EAF rapidly adjusted to the massive influx of Russian-built equipment.[14] Most of the new aircraft came to Egypt by ship before being reassembled and tested at Dākhilah air base near Alexandria. The Il-28 bombers then moved to Cairo West, while the MiG-15 fighters flew to Almāẓah. By early 1956 the EAF had five fighter and ground attack squadrons, though only three were operational. Others were converting to their new Soviet equipment. The EAF's single bomber squadron was similarly out of service as it converted to new aircraft, though the five transport, communications, and navigation-training squadrons were all operational. In addition, there was a jet conversion unit, the Air Academy, plus miscellaneous health and agricultural spraying units. On the other side of the frontier the IAF, at this time, had six operational fighter and ground attack squadrons, another jet fighter squadron being established, one operational heavy-bomber squadron, and three transport and training squadrons.

The process of suddenly converting from essentially British equipment, training, and organization to Soviet equipment and ways of doing things inevitably caused problems. The EAF coped with these, in part because of the high quality of many of its veteran aircrew, and once again the Egyptians demonstrated their adaptability. According to the British air attaché, there were no immediate changes in the command structure, although the EAF did acquire a new Canal Zone HQ once the British had evacuated this area.[15] The EAF merely took over the air defense of the canal, but its ability to maintain the same degree of effectiveness as its RAF predecessors remained to be tested.

Years of British unwillingness to supply new equipment meant that no new squadrons had been formed until the arrival of some additional Meteors and Vampires in mid-1955 permitted a modest expansion of the EAF's front-line strength. This growth was as nothing, however, compared to the massive expansion program that the new Soviet supplies allowed. Some squadrons were changing straight from piston-engined fighters to high performance jets, No. 1, for example finally giving up its very popular Hawker Furies for MiG-15s. Others, such as, No. 30, already had considerable jet experience flying Vampires before converting to MiGs, although it would be interesting to know how the pilots reacted to a sudden change from their forgiving Vampires to the much more demanding MiGs. The conversion of No. 9 Squadron from a heavy bomber unit flying Lancasters to Il-28 jet bombers seems to have caused fewer problems than might have been expected. Though this unit was short of pilots, it was expected to draw some from the transport squadrons. Nevertheless, the British air attaché predicted greater difficulty if the EAF tried to form a second bomber squadron.[16] There was an overall shortage of trained aircrew, yet the shortage of qualified ground staff and mechanics was, as always, much more serious. The work of the Fighter Training Unit (FTU) at Almāẓah was regarded as the most vital during this period of change. Even so, the new high-performance Soviet jets were difficult to master, two MiG-15s and an Il-28 being damaged in accidents at Almāẓah by mid-January 1956.[17] At the same time Egypt's political relationship with other Arab states was considered so important that Syrian as well as Egyptian Air Force personnel were trained there by Czech and possibly also Russian instructors.

## A New Strategy for a New Threat

According to a highly critical report by the British air attaché, many of the EAF's newer pilots suffered from poor flying discipline, bad techniques, and a lack of alertness and were in no position to take on the Israelis. The EAF's main role remained ground support, as it had been during the 1948 war. Meanwhile the weakness of Egypt's air defenses remained a major worry for Egyptian military planners despite the addition of new MiG fighters. The higher echelons of the EAF seem to have been fully aware of their service's limitations, and during 1956 the nucleus of a fighter control system took shape under the command of Air Marshal ʿAlī ʿAṭiyah, previously air attaché in London. For decades the British had been responsible for the defense of the Suez Canal, and Britain had not allowed Egypt to establish an independent network of radars, anti-aircraft guns or fighters.

So far Egypt could only field a few French- and British-built radar sets, but these were not tied together in a network and there were huge gaps in their coverage of the Egyptian frontiers. Under Russian direction the beginnings of a coordinated air defense network, using both existing radars and new systems supplied by the Soviet Union, were gradually put in place. This was to include radar sites located along the Mediterranean coast from Alexandria to Port Said and down the length of the Suez Canal to the Red Sea.[18]

The last British troops left Egypt on 13 June 1956. At the same time a nationwide plebiscite was held to approve a new constitution. The revolutionary government consolidated its position; the Revolutionary Council formed during the coup of 1952 was dissolved, and most of its military leaders gave up their uniforms. Nasser became Egypt's new president, and there was a general feeling that real independence had dawned for Egypt.

In the United States, Egypt's purchase of large quantities of Soviet military equipment led to fears that Egypt was moving into the Soviet sphere of influence, compounded by suspicions about Egypt's policies in the Arab world. As a result, the United States and Britain withdrew earlier offers to help finance the building of the Aswan High Dam, President Nasser's prestige development project. President Nasser's declaration that he would finance the dam by nationalizing the Suez Canal was, in turn, greeted with considerable enthusiasm by the Egyptian people. The canal was, at that time, still owned by a western commercial company.

President Nasser's nationalization of the Suez Canal had grave consequences. Israel had little direct interest in the Suez Canal, and the Egyptian government was correct in assuming that Israel would not go to war over the issue. Cairo was also correct in believing that Israel would not attack Egypt on its own and in expecting the United States to mediate if the dispute between Egypt and Israel did lead to war. What the Egyptian government failed to recognize was the possibility that Israel would form an anti-Egyptian alliance with both France and Britain. This error was based on a misunderstanding of the depth of

French resentment over Egypt's support for the Algerian independence movement, and of British emotional investment in the Suez Canal.

To be on the safe side, Egypt halted Palestinian guerrilla attacks across the Israeli border and moved additional troops into defensive positions in the Sinai peninsula. Half of the Egyptian Army was now stationed in Sinai and the Canal Zone. Reserves were mobilized and plans to change the national guard militia into a better-equipped part-time Army of National Liberation were speeded up. On 15 September 1956 all British and French Suez Canal pilots were obliged to leave the country. President Nasser also delivered a speech to EAF cadets at the Bilbays Air Academy announcing that Egypt would certainly fight if attacked.

For the EAF this crisis could hardly have come at a worse time. It was still in the first stage of a substantial expansion program and, more significantly, in the middle of its transition from western to Soviet equipment, training methods, and tactics. Even without the sudden Suez Crisis of 1956, the EAF would have faced difficulties by expanding and fundamentally changing its sources of supply at the same time. In fact the associated problems would continue long after 1956.

## Quality versus Quantity

The standards of what might be called the pre-expansion EAF were good, though not as high as those of most western air forces. This existing cadre of RAF-trained air and ground crews provided a solid framework for the ambitious expansion program. Different squadrons achieved differing degrees of success, but in general the EAF's transition from piston-engined aircraft to high-performance jets was achieved with an accident rate not so different from that suffered by the USAF, U.S. Navy (USN), or RAF. On the other hand, the accident rate might have been kept down by diluting training schedules in air combat maneuvering and gunnery. Good as they were, the cadre of RAF-trained personnel still had serious limitations, most of which resulted from restrictions that had been imposed by Britain. By and large they were effective when under the direction of ground controllers, yet this did not make Egyptian pilots into effective combat aircrew. They did not, for example, have the dogfighting skills of the RAF or of their immediate IDF/AF opponents. They had inadequate bombing and gunnery training, and the EAF did not properly integrate different missions and aircraft types. As a result the EAF's success in intercepting high-flying reconnaissance Canberras and in several ground attack missions over Sinai during the 1956 Suez War were not repeated when it came to air combat. Such problems recurred in 1967 and even 1973.

The EAF expansion program of the mid-1950s was only a partial success. An existing core of high-quality personnel was diluted by the speed and scope of this expansion because several branches of the service, as well as the army and navy, were all competing for a limited pool of technically competent young men. The demands of size won out over those of quality, and the EAF ended up with a large number of aircraft flown by pilots who had relatively few flying hours and limited tactical proficiency. A comparable situation was seen among support personnel on the ground.

Ever since its creation the Egyptian Air Force had, to some extent, been reliant on foreign advisors and technical support. In the early days these had been British. For a brief period in the mid 1950s the EAF attempted to become genuinely self-reliant. Now the demands of an ambitious modernization and expansion program meant that the Egyptians again had to rely on foreigners, this time from the Soviet Union and other Communist bloc countries.

The years from 1953 to 1956 shaped the history of modern Egypt. During this period the EAF had an importance far beyond that of most other air forces in developed western countries. The EAF became a symbol of Egypt's political aspirations, both at home and abroad, rather than simply being one part of the nation's defensive capability. A large and effective air force was seen as a concrete example of the country's ability to use modern technology. To a lesser extent the EAF was also becoming an instrument for the defense of a new regime as well as of the country as a whole, a trend that would lead to problems in future years. Many of the profound changes in Egypt's political direction were, however, the results of accidents or external events rather than calculated decisions made by the new government in Cairo. President Nasser certainly had no wish for Egypt to become a Soviet satellite, and he had little political or cultural sympathy with what the Soviet Union stood for. In fact his policies within Egypt were strongly anti-communist. Nevertheless, the EAF's need for modern jet fighters contributed to Nasser's decision to embrace Moscow. President Nasser and his government remained above all patriots, at first Egyptian nationalist but increasingly pan-Arab nationalist, though even this pan-Arabism was used as a tool to further Egyptian interests.

The EAF's turning to Soviet equipment and organizational systems resulted from Egypt's inability to obtain what it considered to be its essential defense needs from the West. Years of fruitless efforts to obtain adequate supplies of suitable Western hardware led to such a turning eastward. Once the decision was made it seemed at first to bear considerable fruit. The EAF was able to modernize and to expand at a remarkable rate, but the long-term results would be far less happy. Soviet equipment, Soviet training, and Soviet organization never gave the Egyptian Air Force the success it expected. Meanwhile the political price that Egypt would pay for turning to the Soviet Union would prove very high indeed. In these respects the Egyptian experience was similar to that of many developing countries, particularly in the third world.

## EAF Order of Battle, 31 December 1955[19]

| Unit | Location | Aircraft | Total | Operational | Comments |
|---|---|---|---|---|---|
| **Fighter and ground attack** | | | | | |
| No. 1 (Day-fighter) | | | | | Furies in store, expecting MiG-15s |
| No. 2 (Day-fighter and ground attack) | Kibrīt | Vampire | 18 | 15 | |
| No. 20 (Day-fighter and ground attack) | Dākhilah and Deversoir | Meteor F4 | 10 | 7 | Flights at Deversoir to protect arrival of new aircraft in Alexandria |
| No. 30 (Day-fighter) | Deversoir | Meteor F8 | 10 | 7 | Vampires to FTU, converting to MiG-15 |
| No. 31 (probably ground attack) | Kibrīt | Vampire | 19 | 16 | |
| **Bomber** | | | | | |
| No. 9 (Medium bomber) | Cairo West | | | | Lancasters in store, expecting Il-28s |
| **Transport and Communications** | | | | | |
| No. 3 (Transport) | Almāẓah | C-47 | 11 | 8 | |
| No. 7 (Transport and paratroop) | Almāẓah | C-46 | 19 | 10 | |
| No. 10 (Headquarters communications) | Heliopolis | Bonanza | 2 | 1 | |

*Continued on next page*

*(Continued)*

**Transport and Communications**

| | | | | | |
|---|---|---|---|---|---|
| No. 11 (Communications) | | C-47 | 4 | 4 | |
| | | C-46 | 1 | 1 | |
| | | Mallard | 2 | 2 | |
| | | Sikorski S-51 | 2 | 1 | |
| No. 12 (Airborne forces) | | C-46 | | | Combined with No. 7 |
| **Training** | | | | | |
| No. 4 (Navigational training) | Alexandria | Beechcraft | 7 | 7 | |
| FTU (Jet conversion) | | Meteor T7 | 5 | 5 | |
| | | Meteor NF13 | 6 | 5 | |
| | | Vampire FB52 | 25 | 17 | |
| | | Vampire T55 | 4 | 4 | |
| | | Harvard | 6 | 6 | |
| Air College (basic training) | Bilbays | Harvard | 37 | 19 | |
| | | Chipmunk | 51 | 37 | |
| | | Jumhūrīyah | 13 | 13 | |
| **Miscellaneous** | | | | | |
| Heath and Agricultural Squadron (spraying) | | Vultee BT-13 | 5 | 5 | |
| | | Super Cub | 1 | 1 | |
| | | Magister | 22 | 2 | |
| | | Mraz Sokol | 17 | 8 | |
| | | Morane 502 | 2 | 2 | License-built Fi 156 Storch |
| | | Hiller helicopter | 2 | 0 | |

## Air Defense Structure of Egypt, October 1956[20]

| Unit | Aircraft | Base | Comments |
|---|---|---|---|
| **Eastern zone** | | | |
| No. 30 | MiG-15 | | |
| No. 20 | Meteor | Deversoir | |
| No. 5 | Meteors | Fā'id | |
| No. 31 | Vampires | Kibrīt | |
| Fighter Training | (Mixed) | Fā'id | Meteors and Vampire available for use |
| **Central zone** | | | |
| No. 1 | MiG-15 | Almāẓah | |
| No. 2 | Vampires | Cairo West | |
| No. 8 | Il-28 | Inshāṣ | |
| No. 9 | Il-28 | Inshāṣ | |
| No. 11 | Il-14 | Almāẓah | |
| No. 3 | C-47 | Almāẓah | |
| No. 7 | C-46 | Almāẓah | |
| No. 4 | (Mixed light aircraft) | Dākhilah | |

# 13

# The Other Side of Suez (1956)

Egypt's international relations were decisively altered by the brief Suez War of 1956. In military terms Egypt lost this war. However, diplomatically and politically Egypt, and President Nasser, emerged stronger from the conflict. Israel was forced by international political pressure to withdraw from the Sinai, while Britain and France were compelled to retreat from Egypt in humiliation despite having overwhelmed the Egyptian armed forces. The Suez War was a watershed in the relationship between the Arab states and their former colonial masters. It also confirmed Egypt's leadership role in the Arab world and helped set the stage for further hostility with the West and conflict with Israel. It pushed Egypt into an international alignment with the Soviet Union that was to last for nearly three decades. The conflict also underlined the EAF's importance not only for its country, but for shaping regional and even global events.

## Secret Preparations for War

In the fall of 1956 President Nasser was focused on domestic matters, his new alliance with Syria and Jordan, and the impact of the nationalization of the Suez Canal. Within a short time Egypt found itself in a war with Israel, Britain, and France. In the summer of 1956 political and military leaders from Britain, France, and Israel secretly met to plan a coordinated assault on Egypt. All feared the effects of the Soviet alliance and the rise of Egyptian military power, and wished to strike a blow at Egypt's military and political influence. Each had its own reasons for attacking: Britain, the nationalization of the Suez Canal and its lost influence; France, Egypt's support for the Algerian Revolution; and Israel, the guerrilla groups that operated from the Gaza Strip and Egypt's blockade of the Gulf of Aqaba.

Egyptian military planners were unaware of the tripartite war plans, though they expected raids by Israeli commandos along the border and had considered the possibility of a surprise attack by British forces to seize the Suez Canal. However, the idea that Britain, France, and Israel would join forces and simultaneously invade the Sinai, bomb Egyptian military positions, and assault the Suez Canal was not seriously considered.

France had dramatically increased its military aid to Israel in 1955–56, selling tanks, artillery, ammunition, transport aircraft, and, most significantly, several dozen Ouragan and Mystère IVA fighters. The Israeli Air Force rushed through a training program to convert pilots to these new fighters, several Ouragan and Mystère squadrons becoming operational by October 1956.

During the fall of 1956 Great Britain and France secretly called up individual reservists, assembled an invasion fleet, and flew hundreds of fighters and bombers to bases in Malta and Cyprus for the attack on Egypt. Simultaneously Israel secretly mobilized and deployed to invade Sinai.

To encourage Israeli participation in the attack, France pledged to defend Israel from air attack by Egyptian jet bombers. The emergence of the Egyptian Il-28 bomber threat had caused serious concern at the highest levels of the Israeli government. It had been a key element in pushing Israel toward a security arrangement with France that led to its joining the alliance against Egypt. Here again the EAF had proven to have had a direct influence on the geopolitics of the entire Middle East.

To the Israelis the Il-28s meant that the EAF bombing threat to Israel was serious in a way that the surplus World War II bombers and converted transports could never have been. The Israelis realized that Il-28s operating at 35,000 feet could strike throughout Israel, especially when operating at night, without being intercepted by their Ouragan fighters.

In the longer term, the Israeli answer to the emergence of the bombing threat to Israel was to reorganize its entire air force from an essentially defensive to an offensive counterair strike force intended to preempt this threat, but this could not be accomplished for many years. The immediate answer was to deploy French fighters that could shoot down the Il-28s. Dozens of French Air Force fighters—Mystère IVAs and F-84F Thunderstreaks—secretly flew to Israel in late October 1956 to fulfill French promises to defend Israeli airspace. On landing, the fighters were repainted with Israeli insignia. French pilots and ground crew were issued with Israeli identity cards written in Hebrew. This highly secret affair was code-named "Operation 750."[1]

However, despite their role in bringing France and Israel together, fears of an Egyptian bombing capability against Israel were exaggerated. In 1956 the EAF Il-28 force was more image than substance. Not only were the Il-28s still less than fully operational, but the aircraft was plagued with low reliability and poor performance in the Middle East environment. The EAF was struggling to train air and ground crews in operating the multi-jet Il-28s. The EAF was aware of the limitations of the Il-28, but they were far better than the surplus British Halifax bombers. Air Force leaders planned to create an experienced group of pilots and support personnel that later could be transferred to more powerful bomber aircraft.

Senior EAF commanders had seen the political impact of the limited nocturnal raids performed by both sides in 1948, even if the damage caused by these attacks had been limited. In addition, the RAF influence on the EAF, which included a deep respect for the military and political influence of bomber strikes against countervalue targets, continued to be felt. Meanwhile EAF officers and enlisted men still attended RAF schools until just before the Suez War.

British, French, and Israeli plans called for a four-phase assault on Egypt. First, the Israelis would invade the Sinai on 30 October. Then Britain and France would "impartially" call for Israel and Egypt to pull back from the Suez Canal and issue a deadline for both sides to end the fighting. The safety of the Suez Canal was more of a pretext for the invasion than a real cause for concern. When conflict continued (as anticipated) there would be an excuse to intervene. Phase Two of the Anglo-French plan was four days of air attacks to destroy the EAF. After the air assault would come a British and French paratroop assault and amphibious landings. The final phase called for Anglo-French forces to secure the northern portion of the Suez Canal.

## Transition, Reorganization, and Training

While the army was the largest and politically most powerful military service in Egypt, the EAF had considerable influence. President Nasser, proud of his Russian-supplied air force, frequently employed it as a political and public relations tool. Maj. Gen. ʿAbd al-Laṭīf Baghdādī, the

EAF commander in 1956, was a member of the Free Officers group that had staged the 1952 coup.

The Suez War caught the EAF as it was shifting from a small organization patterned on the RAF and equipped with British aircraft, to a much larger Soviet-influenced air arm. Blessed with a large and well-equipped complex of airfields inherited from the British, the EAF still suffered from a lack of trained and experienced aircrew and support personnel. The transition to the increased sophistication of second generation jets like the MiG-15 and F-86 compared to piston-engined aircraft and early jets hurt the readiness and safety records of even the large and highly experienced U.S., British, and Soviet air forces. Despite many challenges, the EAF was making rapid progress due in part to training assistance from Russian and East European advisors.

By the fall of 1956 many EAF fighter units had transitioned from Furies, Meteors, and Vampires to MiG-15s, and a small number of Il-28s were operational. By 1956 the EAF had divided the country into defensive zones, with the eastern and central being the most important. The eastern zone, with its headquarters at Ismāʿīlīyah, covered the Canal Zone and offered tactical support to the army's Eastern Command in the Sinai as well as reconnaissance along the border with Israel. The central zone, with its headquarters at Almāẓah, was tasked with the defense of Cairo and the Egyptian heartland and management of the EAF's strategic bombers and transports. British, French, and Israeli intelligence sources suggest that some central zone squadrons had deployed to bases in the Canal Zone or had moved separate flights to more than one base by the time the war broke out. According to American, British, French, and Israeli intelligence sources, in October 1956 the EAF had a total of 69 MiG-15s and 24 Il-28s assigned to operational squadrons with many other aircraft in storage and maintenance. These units were backed up by 84 Vampire and Meteor jets and the piston-engined Fury and Spitfire fighters assigned to advanced training squadrons.[2] The Egyptian Air Force also had a personnel strength of 6,400 including some 400 officers, 3,000 enlisted, and 3,000 civilians. However, on the eve of the Suez Conflict some 200 EAF aircrew and technicians were away in Poland and the Soviet Union attending training courses. While the force included some 440 pilots, fewer than 100 were thought to be combat ready.[3]

The primary mission of the EAF was protecting Egyptian airspace and defending army units in the Sinai. Egypt had faced a strong air threat from Israel, and increasing Soviet influence pushed the EAF even more toward a defensive philosophy. In the 1950s the Soviet military was obsessed with air defense. The memories of the 1941 German attack on Russia and threat from American and British nuclear-armed bombers drove the Soviets to create a massive air defense force, the "PVO Strany." This organization became the largest and most important element of the Soviet military after the army.

This essentially defensive Soviet air doctrine was now exported to Egypt, even if it did not fit the conditions and environment of the Middle East. Soviet doctrine emphasized air defense first, followed by air superiority and offensive air support. During the decade following the Second World War only one tactical aircraft, the Il-28, was created and fielded to perform offensive air support. The remaining aircraft, which included the MiG-9, MiG-15, MiG-17, and Yak-15, were interceptors designed to shoot down bombers.[4]

Egypt's air defense network, which provided warning and direction for both EAF fighters and army anti-aircraft guns, was slowly being upgraded with Russian assistance. Since 1955 many new Soviet-built radars and control centers had been established on the coast, in the Sinai and down the Suez Canal and Gulf of Suez. Many of these new air defense sites and radar units were not fully operational in the fall of 1956 as their crews were still in training.[5] Point air defense of ground forces, military bases, and airfields was the responsibility of Egyptian Army anti-aircraft units. Airfields and important military bases were defended by army 20–57 mm anti-aircraft guns.[6]

While still trying to adjust to a new organization and doctrine and to master advanced Russian-built equipment, Egypt's military was forced to fight a war against enemies that were numerically superior and more experienced. In the air the EAF was outnumbered more than three to one in fighters and five to one in bombers. When the impact of the surprise attack plus superior training and experience was added to the equation, the final outcome of combat was not in doubt. Since the aim of the secret Anglo-French-Israeli plan was to break Egypt's military power for years to come and topple President Nasser, major EAF bases and army centers were naturally priority targets.

The British and French air armada included 115 jet bombers, on Cyprus and Malta, and 108 land-based jet fighter-bombers (all flying from the island of Cyprus). On aircraft carriers were a further 117 British and 46 French combat aircraft. Britain had 34 transports and 12 helicopters, while France contributed 36 transports and 4 helicopters to deliver paratroops. Israel entered the battle with 136 combat aircraft including about 60 Mystère, Ouragan, and Meteor jets and 55 Mustang and Mosquito fighter-bombers. Three French squadrons—two with Mystère fighters and one with F-84F fighter-bombers—defended Israel, so the IAF could concentrate on supporting the invasion of the Sinai.[7]

RAF Canberra reconnaissance jets kept watch on the Egyptian armed forces, mostly flying just outside the territorial limits but on several occasions penetrating Egyptian airspace. These flights were noticed, and the Anglo-French buildup was similarly detected by Egyptian intelligence. President Nasser and Chief of Staff Marshal ʿAbd al-Ḥakīm ʿAmr ordered several Egyptian Army units to pull back from the Sinai to protect the Suez Canal. This left some 30,000 troops and about 200 tanks to face Israel's 45,000 troops and more than 300 tanks.

At the end of October, as the crisis deepened, the Egyptian commander in chief, ʿAbd al-Ḥakīm ʿAmr, secretly flew to Damascus to discuss the Egyptian-Syrian military pact. On the evening of 29 October one of the two EAF Il-14 transports that had carried the Egyptian delegation to Syria took off from Damascus and headed out over the Mediterranean. On board were the flight crew plus a number of military officers and Arab journalists. An Israeli Meteor NF13 night-fighter intercepted and shot down this transport: the Israelis thought it was carrying Marshal ʿAmr. However, his plane, piloted by Flt. Lt. Saʿd al-Dīn Sharīf, took off later the same evening and arrived safely.[8]

## 29–30 October: Israeli Surprise Attack

Israel struck the first blow of the Suez War on the evening of 29 October 1956. Several hundred Israeli paratroopers landed near Mitlā Pass while armored columns crossed the border into the Sinai. There were no Israeli counterair attacks in order not to provoke the EAF into retaliating with its Il-28 jet bombers. In any case Israeli military leaders knew that the British and French would soon bomb Egyptian airfields. These Israeli moves were first thought to be another of their frequent commando raids, and Egyptian Army headquarters ordered units to respond. Egyptian military leaders had expected an Israeli strike in response to the new Arab alliance and continued blockade of the Gulf of Aqaba, but not a full-scale invasion of the Sinai.

The next morning (30 October) four RAF Canberra reconnaissance jets flew along the Suez Canal at 12,000 meters to monitor Egyptian reaction to the Israeli assault. Intercepted by EAF MiG-15s, vectored by a Soviet-installed ground control interception (GCI) system, one Canberra was seriously damaged. In Antony Eden's own words, this attack was "a brilliant piece of work by any standards," and led to a rapid change in British air operational plans.[9] No longer could high altitude provide safety in daylight.

Later in the morning four EAF Vampires from No. 2 squadron at Fā'id successfully flew a reconnaissance mission over the Sinai. With their reports the Egyptians realized the size of the Israeli assault. The first Egyptian air attack was made by four MiG-15s; two attacked the Israelis at Mitlā and destroyed an IAF L-8 Cub on the ground, and two successfully strafed an Israeli column. At 11:00 A.M. Egyptian Vampires again attacked the Israelis.

## Air Superiority: The MiG-15 versus The Mystere IVA

In response to these damaging raids, the Israelis started flying combat air patrols with Mystères and Ouragans over the Sinai. Meanwhile the EAF sent in more strikes. On 30 October the EAF flew more than fifty sorties over the Sinai, which disrupted Israeli operations at Mitlā Pass and inflicted casualties on several columns. In return, heavy Israeli attacks caused losses among Egyptian reinforcements coming in from the west, although Egyptian anti-aircraft fire claimed two Israeli P-51 Mustangs.

The first air combat took place at about 3:30 P.M. on 30 October. Two EAF Meteors and six MiG-15s met a patrol of Israeli Mystères near Mitlā Pass. The MiGs engaged the Mystères, allowing the Meteors to make their ground attack runs, and neither side lost any aircraft. A confused engagement between seven MiG-15s and six Mystères late in the afternoon of 30 October near Kībrīt (on the western side of Little Bitter Lake) resulted in the loss of one MiG and damage to one Mystère. The Israelis were thought to have been attempting to attack Kībrīt airfield itself. As the Mystères did not press home such an attack, and the MiG pilots returned to base firmly convinced that they had been fighting Israeli aircraft of whom two had been shot down, the morale of the EAF's fighter squadrons now rose considerably.[10]

These first air battles had demonstrated that Israeli pilots and the new French-built Mystère IVA fighters were superior to EAF MiG-15s. The Mystère was fast, in fact it was supersonic in a dive, but the MiG-15 could climb quicker and was more maneuverable. In air combat, the Mystère's power-assisted controls allowed tighter turns at high speeds. Its two 30 mm DEFA cannon had higher rates of fire and muzzle velocity than the MiG-15's two Nudelman Sunavov NS-23 23 mm and one Nudelman Sunavov NS-37 37 mm cannon. The MiG-15 and Mystère fighters had both just recently arrived in service in the Middle East, and neither the Egyptian nor the Israeli pilots had much experience flying such modern jets.

On the evening of 30 October France and Britain demanded that both Egypt and Israel stop fighting, a demand that was part of the previously arranged war plan. Israel accepted the order, but the army and IAF continued their air and ground attacks. As a result of stiff Egyptian resistance, Israeli forces had not been able to approach the Suez Canal. Egypt rejected the demand since she was defending her own territory. During the night of 30–31 October EAF Il-28 jet bombers claimed to have bombed Israeli military targets at Al-Awjam, ʿAqīr, Ramat Dawid, and Kāstīnā. According to Israeli sources, bombs only fell near Ramat Raziʾel west of Jerusalem, causing no damage.

## 31 October: The Sinai Battle Decided

Egyptian and Israeli ground forces fought intense battles on 31 October, the Israelis pushing back the Egyptians with the help of heavy air support. At dawn on Wednesday, 31 October, four EAF Vampires flying in to strike Israeli forces near Mitlā Pass were bounced by two Israeli Mystères; three were shot down, while one survivor, badly damaged, managed to escape back over the Suez Canal before crash-landing. EAF officers Maḥmūd Wāʾil ʿAfīfī and Baghāt Ḥasan Ḥilmī were killed, and another EAF pilot was captured.[11]

Brig. Gen. Fārūq al-Ghazzāwī (ret.), then a lieutenant, took off from Abū Ṣuwayr on the afternoon of 31 October:

> We were briefed to fly with three sections of two in an open battle formation and to patrol over the Sinai at 6,000 meters (20,000 feet). I picked up an Ouragan target, which was flying low on a reverse course heading southwest.

> I told my squadron mates about the target. They said, "you lead and we will follow."
>
> It was my first engagement with an enemy aircraft. I kept my eye on the enemy aircraft, started to close, and armed the guns. I should have checked whether my leader was following me or not, but all I remember was how excited I was. Then I discovered that I was alone, but I kept my gunsight on the plane and fired one, two, three bursts; he was from time to time changing direction to throw off my aim.
>
> All of a sudden I heard two booms, then nothing. I thought it was a bump from flying through his slipstream. My target reversed his steep turn and then I felt a severe crash. I was at a high angle of attack, perhaps forty degrees, and the stick was jammed. I put on full throttle—there was no afterburner in the MiG-15—and I climbed back to 6,000 meters.
>
> I found myself all alone. It was calm, not a word on the radio. I had taken a bullet through the canopy, so I was losing pressurization. I moved the stick. It jammed in a slight upward position. Luckily I could fly straight and level. I came close to the Suez Canal, losing height. I was worried because there was a big hole in the wing. I was also worried that the enemy fire had damaged the undercarriage. But luckily I came in and landed O.K.
>
> [In the debrief, we figured out what had happened]: I surprised and hit an Israeli Ouragan, and his number two came in and attacked me. Later I found out that the plane I hit had to crash-land.[12]

While strafing a convoy an EAF Meteor was meanwhile lost to Israeli ground fire. EAF MiG-15s also shot down an IAF L-8 Cub over the central Sinai, and a pair of Egyptian Vampires bounced Israeli P-51s strafing a convoy, damaging one that crash-landed in the desert.[13] In several other air battles, one MiG-17 and one MiG-15 were lost to Israeli Mystères. Hit in the wing by cannon fire over the northern Sinai, the pilot of the MiG-15 made a successful wheels-up landing in the shallows of Lake Bardawil near the Mediterranean coast. He escaped, and his MiG was later salvaged by the Israelis.[14]

Pilots of the Fighter Training Squadron at Fāᶜid attacked Israeli forces in the Sinai despite their vulnerability to Israeli jets. Taḥsīn Zakiī an instructor, strafed several Israeli targets with a Fury. (He survived to became the leader of the EAF aerobatic team and command the first squadron of Su-7s in 1967.) One of his students, ᶜAlī Sharmī, made at least one reconnaissance sortie flying a Spitfire F22.

During the second day of fighting, EAF aircraft flew more than a hundred sorties against Israeli forces in the Sinai. French F-84F fighter-bombers based in Israel, leaving the defensive counterair missions to the Mystère squadrons, flew their first raids against Egyptian forces in the Sinai on the afternoon of 31 October. In fact French fighter-bombers attacked Egyptian targets in the Sinai even before the 6:00 P.M. cease-fire deadline that was to be the excuse for Anglo-French intervention.[15] Israeli and French fighters flew more than two hundred sorties over the Sinai on 31 October; in addition to the two planes lost to EAF fighters, several fell to Egyptian anti-aircraft fire.

French transport aircraft, flying from Cyprus, helped maintain the rapid Israeli advance in the Sinai. Named "Operation Archer," this resupply effort was undertaken after dark on 31 October–1 November, without the knowledge or approval of their British allies. Flying at night to avoid interception by EAF fighters, nine French Nord Noratlas transports delivered fuel, food, and ammunition to Israeli paratroopers in the Sinai.[16]

As soon as fighting broke out, Soviet and Czechoslovak instructors and technicians and Syrian pilots had pulled out of Egypt. To keep Syrian jets and nonessential aircraft safe from attack, Soviet, Czechoslovak, and Egyptian pilots also flew numerous MiG-15s, MiG-15UTI trainers, Il-28s, and Il-14 transports to Syria and Saudi Arabia. Egyptian air defenses were put on alert; fighter bases assigned pilots to be prepared for quick alert scrambles, while radar sites and anti-aircraft guns were prepared for action.

## 31 October: The Anglo-French Assault Opens

The British and French assault force included more than 50,000 men, 130 warships and transport vessels (including two French and five British aircraft carriers), and nearly 500 aircraft. More than half of Britain's Bomber Command and a large segment of the RAF's tactical fighter force was flown to Malta and Cyprus to participate in the attack. A Royal Navy task force comprising *HMS Albion, HMS Bulwark, HMS Eagle* (with 117 aircraft), and their escorts operated in the Mediterranean. In addition, eight French Air Force squadrons were committed to the assault along with two French Navy aircraft carriers, the *Arromanches* and *Lafayette* (with 46 aircraft).[17]

To obtain up-to-date intelligence on the afternoon of 31 October, RAF Canberra and French Air Force RF-84 jets penetrated deep into Egyptian airspace. The principal targets of these missions were airfields, and the location of EAF units were confirmed in order to plan air strikes.

Just after dark on 31 October waves of high-flying British bombers began the Anglo-French air offensive, but their targeting was flawed as a result of the EAF's successful interception of RAF reconnaissance Canberras. They were also operating at higher altitude than initially planned due to expected EAF opposition.

The first wave of bombers attacked the wrong target. Diverted in flight from hitting the EAF bomber base at Cairo West, as American civilians were reportedly in the process of evacuating from this airfield, the RAF bombers were sent to strike Almāẓah on the eastern side of Cairo, home of a MiG-15 unit. The RAF instead bombed Cairo International Airport after Canberra pathfinders mistakenly dropped target marker flares, thinking it was Almāẓah. RAF Valiants and Canberras came in and hit the airfield with 225 and 453 kilogram bombs.

From about 10:00 P.M. until early morning, waves of RAF Canberras and Valiants bombed Almāẓah, Bilbays, Inshāṣ, Abū Ṣuwayr, and Kibrīt from high altitude.[18] Detected by Egyptian radar, the high-flying intruders were engaged by heavy anti-aircraft guns and Meteor night-fighters, one forcing Grp. Capt. L. M. Hodges' Valiant to take violent evasive action to evade an attack.[19]

Despite high altitude bombing by the RAF, the EAF's few operational Il-28s continued to fly night raids against Israeli military targets inside Israel and close to the Egyptian frontier. At the same time other Il-28s were dispersed from Almāẓah and Cairo West. According to the ʿAlī Muḥammad Labīb, who was then Director of Operations at Cairo West, the RAF bombing caused little difficulty. The Egyptian bomber crews merely waited for lulls between raids before themselves taking off. Delayed action bombs which the British scattered across Egyptian airfields were, however, a constant threat as these devices could not be located and defused in the dark.[20]

## 1 November: Nasser's Decision—Fight on Our Terms

President Nasser, who was at his house near Almāẓah airfield, heard the RAF bombers and observed the attack from his roof. Leaving for his military headquarters, the president was greeted by crowds chanting "We shall fight!" Calling together his military commanders, Nasser ordered the Egyptian Army to withdraw from the Sinai to defend against the expected Anglo-French invasion while authorizing distribution of weapons to civilians to prepare for guerrilla warfare. Egypt's senior military commanders were surprised by Nasser's decisions, but forces began pulling back from the Sinai by midnight.

Nasser also ordered the EAF not to contest the allied air assault. This act was interpreted by some in the West as evidence of a lack of fighting spirit among Egyptian pilots.[21] However, after the cease-fire Nasser explained: "We have 120 pilots fully trained for combat, and another 250 to 260 still in training. If I sent these to fight against the combined air forces of Britain and France I would be mad. At some stage the British and French are going to withdraw—probably after a month or two. But we are going to be in a state of war with Israel for years, and we shall need all the pilots we can get. Planes can be replaced overnight, but it takes years to train a pilot."[22]

Despite night bombings by the RAF, the EAF continued to focus on events in the Sinai. Major General Shalabī al-Ḥinnāwī (ret.), as a major, commanded the only MiG-17 squadron to serve in the Suez conflict: "At first light on 1 November I led a flight of four MiG-17s in a strafing attack against an Israeli position near Mitlā Pass. The Israelis were dug in and well camouflaged—exactly like sand—and were very difficult to see. We started firing and caused several explosions. We finished our ammunition and turned back."[23]

British and French reconnaissance jets soon discovered that the night bomber attacks had only limited success: some fourteen EAF aircraft had been destroyed, but runways and facilities suffered minimal damage. Information from these early missions altered attack plans for subsequent Anglo-French raids.

## 1 November: The Empire Strikes Back

Soon after dawn, British and French fighter-bombers from Cyprus and aircraft carriers struck all the major EAF bases in northern Egypt. Royal Navy squadrons struck Egyptian bases west of the 32 degree line of longitude, while the French and Royal Air Force jets hit bases located east of this line.[24] Major Shalabī al-Ḥinnāwī had completed his morning strike and was preparing for another mission: "I was giving a briefing to my pilots. At this moment we saw four aircraft—Sea Hawks—come in and strafe. Unfortunately our MiG-17s were sitting in the open. The attack met with very weak anti-aircraft fire. Many of our aircraft were wrecked.

I remember seeing one Il-28 jet blow up. A shell splinter from the strafing hit me in the leg. I was ordered to go to the hospital for surgery, but I refused."[25]

Maj. Gen. Muḥammed Nabīl al-Masīrī (ret.), who was a captain in 1956, described his experiences on the morning of 1 November:

> We were at Cairo West. I was in a state of [cockpit] readiness and could take off and get into action if the enemy were to come. We had four [MiG-15] aircraft near the start of the runway in an open area. However, I was attacked while I was on the ground without any warning. I had only a very narrow chance to jump from my jet and run from the attacking aircraft. The jets were turning to the left toward me. My number two asked me "What should I do?" I yelled "Jump!" I ran to the right to get away from their guns. I got only twenty or thirty meters before my aircraft was destroyed.
>
> I didn't see [my number two] after I ran. I saw that all four aircraft were in flames. After a while I saw this young pilot come out of the flames not touched or burned, and he was still struggling with his parachute which was going left and right . . . it was very funny at the time. No one saw us get out, and they thought we were killed.[26]

Brigadier General Fārūq al-Ghazzāwī (ret.) commented on his experiences: "I was at Abū Ṣuwayr, which was a former Royal Air Force base. It was a lovely place full of beautiful gardens. We were just finishing breakfast, and I was about to go to the squadron when we were caught. I remember seeing four Hawker Sea Hawks come in and strafe our aircraft. I was face down in one of the gardens and looked up and watched the jets hitting us."[27]

EAF pilots and ground personnel wanted to fight back, despite Nasser's orders, but air bases in northern Egypt were under almost constant attack for most of that day. As a result of control procedures long drilled into Egyptian officers (perhaps designed to minimize the possibility of coup attempts), many EAF base and unit commanders waited for orders before making decisions on how to react to the Anglo-French attacks. As a result, opportunities to save aircraft or strike back were lost.

Major General al-Ḥinnāwī remarked: "Some eight of our MiGs covered by nets near the hangers weren't hit in the first attacks. We tried to move the surviving MiGs, but we couldn't get approval from headquarters, and they were destroyed in later raids.[28]

Major General Māsīrī added: "There were repeated attacks, and most of the aircraft at our base were destroyed on the ground. However some of the Il-28 bombers were able to take off between attacks and escaped, but most aircraft were destroyed."[29]

During the next several days, between air strikes, dozens of EAF aircraft were flown out of the country to escape destruction. Most flew to Syria or Saudi Arabia, but some Il-28s went to Luxor air base in the south of Egypt. According to Brigadier General Jawāʿī (ret.), some of the surviving MiG-15s from his squadron escaped to alternate strips. Al-Ghazzāwī and fellow pilots flew a number of missions from the Cairo-Suez road during the next several days.[30] On the afternoon of 1 November, EAF MiG-15s shot up a high-flying Canberra, but due to the ongoing British and French air strikes, the EAF was unable to cover army units withdrawing from the Sinai.

On 1 November Egyptian air bases were hit by 205 Royal Navy, 106 RAF, and 75 French fighter-bomber sorties. These aircraft struck at Abū Ṣuwayr, Almāẓah, Bilbays, Cairo West, Dākhilah, Inshāṣ, and several other bases. During the night Canberra and Valiant bombers again bombed Cairo West, Fāʾid, Kasfarīt, and Luxor, where four Il-28s were destroyed. Several British and French aircraft were damaged by Egyptian fighters and anti-aircraft fire, but none was shot down.

## 2–4 November: Assault and the Truce Talks

By 2 November Anglo-French raids had damaged or destroyed one hundred EAF aircraft and seriously disrupted many air bases. On 2 November French aircraft carriers also launched their first strikes which hit targets near Port Said and Egyptian naval vessels in Alexandria harbor. However, when French pilots detected ships of the U.S. Navy at anchor, they called off their attacks.

Canberras, escorted by French F-84F fighters, struck several Egyptian targets, including Radio Cairo, in daylight raids. Most of the surviving Il-28 jet bombers were then located by Canberras at Luxor and destroyed by French Air Force F-84F fighter-bombers flying from Israel. A total of nineteen Il-28 bombers had been destroyed at this EAF base.[31] The RAF came back again after dark and bombed several airfields on the night of 2–3 November.

Air activity over northern Egypt and the Sinai at this time was described as fantastic, but it was almost all flown by British, French, or Israeli aircraft. British and French aircraft bombed targets near Port Said and the Suez Canal in preparation for the invasion, while Israeli aircraft and French jets based in Israel continued to bomb and strafe Egyptian Army units withdrawing from the Sinai. Several of the attackers were, however, lost to anti-aircraft fire.

Despite heavy damage to airfields and the loss of more than 150 aircraft, the EAF continued to fight back. A MiG-15 attacked and severely damaged a Canberra over central Egypt on 3 November. However, continuing Anglo-French air attacks frequently disrupted EAF repair and counterattack efforts. For example, two Meteors fueling for a mission at Fāʾid on 3 November were destroyed by a flight of Venom fighter-bombers.[32]

For several days the United Nations had been working under heavy pressure from the American and Russian governments to secure a cease fire in Egypt. On 3 November Dag Hammarskjold, secretary general of the UN, approved a plan that called for international peacekeeping forces to rapidly deploy to Egypt. The UN demanded that British, French, Israeli, and Egyptian forces cease hostilities on the evening of 6 November 1956.

Heavy air strikes on 4 November paved the way for the invasion, while Israeli and French aircraft continued to pound Egyptian units in the Sinai from the air. Allied jets maintained their interdiction of EAF airfields, and the EAF lost a number of aircraft during the day to these raids but only one pilot, Flt. Off. ʿAbd al-Munʿim Ḥāfiẓ Muḥammad Iwāsī, was admitted killed during these strikes.[33]

## 5 November: Anglo-French Invasion

On the morning of 5 November, Israeli forces overran Egyptian defenses at Sharm al-Shaykh and also reached the Suez Canal, while British and French troops landed by parachute and helicopter near Port Said to secure the beaches for an amphibious assault. Two Sea Hawks and one Venom, however, were shot down by ground fire in the bitter fighting around Port Said. At dawn on 6 November one of the surviving MiG-15s, operating from a roadway in the Nile Delta, strafed British-held Jamīl airfield. An RAF Venom tried to intercept, but the MiG—the British claimed it had Soviet markings—easily escaped.[34]

Political activities involving the United Nations soon brought an end to the fighting. During discussions the Soviet Union had threatened to send troops to the Middle East and even to use nuclear weapons unless Britain and France immediately ceased their assault on Egypt. American representatives supported the Russian call for an immediate cease-fire. While Israeli forces had secured their objectives and stopped fighting on 5 November, the British and French continued to push down the Suez Canal right up to the cease-fire hour which came into effect at dusk on 6 November.

## Outcome

For the Egyptian Air Force, the conflict had been a disaster. The allied air campaign against Egypt had been the most intense aerial bombardment since the Korean War. During the five-day air campaign, British and French aircraft flew more than 5,000 sorties against Egypt. Royal Navy Fleet Air Arm aircraft flew 1,130 sorties, while RAF jets contributed more than 3,000 strike and support sorties. The French Groupement Mixte No. 1 based in Cyprus had flown 516 sorties, while Aéronavale aircraft performed 186 sorties from the two French aircraft carriers. The number of sorties flown by French squadrons based in Israel has not been revealed.[35]

Five days and nights of air strikes by the Anglo-French air armada had badly damaged airfields and support facilities. Before they withdrew from the Sinai, Israeli forces systematically destroyed Egyptian bases, roads, railways, and other elements of infrastructure.

The EAF was proud of its performance in action against Israel; more than two hundred sorties were flown during the initial phase of combat in the Sinai. Egyptian fighters downed an L-8 Cub and a P-51 and damaged several IAF aircraft, including an Ouragan jet. In air combat with Israeli Mystères, the EAF lost (according to Israeli claims) one MiG-17, three MiG-15s, and four Vampires, while one Meteor was shot down by ground fire.

Following the conflict, the IAF admitted the loss of fifteen aircraft, thirteen of these to Egyptian anti-aircraft fire. British and French sources believed that Israeli losses were actually twenty or more aircraft, if those damaged beyond repair or destroyed in accidents were included. British losses stood at eight aircraft, at least two of which were accidents, while the French lost two planes to ground fire and one to operational causes.[36]

International political pressure, especially the harsh threats from the Soviet Union and lack of support form the United States, forced British, French, and Israeli forces to withdraw rapidly from Egypt. Despite this political setback, the three countries had accomplished one of their goals: the fighting had shattered Egypt's armed forces. British estimates of the number of EAF aircraft damaged or destroyed by air attack in the conflict included 91 MiG-15/17s, 11 Meteors, 30 Vampires, 26 Il-28s, and 63 trainers and support aircraft.[37] This estimate appears to be high, given the fact that many EAF aircraft had flown out of Egypt to safety during the fighting. Only five EAF pilots had died, and about twenty other EAF personnel were killed during Anglo-French air attacks, proportionally much less than the Egyptian Army had suffered.

14

# A Doubtful Anniversary (1956–1961)

The British, French, and Israelis failed in their goal to bring down President Nasser and establish international control over the Suez Canal. While Egypt's armed forces were defeated in battle and the Sinai was seized by Israeli forces, the Suez War strengthened Nasser's political position in Egypt and the Arab world as he was credited with scoring a diplomatic victory. Following the war, President Nasser also consolidated Egyptian control over the strategic Suez waterway and nationalized British and French assets. This defeat was a fundamental blow to British policies in the Middle East. Within a short time the pro-West government in Iraq fell and those in Jordan and Lebanon were threatened, while British forces in Aden and Muscat faced increasing unrest. UN actions, aggressively supported by President Eisenhower and Soviet leaders, forced the British and the French to withdraw from Egypt by 22 December 1956. Israeli forces were also compelled to pull out of the Sinai but delayed their move until March 1957. By then the United Nations Emergency Force was in place to patrol Gaza, Sharm al-Shaykh, and along the Sinai frontier. As the Israelis withdrew, they destroyed Egyptian roads, railways, and military bases. For Israel, however, its military victory bore some fruit. President Eisenhower guaranteed the right of passage for shipping bound for Israel, and the presence of UN forces minimized the threat of attack by Arab guerrilla forces. Egyptian leaders still wanted to improve relations with the United States, and diplomats met with senior representatives of the State Department. However, when these meetings produced no results, Egypt again turned to Russia. As a result, the Soviet Union solidified its position in Egypt through the rapid delivery of military supplies worth more than $150 million. Along with this military aid, the Soviet Union agreed to increase its funding for the Aswan Dam project and other pressing infrastructure projects. Egypt was allowed to pay for this assistance with raw materials such as cotton and long-term, low-interest loans.[1]

## President Nasser's Military Expansion Plan

President Nasser and his advisors reviewed the lessons of the Suez War and developed a strategic plan to guide future military and political activities. Objectives of the new plan included: (1) confronting Israel; (2) deterring western aggression; and (3) supporting Egyptian efforts to lead the pan-Arab movement.

To achieve these general goals a long-term plan for the expansion of the Egyptian armed forces was created. Air power was given heavy emphasis in the new development plan since the air force could both threaten Israel and the West. Air, mobile forces, and armored units also received priority funding. President Nasser later admitted in a speech that not until after the 1956 War did Egypt create a well-thought out plan to guide military development.[2]

The Anglo-French air bombardment caused limited EAF casualties. Nevertheless, hangers, runways, and other facilities at dozens of air bases suffered extensive damage, and most of the EAF's inventory of aircraft had been destroyed. Following the withdrawal of enemy forces, the Egyptians began repairing damaged facilities and rebuilding their armed forces.

The USSR promptly responded to Egypt's requests for assistance, and in March 1957 three Rumanian ships delivered fifteen crated MiG-17s, ten disassembled Il-28s, and other military equipment. This massive Russian resupply program quickly restored the military balance between Egypt and Israel. More than 650 Soviet, Polish, and Czechoslovak military officers and technicians arrived with the new weapons. Foreign "advisors" serviced Russian-built equipment, translated Soviet training manuals into Arabic, and encouraged Egypt's armed forces to reorganize along Russian lines.[3] The infusion of new aircraft and intensified training also allowed the EAF to standardize on fewer types of aircraft and streamline its maintenance organization.

UN forces in the Sinai separated Egyptian and Israeli forces, reducing the opportunity for conflict and thus enabling Egypt to concentrate on domestic matters as well as the task of rebuilding its military. Thousands of officers and enlisted men were recruited and existing staff retrained.

Hundreds of EAF aircrew and ground personnel were sent to the Soviet Union, Poland, Bulgaria, and Czechoslovakia for training. Selected officers attended the Soviet Frunze General Staff Academy.[4] The high degree of coordination between Egypt, the Soviet Union, and its allies came to light on 14 May 1957 when Pilot Off. ʿAbd al-Munʿim al-Shinnāwī inadvertently landed his Czechoslovak Air Force MiG-15 at Schwechat airport outside Vienna.

Egyptian morale soared as new jets and other weapons poured into Egypt. By June 1957 the EAF had received one hundred MiGs and forty Il-28 bombers. On 23 July 1957 the EAF staged a massive air show over Cairo to demonstrate its air power. Air Vice Marshal Maḥmūd Ṣidqī Maḥmūd, the new head of the EAF, delivered a fiery speech as twenty-one MiG-15s, eighteen MiG-17s, and a dozen Il-28s flew overhead. To celebrate the EAF's twenty-fifth birthday on 19 December 1957, an impressive static display of aircraft was highlighted at Almāẓah, while flights of MiG-17s and Il-28 bombers buzzed over Cairo.

## The Challenge of Change

Egypt was clearly undergoing great political and social changes in the late 1950s. A new class of merchants, bankers, government bureaucrats, and military officers was taking power from those who held power during the monarchy. It was not an easy transition; liberal political factions demonstrated for radical change, while more conservative groups called for the strict following of Islamic principles. President Nasser embraced a strong Arab nationalist political position and was implementing his doctrine of Arab socialism through government domination of economic enterprises and activities. Increasingly Nasser voiced an anti-western and anti-imperialist line in foreign affairs while adopting socialist and authoritarian policies within Egypt.

Slowly Nasser increased his control over all aspects of Egyptian government. He did not tolerate groups or individuals that challenged his authority. Those who opposed his regime were denied employment, jailed, or deported. The political and social tensions of the time had an impact on the armed forces. Late in 1956 an Egyptian Air Force intelligence officer named ʿIsām al-Dīn Maḥmūd Khalīl discovered that members of the royal family and other individuals who had lost out in the 1952 Revolution were plotting an anti-Nasser revolt. Khalīl was contacted by Murtaḍá al-Marāghī, Fārūq's one-time minister of the interior, and Ḥusayn Khayri, a relative of the king, who offered him a large sum of money to help organize a coup d'état. Maḥmūd Khalīl then played along with the plotters and assisted the Egyptian police in arresting the conspirators.[5]

## The EAF and the United Arab Republic

In an attempt to expand Egyptian power and influence, President Nasser formed a political union with Syria and Yemen. The confederation with Syria was known as the United Arab Republic (UAR). Syrian leaders joined this confederation to help consolidate their power and head off a feared military coup. The imām of Yemen agreed to ally with Egypt to deflect British and Saudi pressure. President Nasser stressed that the Arabs were one nation that had been divided by the imperialists to keep them under control. By working together under his Egyptian leadership, President Nasser said that the Arabs could again be strong, an idea that had strong appeal throughout Arab countries.

The political union between Egypt and Syria was announced in February 1958. Yemen formed a confederation with Egypt, named the United Arab States, a month later. Consequently President Nasser's prestige soared within Egypt and the Arab world. While a lofty goal, political unification and military cooperation were difficult to achieve. Plans were drawn up for a united military command, and the Egyptian, Syrian, and Yemeni armed forces began to coordinate their activities. Egyptian military and political missions were sent to Yemen to cement relations, while the aircraft of the three countries were repainted with a common United Arab Republic Air Force (UARAF) insignia.

The air arms of the three nations were at different stages of development. Egypt had the largest and most experienced air arm of all the Arab nations. The Syrians had a capable air force, but it had been badly affected by a series of coups the country had experienced (see Appendix 3). Meanwhile Yemen fielded a very small and inexperienced air arm that relied heavily on the Soviet Union and East bloc military advisors.[6]

The Egyptian and Syrian air forces faced similar challenges and the same Israeli adversary. President Nasser hoped that closer coordination among Egyptian, Syrian, and Yemeni armed forces would increase Egyptian prestige and pressure Israel into a defensive situation. The confederation with Yemen also gave Egypt a vital bridgehead at the southern end of the Red Sea. However, this link would also bring Egypt into the savage and prolonged Yemeni civil war.

A key long-term military goal of the UAR was the expansion and modernization of its air force. Egypt had a long-term plan to expand the combined air force to thirty front-line squadrons, but in 1958 many of these new units existed only on paper.[7] Massive Soviet training and material support were essential to the success of this plan. The Egyptian, Syrian, and Yemeni elements of the UARAF shifted away from years of British and French influence and moved toward Soviet-style ranks and organization. Major changes were made in training techniques as well as command and control. Nevertheless, the formation of the UARAF did mean that Israel had to be prepared to counter a combined air attack from both Egypt and Syria.

As the armed forces of the three countries began to coordinate their activities, Egyptian officers received most of the top command and advisory positions. The formation of the UAR increased the responsibilities of the Egyptian armed forces, with both officers and enlisted men gaining valuable experience on postings to Syria and Yemen. On the other hand, many long serving Syrian and Yemeni officers were upset at this situation.

While on paper the UARAF appeared an imposing threat, the force had limited experience, a lack of technical expertise, and poor training. Soviet advisors rooted out all British influence, and any Egyptian who did not embrace Russian concepts was hurting his career. Soviet and east European training focused on how to fly safely, maintain aircraft, and follow commands. Senior officers and ground controllers gave the orders, leaving pilots little initiative. In fact, training limitations and this "follow orders" philosophy imposed by Soviet advisors had a dampening impact on Arab armed forces, the effect of which was later revealed in future conflicts.[8]

## Border Battles with Israel

Because of the presence of UN peacekeeping forces, the Egyptian-Israeli border and the Gaza Strip remained relatively quiet following the 1957 withdrawal of Israeli forces. However, this was not the case on the Golan Heights. Israel and Syria each had substantial fortifications along the border, and both sides aggressively patrolled the area with tanks and infantry. In the late 1950s Israel constructed a canal to divert water from the Jordan River to the Negev Desert for irrigation. Syria and Jordan strongly objected to this project. Tension along these borders increased, and in 1958 Syrian and Israeli forces frequently clashed. Both sides traded artillery fire and sent in commando raids, while Syrian and Israeli jets confronted each other above the frontier.

To support Syria, Egyptian jets began flying aggressive patrols along the Sinai border. Egyptian aircraft flew over several Israeli villages on the night of 14 December 1958, dropping flares. A week later Israeli Mystères ambushed Egyptian jets flying near al-ʿArīsh. One EAF MiG-17 crashed at Biʾr Lahfān, 65 kilometers inside Egyptian territory. Tensions in the region remained high, and in January 1960, following an Israeli raid on the Syrian village of Tawāfīq, both sides massed troops and armor along the border. Then, on 14 February 1960, an Israeli Mystère was shot down by a UARAF MiG-17.[9]

Substantial Soviet military and economic assistance and the formation of the United Arab Republic had by now made Egypt the dominant Arab country in the Middle East. President Nasser did not miss an opportunity to exert his political and military might, even against fellow Arabs. In 1960 he threatened to use force unless the leaders of Sudan backed down when questions arose over the construction of the Aswan Dam.

Israel, however, remained the focus of Nasser's greatest concern. The rapid resupply and expansion of the Egyptian armed forces had seemingly restored the military balance between the two countries. By 1959 Egypt fielded 20 percent more combat aircraft and a much larger transport and training force than Israel. When Syrian aircraft were added, the UARAF outnumbered the IAF by more than two to one. Soviet rearmament enabled the Arabs to match Israel in quantity and quality, while the new MiG-17 fighters, which served with the Egyptian and Syrian elements of the UARAF, outperformed Israel's French-made Ouragan and Mystère jets.

The UARAF's force of eighty Il-28 bombers was one area in which the IAF clearly could not compete, as Egypt and Syria could threaten strikes against Israeli cities as well as military targets in the event of war. The Il-28 could carry a 3,000 kg bomb load over a range of more than 950 kilometers. With its radar and navigation systems this bomber could also effectively strike large targets from high altitude at night.[10] In the years since the 1956 War, Egyptian and Syrian Il-28 crews had, of course, become more proficient and experienced.

This counterforce doctrine was an outgrowth of the long relationship between the British Royal Air Force and the EAF and not the product of Soviet teachings. The EAF had always believed in bombers, and these were among the first weapons to be used, with considerable impact, against Tel Aviv during the 1948 War. This countervalue threat was taken seriously by Israel.

## Israel's Response to the Arab Air Threat

Israeli leaders learned a powerful lesson from the events following the 1956 War. UN pressure forced Britain, France, and Israel to pull out of Egypt and eliminated the benefits of the allied military victory. Israeli political and military leaders vowed that this would not happen again. In the event of a future war they felt they could not count on external assistance like that provided by Britain and France in 1956. The formation of the United Arab Republic and United Arab States increased Israeli paranoia about being surrounded and outnumbered.

To overcome disadvantages in geography and manpower, Israel developed an offensive military doctrine. The primary weapons of this new strategy were the tank and jet fighter. The IAF was strengthened through the purchase of large numbers of Ouragan and Mystère IVA jets. Israel also received the twin-engine Vautour fighter-bomber in 1958.

A year later France sold its top fighter, the Super Mystère B2, to Israel. The first European fighter capable of sustained supersonic flight, this new jet was considerably faster than the MiG-17 and could perform both fighter and attack missions.[11] As a result of very demanding selection criteria and intense training, the IAF also had a sizable lead over the UARAF in pilot experience and tactics.

On 28 September 1961 Syria withdrew from the United Arab Republic following a coup d'état that toppled the government. The United Arab Republic's name continued to be used by Egypt for years, yet the political and military unification of the three countries never materialized, and in November 1961 Yemen withdrew from the federation with Egypt. The termination of the UAR left a legacy of mistrust that was to have an effect for many years to come.

Following the end of the UAR, President Nasser concentrated more on domestic activities. Since the 1952 coup, the military pervaded all aspects of Egyptian life. As part of his move to a more socialist structure, President Nasser reorganized the Egyptian military in the early 1960s. The officer corps was split into two groups: professional soldiers and a civil bureaucracy that performed political and administrative roles.

The separation of political/administrative and military roles was not, however, carried out completely, and many uniformed officers continued to be involved in the political process. For example, Marshal ʿAbd al-Ḥakīm ʿAmr, a close friend of President Nasser, was very active in political activities. ʿAmr had been "viceroy" of Syria and UAR's deputy supreme commander during the years of political and military union. This gave him effective control of the United Arab Republic's armed forces and considerable political power.

## Egyptian Defense Industry

Egypt's defense industries can trace their beginnings back to the 1820s, but it was not until after the 1948 War that the government sponsored private sector defense manufacturing. The Egyptian aviation industry began in the 1920s with the development of several domestically designed aircraft. During the late 1950s and early 1960s Egypt pursued an aggressive defense industry development program.[12]

During World War II a pool of skilled Egyptian workers was created by RAF and U.S. policies that awarded maintenance and overhaul contracts to private firms. Egyptian authorities further encouraged the development of indigenous arms and aircraft industries beginning in 1949.[13] Initially Egypt was willing to work with any group or country to obtain technology and expertise. The first industrial agreements were reached between Egypt, the British Bristol Aircraft Company, and America's Pratt and Whitney company in 1949 for the overhaul of civil and military aircraft and engines.

Several West German companies built a factory to produce Bucker Bu 181D Bestmann trainers in 1950. The first Egyptian-built aircraft entered service with the EAF in January 1952.[14] Eventually more than three hundred of these aircraft were built at the Egyptian Aircraft Construction Factory, located in Heliopolis. Known as the Jumhūrīyah (Republic), they remain the EAF's primary trainer. Eight versions of the aircraft were built, and in the 1970s all the surviving aircraft were remanufactured and brought up to a common configuration with a U.S.-built Lycoming engine and a bubble vision canopy.

On 5 July 1950 Defense Minister al-Fāriq Ḥaydar Pasha announced plans for the production of a jet fighter in Egypt. British firms then built and equipped a factory at Ḥulwān to assemble the De Havilland Vampire jet fighter, but increasing Anglo-Egyptian friction over the continuing British presence in the Canal Zone slowed this project. Then during the height of the crisis in 1951, Egypt dismissed the British personnel working at the factory.[15]

As the new government was consolidating its power following the 1952 Revolution, no progress was made in developing an Egyptian defense industry. By 1955 Egypt was no longer interested in building first generation jets like the Vampire, the Egyptian government reportedly attempting to sell the jigs and other specialized tooling at the Ḥulwān factory back to De Havilland. Continuing friction with Britain prompted Egypt to turn to other countries.

German technicians were invited to support Egyptian defense industrialization projects.[16] President Nasser decided to embark on a massive effort to establish domestic jet fighter, missile, and other defense programs. In the late 1950s Colonel Maḥmūd Khalīl gathered together a team under the direction of the famous German aero-engineer Willi Messerschmitt. Their first project was production of the Hispano Ha-200 jet trainer, originally designed for the Spanish Air Force. Starting in 1959, Factory 72 at Ḥulwān was retooled to build the new jet, and in July 1960 four Spanish-produced Ha-200 jets in UARAF markings flew over Cairo to announce the project. By the late 1960s sixty-three Ha-200s, known in Egypt as *Al-Qāhirah,* had been built. However, after several years of service, they were retired after their engines proved to be unreliable due to erosion in the sandy Egyptian environment.[17]

Egypt also committed itself to an ambitious project to develop and produce a supersonic fighter. Again with Spanish assistance, Willi Messerschmitt and his design team conceived a lightweight supersonic fighter. When Spain ended its support for this project in 1960, the entire Messerschmitt team was moved to Egypt. The Austrian jet engine expert Ferdinand Brandner was also invited to Egypt to develop a new turbojet for this small fighter.

Test facilities and workshops for the new fighter and engine were built at Ḥulwān. Brandner's jet engine ran for the first time in July 1963, while the Ha-300 prototype made its first flight on 7 March 1964. India helped fund development of the E-300 engine in order to acquire a new power plant for its HF-24 Marut jet fighter, a similar program.

Two Egyptian pilots were sent to India in 1964 and attended the Indian Air Force test pilot school to prepare for the Ha-300 flight development. Maj. Zuhayr Shalabī demonstrated exceptional flying ability, while Maj. Ṣaḥbī al-Ṭawīl was both a pilot and engineer. Both of these officers completed their courses successfully.[18] India also provided a Marut test aircraft to support the E-300 engine test program. Group Capt. Kapil Bhargara of the IAF flew more than one hundred flight hours with test versions of the E-300 engine, but development problems and the 1967 War eventually led Egypt to cancel this ambitious fighter development program.[19] Soviet MiG-21s sold on credit also undercut this national development project. After the 1967 defeat, all available funds naturally went toward expanding the Egyptian armed forces to confront Israel.

## Egyptian Missile Programs

Following Arab defeat in the 1948 War, the Egyptian government invited many firms and skilled individuals to the country to provide training for the army and assist in setting up an arms industry. Included in this program were a number of former German Army officers and rocket scientists. A small team headed by Dr. Wilhelm Voss began operating in Egypt in 1951. They launched their first test rockets in 1952. In 1953 the Ṣaqr factory was set up to serve as the center for Egyptian rocket development. While several rocket designs were created, technical difficulties and the new alliance with the Soviet Union gradually led to the termination of this program.

While the Soviets and their east European allies provided the Arabs with MiGs, jet bombers, and many other weapons, they refused to supply missiles. As a result, in the late 1950s Egypt renewed its missile development program. A design team of foreign and Egyptian technicians based at the ultrasecret Factory 333 (also known as the Ṣaqr factory), at Ḥulwān, developed several different missile designs.

When it was discovered that Israel had successfully test-fired a solid fuel rocket in 1961, both sides intensified their efforts to win this new missile race. France assisted Israel, and as a result it rapidly moved ahead to field an operational missile.

Nevertheless, fear of the threat from the German-Egyptian team reportedly prompted Israel's secret service to launch a campaign of terrorism against Factory 333 and its personnel. Dr. Emil Kleinwachter was machine-gunned in his car, Dr. Ḥasan Kāmil died when his plane crashed, and a bomb blast at Ḥulwān killed five workers. Within weeks three other personnel died in unexplained accidents, and a woman secretary was blinded by a mail bomb.[20] These attacks prompted most of the foreign members of the Egyptian missile development team to flee the country.

Egyptian scientists continued their development work and constructed three types of missiles: the Qāhir (Conqueror), the larger Ẓafar (Victory), and the two-stage Al-Rāʾid (Leader). All of these missiles were test-fired, but they suffered technical difficulties, and despite years of effort and a substantial investment, none of the missiles ever entered operational service.[21] In the 1960s, however, Egypt was finally allowed to purchase FROG-7A rockets from the Soviet Union.

## Growing Pains

Following the end of the political union with Syria and Yemen, President Nasser went ahead with his ambitious military expansion plan. Because of its limited supply of experienced personnel, this rapid expansion of the EAF forced Egypt into the classic trap of sacrificing quality for quantity. For several years it suffered high aircraft and aircrew loss rates and had to struggle with low aircraft readiness.

However, the massive investments of the late 1950s and early 1960s in training, facilities, and equipment eventually bore fruit. Egyptian pilot candidates, aircrew, and technicians received training from Polish, East German, Czech, Indian, and Russian instructors. A new generation of air force personnel educated in the Soviet Union, eastern Europe, and in new Egyptian schools such as the Air Force Technical Training Institute and Bilbays Air Academy began to demonstrate their abilities. For new pilots the accident rate in training, which had reached high levels in the early 1950s, had been reduced to the same level experienced by the Soviet Air Force at that time.

By the early 1960s the UARAF's efforts had produced a large force of skilled Egyptian aviators and technicians. However, the training given by Soviet and east European instructors did not properly prepare

Egyptian aviators for the war they would face in the future. Maj. Gen. ʿĀdil Naṣr, who served as the EAF director of operations in the 1980s, commented on Soviet training: "When the Russians came, they emphasized training to make us staff and general officers. They didn't teach us tactics, but they succeeded in teaching us to think in a proper and organized manner." He added, "When I was in the Soviet Union, I had many relations with the training department because of my job. I was discussing with them how to train our troops. I discovered that they gave us a course in elementary training, but they didn't teach tactics. They had their own tactics, but they wouldn't be good for us because they depended upon massing and the use of large numbers that were not available to us."[22]

Maj. Gen. Nabīl Shuwakrī (ret.), a former chief of staff of the EAF who trained with both Russians and Americans said: "In the 1960s we sent young men to take their courses in the Soviet Union. When they came back they had unbelievable safety measures. They were afraid of the aircraft. They were told, 'if you do this, you will spin and die, and if you do that, you will crash.' When we took back these pilots from the Russians and they returned to Egypt, we gave them a refresher course to get rid of that conservativeness."[23]

## UARAF Order of Battle 1958[24]

| | Fighter (Jet/Prop) | Bomber (Jet/Prop) | Transport (Jet/Prop) | Other (Jet/Prop) | Personnel |
|---|---|---|---|---|---|
| Egypt | 178/15 | 70/3 | —/48 | 10/70 | 4,375 |
| Syria | 50/16 | —/— | —/10 | 18/106 | 1,259 |
| Yemen | —/30 | —/— | —/3 | —/12 | 400 |

15

# Wider Horizons (1962–1967)

The early 1960s were a time of change and growing confidence in Egypt. President Nasser pushed forward with his doctrine of Arab socialism which nationalized industries and redistributed land to wider ownership. The country played a leading role on the international scene, particularly in the nonaligned movement. Egypt maintained a strong military to maintain internal order, deter Israeli and western aggression, and support Egypt's leadership in the Arab community. In the years following the Suez War of 1956, Egypt, Syria, Israel, and the other Arab nations of the Middle East were locked in a struggle to expand their military power. The leading players in this arms race were Egypt, still officially called the United Arab Republic (UAR), and Israel. With help from France and West Germany, Israel's air and ground forces expanded at a rapid rate. Egypt matched this buildup with aid from the Soviet Union. During this same period Syria, Jordan, Iraq, and other Middle East nations expanded their armed forces. Egypt was usually the first to receive the newest Soviet systems and always hosted the largest group of Russian advisors.

## Low-Level Tensions

Despite President Nasser's efforts to remain independent, in the eyes of the West, Russian assistance placed the country in the Soviet sphere of influence. Nasser tried to demonstrate that Egypt was an independent regional power by engaging in combat in Africa and Yemen and by aiding many Arab countries while still confronting Israel. In fact, Arab efforts to isolate and threaten Israel created an atmosphere of paranoia in Tel Aviv.

This situation led to clashes along the Sinai border. Israeli pilots, eager to demonstrate the superiority of their new Super Mystère fighters, attacked Egyptian jets in November 1959 and May 1960. While neither side suffered losses in these dogfights, two MiGs were damaged. On 28 April 1961 a pair of EAF MiG-17s flying along the border were intercepted by two Israeli Super Mystères. One of the Egyptian pilots went into a spin trying to evade the Israeli attack and was forced to bail out of his aircraft.[1]

Brig. Gen. Tamīm Fahmī ʿAbd Allāh (ret.), an EAF pilot, described his experiences flying along the Egyptian-Israeli border during this period:

> In September, 1962 I was sent to No. 2 Squadron at al-ʿArīsh flying MiG-17s. We made many patrols along the border and were sitting in readiness much of the time. We patrolled and played many games with Israeli jets. but my squadron never fought with the Israelis at that time.
>
> We flew a lot of patrols and feints but not much real fighting. We would fly along the border at 13,000 meters and look into Israel, not a reconnaissance mission, just for training. The theme at that time was fly high, which was wrong, but we did not know it until later. We were young and had to obey orders, so that is what we did. It was exciting because it was my first experience with a real fighter squadron, and we were just 40 km from the border which was hot at that time.[2]

## Russian Training, Doctrine, and Aircraft

Combat lessons of jet age air combat had changed Russian fighter doctrine dramatically. Instead of the emphasis on low altitude air combat and ground attack seen in World War II, Russian fighter pilots in the 1950s were trained to defend against American and British bombers. Most air engagements during the Korean conflict and the cold war began at medium and high altitudes. Ground attack training for Soviet aviators emphasized delivery of tactical nuclear weapons, with conventional munitions being a lower priority. These tactical concepts were exported to Egypt and other countries that the Russians assisted.

The anti-bomber emphasis deeply influenced Russian fighter design and led to the MiG-19, MiG-21, and many supersonic interceptors produced for the Soviet Air Defense Force (PVO Strany). The MiG series of aircraft was supplied to Russian Frontal Aviation tactical squadrons and the air forces of client states such as Egypt. The Soviets never exported sophisticated all-weather interceptors such as the Yak-28 and Su-11/13/15. Russian strike aircraft also reflected the cold war emphasis. The Il-28, Tu-16 bombers, and Su-7 fighter-bomber were developed to deliver nuclear weapons. As a result, the range and payload of these jets with conventional bombs was greatly reduced.[3]

## The EAF Goes Supersonic: MIG-19 Farmer

When the supersonic MiG-19 interceptor was offered to Egypt in 1960, the EAF was quick to order the jet. By this time Israel already had a squadron of supersonic Super Mystère fighters. The first batch of MiG-19s arrived in Egypt during the summer of 1961. Eventually some eighty of these Soviet-built jets served with the EAF.

While it had high performance for its day, the MiG-19 was not liked by Egyptian pilots and support personnel. Compared to the older MiG-17, which the Farmer replaced in the intercept role, the new fighter was difficult to fly and maintain. In addition, the twin-engine MiG-19 had limited range and a modest payload. Egyptian MiG-19s first fired their guns in anger in the final week of July 1963 when Flt. Lts. ʿĀdil Aḥmad and Muḥammad Kāmil al-Muwāwī fought with four IAF fighters that had crossed into the Sinai. Neither side achieved any victories during this brief encounter.

## MiG-21: The EAF Joins the Mach Two Club

In 1961 the EAF ordered the MiG-21 Fishbed, the first Soviet fighter capable of achieving Mach two. This aircraft would form the backbone of the EAF fighter force for nearly two decades and replaced the older MiG-19 and MiG-17 in the intercept role. The new MiG-21's performance was superior to that of the IAF's Super Mystère and roughly equal to that of the Mirage III, though Israeli jets had better armament and IAF pilots were generally more highly skilled in air combat.

A small and relatively simple high-performance interceptor, the MiG-21 was developed to work in cooperation with a ground-based radar and surface-to-air missile network to intercept and shoot down bombers. The first generation MiG-21F-13 was equipped with a simple ranging radar, a gunsight, and an armament of one 30 mm cannon plus two Atoll infrared-homing air-to-air-missiles. In the mid-1960s the EAF received the MiG-21FL, a night-fighter version.

This jet featured a higher thrust engine and improved radar, however the Russian designers had removed the cannon to save weight.

Brig. Gen. Fārūq al-Ghazzāwī (ret.) recalled his experiences with this new Soviet fighter: "In 1961 I was among the first group to convert to the MiG-21 in the Soviet Union. Our (former) Air Force Commander Lt. Gen. ʿAlāʾ Barakah led this group. It was wonderful to fly supersonic in the MiG-21F-13. I flew many hours in the MiG-21. Later I commanded a squadron of MiG-21FLs."[4]

Brig. Gen. Tamīm Fahmī ʿAbd Allāh (ret.) also remembered his first experiences flying the MiG-21: "In November 1963 I joined a squadron at Cairo West flying MiG-21F-13s. It was a beautiful aircraft. I enjoyed it very much."[5]

Maj. Gen. ʿAwaḍ Ḥamdī (ret.) discussed his view of the MiG-21:

> I believed that flying a MiG-21 against any other fighter at that time I was safe. There is not yet a fighter tailored for efficiency in aerodynamics like the MiG-21—it was a beauty. The best aerodynamic/engine mix from zero height to 20 km, almost 80,000 feet, is the MiG-21. The MiG has an automatically adjusting shock cone that works better than any other fighter I have seen, including the American F-16. You can maneuver the MiG-21, pitch it over at 70,000 feet, and it will work.
>
> However, the fire control system is not good—it can't do much. We were also short of fuel and had limitations on external fuel tanks. With the MiG-21F-13 you could only carry one tank and could not drop it at high speed. In addition to being short of fuel, the MiG-21 was underpowered. If you engaged in a close combat with a MiG, you lost speed fast. This was all the time a frustration because you did not have enough fuel, range, or weapons. We had two very limited R-3S Atoll missiles and a single (30 mm) cannon. In the MiG-21F-13 we had only sixty rounds of ammunition, so you were on the defensive most of the time. However, it was a reliable aircraft, and we had faith in the MiG-21 regardless of its deficiencies.[6]

Brig. Gen. Tamīm Fahmī ʿAbd Allāh commented about a later version of the MiG-21 that the EAF bought for the night intercept mission:

> In 1966 I transitioned to the MiG-21FL. Our squadron was at Inshāṣ. At the time we were adding four additional MiG-21 squadrons, so we were converting a lot of pilots and support crews to the aircraft. . . . The MiG-21FL was a good airplane. It was smooth, easy to fly, but as a fighter aircraft it was very limited. We had just two Atoll [air-to-air] missiles, no cannon, and a lousy

> radar. It was a very maneuverable airplane, especially in rolling, but it had no gun and no camera. The only nice thing about the FL's avionics was the autopilot. You had to be directed to an intercept by ground controllers, aim, shoot and that's it. Soviet philosophy worked with mass waves of aircraft. They would flood the area with aircraft, but in Egypt we had only a limited number of fighters.[7]

The R-3S Atoll air-to-air missile homed in on the infrared emissions from a target's engine and airframe. The missile was on its own once fired and could fly over 3 kilometers. However, the missile's infrared seeker had a very narrow field of view and was susceptible to being decoyed by infrared sources of higher intensity than the target, such as the sun or hot desert terrain. This required an EAF pilot to fly below and directly behind his target and wait several seconds for the missile to lock on before firing. In air combat over North Vietnam the similar American AIM-9B missile achieved a success rate of only 15 percent.[8]

## Tu-16 Badger: The EAF's Big Stick

During the early 1960s Egypt fielded two squadrons of Tu-16 "Badger A" bombers. This large, twin-engine medium bomber had twice the range of the older Il-28 while carrying three times the bomb load. A small number of Tu-16 "Badger G" aircraft equipped to carry AS-1 Kennel air-to-surface missiles were also provided to Egypt. These missiles could be fired at both ship and land targets.

Egyptian Tu-16 and Il-28 jet bombers were Egypt's first-line strike force. Egypt's seventy-five jet bombers could attack any target within a range of 700 miles. Flying at night or with a strong MiG escort, most of the bombers were expected to reach their targets despite Israeli or western defenses. Husni Mubarak, later the president of Egypt, served as a bomber pilot and later commanded a squadron of these Tu-16s during the 1960s.

## Su-7 Fitter: Disappointing Performance

Another new Soviet aircraft that entered service in the 1960s was the Su-7 Fitter, to replace the MiG-17 in the fighter-bomber role. A large and powerful jet, the Su-7 had a maximum speed of more than 1,600 kph (Mach 1.5) at high altitude.

Col. Taḥsīn Zakī (ret.) remarked about his first experience with the Su-7: "In Russia, late in 1966, I was with some Egyptian pilots who were sent to train on the Sukhoi Su-7. I noticed after flying a couple of sorties in the aircraft that it had lots of defects with little fighting capability plus a small ordnance load. Worst was its short range! Therefore I put it all in my report to the supreme commander of our forces, Marshal ʿAmr. When he arrived in Russia, I was called to discuss the matter and later, after reading my report, we decided to buy only thirty-six and cancelled the rest of the order."[9] However, Egypt eventually bought more Fitter fighter-bombers because of Soviet pressure and the need to replace aging MiG-17s.

## Support for Egyptian Political Initiatives

President Nasser often used the EAF to further his political goals. The earliest examples date to 1954 when Nigerian students entered the Bilbays Air Academy. In 1958, following the overthrow of the Iraqi monarchy, the Iraqi Air Force received its first MiG-15s and support from an Egyptian training mission. In 1959 the Libyan Air Force was created with the delivery of a pair of Jumhūrīyah trainers from Egypt.

The Algerian Air Force similarly acquired its first MiG-15 jets and training support from Egypt. Until 1965 all Algerian pilots received their advanced training in Egypt. Egyptian support for Muslim countries extended to Asia, and in the late 1960s Indonesian Air Force pilots were also trained by the EAF.

Egypt under President Nasser played a leading role in Arab and international affairs. This was particularly true on the African continent, and the air force was called upon to support the country's involvement in a series of crises that followed the independence of the former Belgian Congo (now Zaire) in 1960. Egypt contributed 1,000 paratroopers to the UN peacekeeping force. Over the next several years Egypt covertly supplied arms and support to the new government in the Congo in exchange for gold. Egyptian C-47 and Il-14 transports maintained an air bridge carrying material and arms between Cairo and the Congo.

## Yemen: A Test of Wills

Egypt's involvement in Africa was minor compared to its investment of support during the Yemeni civil war. Yemen is a rugged country located at the southern end of the Arabian peninsula bordering Saudi Arabia, the Red Sea, the Arabian Gulf, and Oman. The area was under the influence of Great Britain from 1839 to 1967.

Aden has long been an important refueling stop and supply depot for Britain's Royal Navy. To secure the safety of the port and surrounding territory, Britain divided the region into the southern Aden Protectorate and independent Yemen, while signing treaties with local sheikhs and using military force to keep the peace. Imām Aḥmad, the ruler of Yemen from 1948 to 1962, hated the British presence in the neighboring Aden Protectorate, and he also had a continuing territorial dispute with Saudi Arabia. In 1958 Imām Aḥmad joined with Egypt and Syria in the Union of Arab States, mostly to distance his country from British and Saudi influence. However, Yemen withdrew from the union in 1961, a slight that President Nasser never forgave.

On 19 September 1962 Imām Aḥmad died, and his heir, Imām Muḥammad al-Badr, assumed leadership of the country. A week later a group of Yemeni officers staged a coup and forced out the new imām. The Egyptians had secretly trained and supplied this group of Yemeni military officers, and Egypt was also supporting rebel forces fighting a guerilla war to drive the British out of Aden to the south.

During the first week of October 1962, EAF transports flew almost nonstop carrying arms, troops, and advisors to the Yemeni capital of Sanʿāʾ. At first it appeared that the imām had been killed. However, the imām survived the coup and rallied the support of many northern tribes to his cause.

By late October the battle lines had been drawn. The Republican government, headquartered at Sanʿāʾ, was backed by tribes from southern and western Yemen plus the small Yemeni Army and an Egyptian expeditionary force. Most tribes located in the northeastern portion of Yemen supported the imām. Jordanian, Saudi, and British advisors established camps just over the border inside Saudi Arabia to arm and train men from these tribes to fight for the imām.

As the fighting intensified the Egyptians sent additional air and ground forces to Yemen. President Nasser eventually admitted that his intervention was a "miscalculation." The vital role of air power became evident early in the conflict, as resupply missions using C-47 and Il-14 transports flown by Egyptian and Yemeni crews helped save several isolated outposts from being overrun.

Yemen's small air force, whose pilots and crews had been trained by the Soviets and Egyptians, sided with the Republicans. Based at a single airfield near Sanʿāʾ, the Yemeni Air Force flew a diverse collection of Russian and western aircraft including Zlin 126 trainers, Il-10 Sturmovik ground attack aircraft, York transports, and Mi-1 helicopters. Egypt gradually replaced these worn-out aircraft with C-47 and Il-14 transports, Mi-4 helicopters, and armed Yak-11 aircraft. Using guerrilla tactics, the Royalists inflicted serious losses on the Yemeni Army and the Egyptian forces. In response Egyptian and Yemeni aircraft bombed and strafed Royalist convoys, base camps, and villages.

The Egyptian Army and Air Force had been trained to fight Israel, not a guerrilla war in difficult terrain against a vicious foe. Both services suffered losses and were forced to adjust quickly to a new way of warfare. The first aircraft used in combat over Yemen were Egyptian Yak-11 trainers armed with machine guns and rockets. This was not a new idea; the RAF employed similar armed T-6 trainers over Kenya a decade earlier, and Belgian T-6s fought in the Congo. Later, when Royalist anti-aircraft defenses improved, higher-performance MiG-15 and MiG-17 fighter-bombers as well as Il-28 bombers took over the ground attack mission.

EAF transports and helicopters played a vital role in the deployment and resupply of Republican forces. Transports were used to resupply remote outposts and fly reconnaissance missions over Royalist territory. When a caravan or base camp was spotted, fighter bombers, transports, and helicopters were called in. Egyptian Army General Saʿd al-Shāzlī often flew in specially trained paratroops by helicopter and transport to fight Royalist units.[10]

By early 1963 Egypt had sent more than two hundred aircraft to Yemen to assist the Republicans and support the 75,000-man Egyptian expeditionary force. Egyptian MiG-15s, armed Yak-11s, plus Il-14 and C-47 transports flew from the Sanʿāʾ airfield and nearby Raḥabah airstrip. At times both of these air bases came under Royalist mortar attacks.

Brig. Gen. Tamīm Fahmī ʿAbd Allāh (ret.) described his experiences in this conflict: "I went to Yemen for three months in 1963. At first I flew the Yak-11. We flew mostly armed reconnaissance missions. Our Yaks in Yemen were armed with four 68 mm rockets and machine guns. In two months I flew dozens of reconnaissance and attack sorties. We took some hits because the Royalists were born with a rifle and could shoot very well. Then I was transferred to another base in Yemen where there was a MiG-15 squadron. I flew many ground attack missions. The cannon of the MiG-15 [the two 23 mm and one 37 mm] could do plenty of damage."[11]

Aggressive Republican ground offensives, which included massive air support, gradually forced Royalist forces into the mountains, and even there the Royalists could only move freely at night. Even with heavy bombing and aggressive patrolling, Republican and Egyptian forces could not defeat Royalist resistance. By late 1963 the Yemeni civil war settled down to a savage stalemate. Warriors from the northern tribes continued their guerrilla raids and fought in a savage fashion. The Republicans and Egyptians struck back with commando raids and bombed Royalist villages to rubble.

At the end of 1963, EAF planes were bombing Royalist camps in Saudi Arabia and suspected camps in the British-ruled Aden Protectorate. Egypt also threatened to send ground forces across the borders unless Saudi Arabia and Britain ended their aid to the Royalists. British Royal Air Force Hawker Hunters flew aggressive patrols to deter such raids and struck at targets in Yemen in retaliation for Egyptian attacks.

President Kennedy demonstrated American support for Saudi Arabia by deploying USAF F-100 fighters to the kingdom. In "Operation Hard Surface," the American fighters flew along the border, bringing an end to the Egyptian raids on Royalist camps in Saudi Arabia.[12]

As the civil war dragged on, the EAF was assigned increasing responsibilities. Replacement aircrew and whole Egyptian squadrons were deployed to Yemen for short tours of duty (3–5 months) to gain operational experience. By early 1964, five EAF squadrons flying Yak-11s, MiG-15, MiG-17s, and Il-28s, plus Il-14 transports and Mi-4 helicopters were operating in Yemen. Tu-16 bombers flew regular strikes against Royalist targets from bases in Egypt.[13] Egypt also supplied, trained, and supported the Yemeni Air Force. There is also some evidence that Russian aircrew participated in the fighting. A Yak-11 that was shot down in Yemen near the Saudi border was found to contain a Slavic-looking pilot, though, of course, fair-haired Egyptians of western, Bosnian, or Albanian family backgrounds have always served in the Egyptian military.

In 1963 Royalist fighters and western visitors reported attacks with gas bombs on remote villages, killing many civilians. Evidence of air attacks involving gas increased in 1965, 1966, and 1967. An article in the St. Louis Post Dispatch stated: "As early as 1963 the Egyptians fighting in Yemen against the Royalists supported by Saudi Arabia have used chemical weapons dropped from airplanes, fired from artillery and in land mines. In the first phase this was tear gas. In March of 1965 mustard gas was used and this has been repeated in July of that year. On January 4 of this year [1967] nerve gas was used for the first time in attacks on the villages of Hadda and Kitaf. In Kitaf 200 civilians were killed. The number of casualties in Hadda was not determined."[14]

Egypt denied that lethal gas was used in the war, and a long-term investigation by the International Red Cross failed to prove beyond doubt that lethal chemical weapons had been used in Yemen. Nevertheless, the organization asked Egypt to give its solemn pledge "never again to use, under any circumstances, poisonous gas or any other poisonous substance."[15] The Soviet Union had supplied Egypt with several types of chemical munitions including shells and aircraft bombs plus decontamination and protective equipment.

After a series of successful but costly ground offensives in 1964 the Egyptian expeditionary force in Yemen was reduced in size, and many Egyptian aircraft brought home. Even so, the fighting continued, and on the eve of the 1967 War, 15,000 Egyptian troops and fifty EAF aircraft still remained in Yemen. The Egyptian involvement in Yemen dragged on much longer than President Nasser had expected and was very costly in terms of lives lost, money spent, and negative political impact.

Meanwhile the Soviet Union continued to supply Egypt with weapons and assistance. This allowed Egypt to support the Royalists in Yemen while at the same time continue its military expansion. During operations in Yemen the UARAF did gain valuable experience, air force support personnel working hard and maintaining a high operational rate of some 95 percent despite primitive working conditions and logistical problems.[16]

While air operations in Yemen boosted Egyptian experience and morale, it did not give the UARAF any air combat experience. This was the area in which the IAF placed considerable training emphasis. Limitations of Soviet training had been revealed during the fighting in Yemen, however, few changes were made to the UARAF training program because of the influence of Russian and east European advisors.

Brig. Gen. Tamīm Fahmī ʿAbd Allāh (ret.) commented about Russian training concepts during this time:

> I didn't like the kind of training we were getting because it was very hypothetical. We were not unique, probably the Soviet Union and all Soviet bloc

countries trained the same way, but I will never forgive us for following this same path. We were in combat a lot in Yemen, so we should have known better. But we got our equipment from the Soviets and we believed them, we didn't believe ourselves. We were flying all the time at high altitudes, high speeds—supersonic much of the time. Anyone who would fly low could get courtmartialed. I was once in big trouble because I flew low. We did so in Yemen a lot, but after we came home it was forgotten.[17]

## Air Tag along the Border

On the important Sinai front the UARAF found that the Israelis were growing bolder. To test Egyptian defenses Israeli jets flew an increasing number of feints and repeatedly violated Egyptian airspace. Maj. Gen. Muḥammad Nabīl al-Masīrī (ret.), a former chief of staff of the EAF, talked about one of his missions over the Sinai in 1965:

We had two MiG-15s, I was lead with one other pilot. We got in a combat for sixteen minutes flying MiG-15s against Mirages. We stayed in a battle for ages with the six Mirages.

We started from Kibrīt and went a long way into the Sinai and were flying up to al-ᶜArīsh on a training mission. We heard on the radio that enemy aircraft were in the area. While we were approaching al-ᶜArīsh from the south, we were attacked. We maneuvered with two Mirages, then four and then six. I remember turning and turning, and I got in behind one Mirage that was diving down in front of me. I thought it was a trap—they used to do that a lot—put one aircraft in front in a dive, and if you chased it, another would pop up and get you. So I started looking on both sides and behind and saw nothing. At that time my fuel indicator was flickering, and we had to be on the ground in ten minutes.

I called my wingman who had remained with me throughout the combat. He was great, he would call, "they are coming in—they are firing," and we would make very hard maneuvers never to give them the opportunity to get behind and get a good shot. We would turn into the Mirages and fire at them. I think that we hit them, and they withdrew because they were low on fuel. After that I started to be afraid. We made it to al-ᶜArīsh but all the way I kept looking for other Mirages. . . . What was very astonishing was that these fools didn't win—I say fools—because if they just flew over us and not let us go back, we would have run out of fuel and had to eject.[18]

Despite many air engagements with Israeli jets, neither side lost any aircraft.

Colonel Tazḥsīn Zakī (ret.) also discussed his activities during this time:

> I was commander of the 2nd Air Wing of MiG-17s during 1965 which included a squadron of MiG-17s stationed permanently at al-ʿArīsh. Our orders were not to fly near the Egyptian-Israeli border, but the Israelis often penetrated our airspace, flying very low, even right over al-ʿArīsh airport. This made my pilots feel very bad because of the restrictions the high command had put against penetrating the Israeli border.
>
> To boost the morale of my pilots, I allowed them, at times, to penetrate at low level until Beersheba. In one of these penetrations by two MiG-17s, one of the pilots, the late Ṣalāḥ Mansūr, reached the area of Dimona to see a very strange kind of construction taking place. He thought from the shape of it that it was a nuclear reactor. He immediately informed me, and I decided to fly with him again for confirmation. There was no doubt about it, it was a nuclear reactor under construction. I contacted Gen. ʿAbd al-Majīd al-Rifāʾī, commander of the Eastern Region at that time, and confirmed what I saw, risking to be courtmartialed for penetrating deep into Israel. However, the commander promised not to ground me and my pilots and said well done! It was the first time the Egyptian authorities came to know directly that there was a nuclear reactor under construction in Israel.[19]

The increase in tension along the frontiers between Egypt, Israel, and Syria prompted Syrian leaders to seek assistance from Egypt. In 1966, following a series of meetings of senior leaders of the two counties, Egypt and Syria signed a defense cooperation treaty. As a sign of political support, a squadron of Egyptian MiG-17s was sent to Damascus' Mazzah airfield to bolster Syria's air defenses.

Egypt and Syria responded to Israeli pressure with artillery fire, air patrols, and commando raids. Israeli jets downed two Egyptian MiG-19s that had flown into Israeli airspace on 29 November 1966. One of the jets was shot down by cannon fire and the other destroyed with a Matra 530 air-to-air missile, the first time missiles were used in Middle East air combat.[20]

## The Egyptian Air Force: Confident but Untested

By mid-1967 the Egyptian Air Force, or UARAF as it was still called, had grown to 450 combat aircraft and 350 support planes, staffed by 11,000 uniformed personnel and 5,500 civilians. This force included some 700 trained pilots and 150 experienced navigators. A further 300 pilots, navigators, and foreign students were in training. About a third of Egyptian aircrew had seen service in Yemen. EAF support crews were fairly well trained, though their average time of two hours to prepare a jet for combat was slow compared to that regularly demonstrated by their Israeli counterparts.

Egypt's large air transport fleet had contributed to successful operations in Yemen, in Africa, and in domestic operations. The UARAF fielded ninety An-12, Il-14 and C-47 transports, and several squadrons with sixty helicopters including six giant Mi-6s which could each carry seventy troops or 12,000 kilograms. of cargo. This sizable fleet could air-drop 6,000 paratroopers and over 600,000 kilograms of supplies in an hour. In addition, more than one hundred trainer aircraft were in service.[21]

To deter Israeli or western attack, Egypt threatened counterattack by the EAF's bomber force. A 1967 American Intelligence report stated:

> The (Egyptian) bomber force poses a real threat to Israel. Equipped with Il-28 and Tu-16 aircraft, both within sufficient range of Israeli targets from any base in the Nile Delta, the EAF is capable of launching an initial surprise attack, if executed at night. Some of the Il-28s are equipped for photoreconnaissance, and the Tu-16s have a good tactical reconnaissance capability. The EAF has a flight of Tu-16 bombers equipped to carry KENNEL AS-1 air-to-surface missiles. The remaining Tu-16 bombers and Il-28s have a bomb load capability of up to 6,600 pounds. Aerial bombs used include FAB 5000, 3000, 1500, 1000, 500, 250 and 100.[22]

Israel was very concerned with this threat of attack by Egyptian bombers armed with conventional or chemical bombs. A single bomber armed with chemical weapons could kill thousands of civilians. Egypt's apparent willingness to use lethal chemical weapons in Yemen, despite the negative international reaction, further alarmed Israeli military planners.

Following the American intervention to support Saudi Arabia during the Yemeni civil war, the American-Saudi relationship was strengthened. Israel also asked for defensive systems to protect against Egyptian bombers. In 1964 the United States delivered Hawk anti-aircraft missiles to Israel. This system was, at the time, the world's most advanced air defense weapon, and the Hawk missile system was specifically designed to shoot down Russian aircraft like Egypt's Tu-16, Il-28, and MiG-17. This was the first time the United States had provided any type of advanced weaponry to Israel.[23]

Displays of Egyptian military might and optimistic reports about victories in Yemen created overconfidence in Egypt. In reality the Egyptian armed forces had not yet achieved the level of expertise they were credited with. The rapid expansion of the last decade had also diluted quality. The UARAF had been taxed by the creation of many new squadrons, assignment of personnel to Soviet and east European training courses, foreign deployments, and the rapid introduction of many new types of aircraft.

Conversion of squadrons from the relatively simple MiG-17 and Il-28 to the much more complex MiG-21 and Tu-16 was a challenging task. Often it took more than a year for the personnel of the squadron to adjust to this type of upgrade and resulted in reduced operational readiness as well as higher accident rates. Nevertheless, the good Egyptian weather and the professional attitude of Egyptian crews enabled UARAF units to convert to new aircraft more smoothly than many of their Soviet or third world counterparts.

According to a 1967 American intelligence assessment: "The [EAF] fighter force is equipped with MiG-19, MiG-17 and MiG-15 fighters used for ground attack and in close army support roles and the MiG-21, a point defense interceptor, which probably also has a ground attack capability. The Su-7 is a ground attack day fighter. The MiG-17 is equal to the Israeli Mystère, and the MiG-19 is better, but the MiG-21 is inferior to the Israeli Mirage IIIC."[24]

In addition, a good portion of Egypt's air fleet was not ready to fly at any given time. The operational readiness rate for MiG-21s and Tu-16s was estimated to be 60–65 percent, but older aircraft like the MiG-17 maintained a higher rate, with probably 75–80 percent able to fly at any one time.[25] This operational readiness rate, however, was good for a third world air force and not far behind that of the USAF and Soviet Frontal Aviation units. Egypt's problems could be attributed to a shortage of experienced technicians and the impact of the hot, sandy climate which was hard on Soviet-built aircraft and systems designed for colder climates.

The UARAF and the army were jointly responsible for the air defense of Egypt. In 1967 there were four main air defense zones, each with its own control center (Alexandria in the northern region, Cairo/al-Zamālik in the central, al-Firdān in the east, and al-ᶜArīsh in Sinai). The country was also ringed by ground observers and radar warning sites. Many military bases were protected by anti-aircraft guns ranging in size from 20 mm to 100 mm. Alexandria, Cairo, the Aswan Dam, Ṭanṭá, Port Said, Ismāᶜīlīyah, Suez, and many military bases were similarly ringed by SA-2 anti-aircraft missiles.[26] Such weapons had demonstrated their effectiveness during this time period against American aircraft in the skies over North Vietnam. This Soviet-built missile system was capable of hitting aircraft flying at an altitude of between 1,000 and 3,000 meters out to a range of 30 kilometers.[27]

While impressive on paper, Egypt's air defenses had many limitations. Colonel Taḥsīin Zakī commented on the effectiveness of the Egyptian air surveillance network:

> The Egyptian Air Force held maneuvers to test the efficiency of our air defense against low-flying aircraft. The air defenses failed completely to spot any aircraft flying below 400 meters because of the outmoded Russian radars that were incapable of spotting any aircraft flying low. A meeting was held after the maneuvers at the Institute of War Studies at Almāẓah which was attended by the supreme commander of the Air Force. This meeting ended after one of the Russian advisors said that the air defense system of the UAR was sound but needed some minor modifications of the SAM sites north of the Canal Zone—which surprised me a lot![28]

Air force effectiveness was hampered by Egyptian command and control limitations. There was little communication between the Egyptian Army and the Air Force below top levels. Orders for air operations came solely from the Supreme Command Council via the chief of operations at GHQ and the chief of air staff. These orders were directed through channels to the various operational squadrons. Only general plans had been prepared for the air force to support the army in the event of a war with Israel. There were no coordinated procedures to allow Egyptian aircraft to fly safely in areas protected by army anti-aircraft guns and the new SA-2 air defense missiles.

## Blundering into War

On 7 April 1967 Israeli and Syrian jets clashed over the Golan Heights, and the IAF won a resounding victory. Seven MiG-21s were shot down, and Israeli jets then flew in triumph at supersonic speed over Damascus. Three days later the air force chief of staff, Gen. Maḥmūd Ṣidqī Maḥmūd, and other senior Egyptian officials flew to the Syrian capital for meetings on the military situation.

In early May 1967 senior Soviet advisors informed Egyptian and Syrian leaders that Israeli troops were massing along the Syrian border. After the air clash over the Golan Heights, Israel in fact did send additional forces to the Syrian front. President Nasser and Syrian leaders, however, reacted with alarm. On 14 May 1967 the UARAF was put on alert. At most bases, interceptors were ready to scramble at any time during daylight hours. Attack aircraft and bomber units were also prepared. Pilots and aircrew not already assigned to combat, support, or vital training units were sent to operational squadrons to hone their skills and provide a reserve.[29]

The Egyptian Army air defense units increased their readiness, and ground force reserves were mobilized. Egypt's plans for the defense of the Sinai and confrontation of Israel had the code name Qāhir (Conqueror). These plans called for a Soviet-style multilayer defense supported by a mobile reserve of armored forces, yet the plan had not been fully implemented when war broke out.

Normally only a single Egyptian infantry division, plus supporting armor, was deployed in the Sinai. In April and May 1967 President Nasser ordered a total of four infantry divisions, one armored division, and several independent brigades into the Sinai. These units were moved well forward along the border with Israel as a political gesture. As a result, when war came and Israeli forces quickly penetrated the front lines, many Egyptian units were rapidly cut off.[30]

The UARAF had more than two regiments of interceptors and fighter bombers based at al-Mulayz (Biʾr al-Thamādah), al-ʿArīsh, Biʾr al-Jifjāfah, and Canal Zone airfields such as Fāʾid. These squadrons were responsible for providing air cover and ground attack support for Egyptian Army units.

In a speech on 17 May President Nasser demanded that UN forces stationed in the Sinai be withdrawn, but when the UN did start to pull out its forces, this put Nasser in a tough position. He could not now back down without losing face. On 22 May, President Nasser announced a blockade of the Strait of Tiran. This was the same move that helped initiate the Suez War back in 1956. Further aggressive statements by President Nasser and military action on the part of Egypt and Syria heightened tensions with Israel. The blockade also had given Israel a strong reason to initiate hostilities. The announcement on 30 May that Egypt, Syria, and Jordan had signed a joint defense agreement was the last straw in Tel Aviv.

Israel quietly started calling up its reserves and set into motion plans for war. During the last days of May and first week of June, the Israelis flew dozens of flights into the Sinai to test Egyptian air defenses.

Brig. Gen. Qādrī al-Ḥamīd (ret.) remarked about operations during this period: "We thought that we had superiority over the Israelis since we had three wings of MiG-21s, one wing of MiG-19s, and the new Su-7s versus one wing of Mirages and their older jets. Of course we were lacking in the theory of air combat; the Russians had given us training but not good tactics. They trained us to fly at Mach-two and do high-level intercepts, night fighting, and all that didn't happen during the 1967 War with Israel—it was all fought on the deck. You train for something, and if it doesn't happen that way, you aren't prepared."[31]

Egypt and Syria relied on the MiG-21 and older MiG-17 (plus antiaircraft weapons) for defense against Israeli air attack. The MiG-21 outperformed the Israeli Mirage III at high altitude, and the MiG-17 was more maneuverable. However, Israeli pilots knew exactly how to counter the MiG-21 and MiG-17. In 1965 a Syrian MiG-17 was forced to land at an IAF base, and a year later an Iraqi pilot defected to Israel with his MiG-21F-13. Israeli pilots flew their Mirage, Super Mystère, and Mystère fighters in mock dogfights with the MiG-21 and MiG-17, which gave the IAF an opportunity to learn the strengths and weaknesses of these top line Soviet fighters.[32]

Brig. Gen. Qādrī al-Ḥamīd added:

> I was with Squadron No. 45 at al-Mulayz in the central Sinai. We used to fly above Israel and do reconnaissance at a height of 18 km. They shot at us with their Hawk missiles, but because of our height they didn't hit us. We were flying over Israeli territory but stayed over it just a short time, so the Mirages couldn't catch us. None of us thought that we would really fight with Israel but we felt that we were very good. We were scrambling two, three, and sometimes four times a day to catch their reconnaissance missions. They were flying on the deck, and our troops were telling us to try and catch them. We couldn't catch them because they flew below our radar detection network. I believe they were flying the same routes they used in the 1967 War to see if we could catch them.[33]

These Israeli probes prompted the UARAF to prepare for a possible air attack down the Gulf of Aqaba and across the Red Sea, so a squadron of MiG-19s and eight MiG-21s were sent to Hurghada (al-Ghardaqah) air base on the Red Sea to protect this area.

With the threat of war now growing serious, senior Egyptian military leaders warned of the danger of an Israeli first strike. President Nasser came under increasing pressure from his military chiefs to strike first. The Egyptian leader dismissed the possibility of a surprise Israeli assault and was confident that his army and air force could repel any attack. At a military conference on 2 June 1967, the UARAF chief of staff warned that an Israeli preemptive strike could prove fatal. President Nasser ignored the plea to strike first and ordered additional Egyptian Army units in the Sinai to near the frontier. This was a political move to show that Egypt was ready to fight if necessary.

President Nasser aimed for a diplomatic showdown with Israel, but not a war. However, he had given Israel reason and an opportunity to destroy Egypt's military strength. In the court of world opinion, Israel cast Nasser as the aggressor and quietly prepared to strike.

## Last Opportunity

From 4:00 to 8:00 A.M. on the morning of 5 June 1967, several squadrons of Egyptian jets flew their regular morning patrols along the border. At 8:05 A.M., three Egyptian Il-14 transports took off from Kibrīt. The Egyptian chief of staff, Marshal ᶜAbd al-Ḥakīm ᶜAmr, the UARAF commander, Gen. Maḥmūd Ṣidqi Maḥmūd, and several other senior military officers were on one Il-14. The two other transports carried the vice president of Egypt, Ḥusayn Shāfiᶜī, the prime minister of Iraq, senior Iraqi officials, and Arab journalists who were planning to visit Iraqi troops stationed near the Suez Canal. Egyptian fighters and anti-aircraft gun and missile sites were instructed not to fire on any aircraft over the Sinai between 8:00 A.M. and 9:00 A.M. to ensure the safety of these aircraft. The three Egyptian transports were watched on radar both by Egyptian ground controllers and the Israelis who were about to spring a major surprise.

## UARAF Order of Battle[34]

UARAF had 18 operational air bases: 4 in Sinai, 3 in Suez Canal area, 6 in Nile Delta, 5 in Upper Egypt, each with a base number.

Nos. 5, 7, and 9 Interceptor Regiments (9 squadrons; nos. 22, 26, 40, 41, 42, 45, 46, 49, plus one other) with approximately 200 MiG-21F13 and MiG-21PF fighters.

Nos. 2 and 5 Fighter-Bomber Regiments (6 squadrons). No. 2 Air Regiment with MiG-17Fs and MiG-19s consisted of Nos. 16, 18 (MiG-17), 31

(MiG-17), and 24 (MiG-15) squadrons, although No. 31 may have been moved to No. 61 light bomber Air Regiment.

No. 15 Air Regiment included No. 20 (MiG-19) Squadron and 21 Squadron (MiG-19) perhaps plus another unit, based at Ḍumayr near Damascus in Syria.

No. 12 Fighter-Bomber Air Regiment, consisting of five fighter-bomber squadrons with Mig-17Fs and MiG-15bis, including elements of Nos. 24 (MiG-15), 25 (MiG-17), and 30 (MiG-15) was placed under direct army control.

No. 1 Light bomber Air Regiment with Su-7s and MiG-17s consisted of No. 55 Squadron (probably Su-7) and another unit.

No. 61 Light bomber Air Regiment had approximately 35 Il-28s of Nos. 8, 9 plus one other Squadron, possibly also elements of No. 25 Squadron (MiG-17).

No. 65 Strategic bomber Air Regiment had about 30 Tu-16s, some armed with Kennel anti-shipping missiles, of No. 95 Squadron plus one other.

Air transport was provided by No. 16 Squadron (An-12), two other squadrons (Il-14), No. 11 Squadron (C-47) Dakota), No. 7 Squadron (Mi-6), No. 12 Squadron (Mi-4), probably No. 43 Squadron (Mi-8).

EAF units were located at the following bases:

Cairo West: No. 65 Regiment.

Cairo International: MiG-21s of an unknown unit.

Kibrīt (base no. 228): elements of No. 25 Squadron, No. 12 regiment, No. 31 Squadron/No. 2 Regiment.

Abū Ṣuwayr (base no. 229): elements of No. 25 Squadron/No. 61 Regiment, Il-28s of an unknown squadron from No. 61 Regiment, MiG-21s from No. 45 Squadron.

Fāʾid (base no. 233): elements of No.55 Squadron/No. 1 Air Regiment.

Biʾr al-Jifjāfah (base no. 244): Il-14s and Mi-6s of unknown transport and helicopter units, plus elements of No. 45 Squadron.

Jabal Libnī (base no. 248): No. 24 Squadron/ No. 2 Regiment, elements of No. 25 Squadron/No. 12 Regiment, No. 16 Squadron/No. 2 Regiment

al-ʿArīsh(base no. 259): No. 18 Squadron/ No. 2 Regiment, elements of No. 25 Squadron/ No. 12 Regiment.

Biʾr al-Thamādah (base no. 260): elements of No. 25 Squadron/No. 61 Air Regiment, elements of No. 55 Squadron/No. 1 Air Regiment, plus some An-12s of an unknown transport unit.

Inshāṣ: Nos. 40, 45 plus one other Squadron/No. 9 Air Regiment (MiG-21s).

Hurghada: No. 20 Squadron/No. 15 Air Regiment (MiG-19), plus MiG-21s and MiG-19s of unknown units.

Ra'ʾs Banās: Il-28s of an unknown unit/ No. 61 Air Regiment.

Luxor: No. 31 Squadron/ No. 2 Regiment and unknown Il-28 Squadron/ No. 61 Air Regiment.

16

# The Surprise Assault (1967)

The 1967 War was a turning point in the Middle East. This conflict still evokes questions that are hard to answer. Why did the Soviet Union send Egypt information about an impending Israeli attack on Syria? Did senior Israeli government officials believe that Egypt, Syria, and Jordan were really about to attack? Or did events get out of control and give the Israelis an opportunity to strike first? How were President Nasser and the Egyptian military caught by surprise, given the level of tension leading up to the war? Despite Nasser's aggressive rhetoric and deployment of troops into the Sinai, most Egyptian political and military leaders did not want to go to war. Arab leaders realized that the Arabs had little chance of defeating Israel. But if a conflict were to occur, joint military plans called for Egypt, Jordan, and Syria to attack in a coordinated manner. The goal was to seize some Israeli territory and cause enough casualties to force a cease-fire quickly. The degree to which these hopes were ill-founded became all too clear on that first day of the 1967 War. Because of Egyptian government misjudgment, the country lost the Sinai province, nearly the whole army and air force, as well as the lives of thousands of soldiers, sailors, and airmen. For the country and the personnel of the UARAF who survived the catastrophe, June 1967 lingers as a nightmare that has a place in Egypt's memory comparable to that of the attack on Pearl Harbor in America.

## Operation Focus: The Israeli Attack Plan

Israeli operational planners demonstrated a greater grasp of the potential impact of air power than did the Egyptians. Taking a great risk, Israel committed its total air force to an overwhelming initial air strike and then shifted to providing tactical support for the ground forces. Egyptian war planners followed a more conservative course as a result of Soviet influence and Arab estimates of anticipated Israeli actions. In acknowledgment of Israeli superiority in the air, a large portion of Egypt's fighter force was planned to be held back to protect the Egyptian Army in the Sinai and defend home territory in the event of war.

For years the IAF had planned and practiced a campaign to destroy Egypt's air power, which on paper was greater than Israel's. When the air arms of Syria, Jordan, and Iraq were added, the Arabs had what appeared to be an overwhelming edge. According to the Military Balance assessment released by the International Institute of Strategic Studies, on the eve of the conflict the Arab states surrounding Israel (Egypt, Syria, and Iraq) had a five-to-one advantage in bombers and three-to-one superiority in fighters.

Yet the Israelis never intended to take on all the Arab states at once. The IAF planned to knock the UARAF out of action at the outset of a conflict. Only when Egyptian air power was sufficiently battered would the IAF take on the other Arab air forces. The Israeli name for this well-rehearsed plan was Operation Focus. For this reason, it was only the Israeli-Egyptian balance that mattered at the start of the 1967 War.

To face Israel's 270 combat aircraft plus 100 support aircraft, Egypt fielded some 330 combat aircraft and about 200 transports, helicopters, liaison aircraft, and trainers.[1] However, this total included two fighter squadrons stationed in Syria and several squadrons located in faraway Yemen. Thus the balance of forces in the immediate combat zone (Upper Egypt, the Canal Zone, and the Sinai) was very close.

Qualitative factors in servicing and turnaround time enabled the IAF to bring far more air power into action than Egypt and other Arab air forces. With a wartime manning level and maximum effort, the IAF could fly and maintain about 90 percent of its combat aircraft in action (a level well above that maintained at the time by the U.S. and Soviet air forces). At any one time the UARAF had an average of only 70 percent of its aircraft ready for action, and even this compared favorably with some of its allies. Even more beneficial was the Israeli ground turnaround time, which was as short as ten minutes for refueling, rearming, maintenance, and pilot briefing. It meant that an Israeli pilot could fly up to eight sorties in a single day, while Egyptian pilots were expected to fly no more than two missions.[2]

Israel also had the significant advantage of striking the first blow. By making their initial strike at 8:45 A.M. Cairo time, the Israelis struck at a moment when Egypt's regular early morning period of alert had been reduced. Traditionally the hours before dawn were the most likely time for a surprise attack. By 8:45 A.M. the morning patrols of Egyptian fighters had landed and their pilots were at a reduced state of readiness.

Flying in at less than 30 feet over the sea and following routes through unpopulated desert areas, Israeli pilots eluded the Egyptian radar surveillance network and remained undetected. Israeli jets going in to attack targets in the Sinai flew directly to these targets but flew at low altitude and took advantage of hills to shield themselves from Egyptian radars. IAF C-47s dropped a chaff barrier along the border during the early morning of 5 June, and other electronic deception was used. UARAF and Army air defense radar operators were not trained to deal with such jamming and other Electronic Counter Measure (techniques) employed by the Israelis.[3] To this day it has not been revealed whether or not the Israeli attackers were picked up by the radars of American and Russian ships in the Mediterranean.

Another major Israeli advantage was accurate intelligence on the UARAF. The IAF flew dozens of reconnaissance missions over Arab airfields before the conflict. War planners also had excellent information collected by the Israeli intelligence services. This enabled the IAF to target high-priority air bases, radar sites, and other positions and thus not waste sorties. Taking into account the element of surprise, greater sortie rates, and pilot training advantages, the real force ratio of air power between Israel and Egypt on 5 June 1967 was very much in favor of the IAF.

## Missed Opportunities

The Egyptian military was caught off guard despite repeated warnings of an impending Israeli attack by intelligence sources. The country's political leadership, especially President Nasser, failed to appreciate the risks of his threats and saber rattling. Another opportunity for warning of the Israeli strike was missed. After Egypt and Jordan signed a defense cooperation treaty in early 1967, the Egyptian military was supplied with intelligence information from a radar station 1,350 meters up on the mountains near ʿAjlūn in northwestern Jordan. As this radar station could monitor almost all air activity in Israel, it was agreed that warning of any unusual IAF air activity headed toward Egypt would consist of the code word *ʿInab* (Grapes). Radio links and ground lines between ʿAjlūn and Cairo were both reportedly operational on 5 June.

Unfortunately three days before the outbreak of hostilities Egypt's war minister, Shams al-Dīn Badrān, had ordered that all communications with Jordan be routed through his office. A warning of unusual activity was in fact sent on the morning of 5 June, but this new order, as well as other problems, prevented the urgent warning from reaching the UARAF and army air defense commands in time to react. An officer at the War Ministry, who had heard the initial warning, later telephoned the Air Defense HQ to check and see that the message had been received. The reply he got from Col. Ibrāhīm ʿUmar at the HQ was curt and to the point: "To hell with it, what grapes and what onions? They are all over us!"[4]

## The Initial Israeli Strike: 8:45 A.M., 5 June 1967

8:45 A.M. (Cairo time) Cairo West air base. Israeli jets first blasted the runways with dibber bombs that dug large holes in the concrete. Then they bombed and strafed the Tu-16 bombers, which exploded and burned in their revetments. Egyptian soldiers and airmen were totally stunned.

Maj. Gen. Muḥammad Nabīl al-Masīrī (ret.) said: "When the war started, I was on the tarmac. I heard explosions and looked up. I saw a Mirage coming out of a dive firing its cannon. They were shooting at the Tu-16s which were in their [open-topped] revetments, but these offered no protection. You won't believe it, but at that time I saw the [anti-] runway bombs still hanging from their parachutes. And then they hit and destroyed the runway intersections."[5]

8:45 A.M., al-Mulayz (Biʾr al-Thamādah) air base in the central Sinai. Brig. Gen. Qādrī al-Ḥamīd (ret.):

> I was in the tower, near the radar and saw the whole scene. I heard the sound of many aircraft. I thought that the MiG-19s based at Hurghada were coming back to visit us again: head on the MiG-19 and Ouragan look alike, especially in the sunlight. The MiG was grey and the Ouragan a grey and brown.
>
> Then I saw the bombs falling from four Ouragans, two from each plane. Two Mirages then came in, popped up to 200 meters, and dropped parachute anti-runway bombs. Then the Israeli jets started to strafe. Our first-stage alert

fighters needed about two minutes to scramble. The Israelis attacked them with cannon. They hit two of these MiG-21s and set them on fire, but the other two were able to take off. One of the Ouragan pilots turned and attacked a MiG-21 that had just left the ground. It was a quarter attack, the Israeli hit the MiG, and the pilot ejected. The Ouragan pilot couldn't attack the second MiG, and so he was able to go up and engage the Mirages.

One of our pilots was also able to take off in a third wave [an aircraft not on immediate alert but ready for flight] MiG-21. He was hit in the rudder by cannon fire from an Israeli jet but was able to keep flying. He fought with the Israeli jets, but the ground crew forgot to remove the safety pins from the missile launchers so he could not fire his Atolls. The pilot continued to fight and didn't know that he was damaged until he landed at Cairo West later.[6]

Fāʾid air base, Canal Zone, 8:45 A.M. A Coptic Christian officer, Brig. Gen. Samīr ʿAzīz Mīkhāʾīl (ret.) (then a lieutenant), had just finished sitting first-stage alert and had gone off duty:

We finished our duty at 8:30 A.M., and I went to shave. When I was shaving, I heard the strafing. We didn't imagine that the Israelis would attack us. We took our jeep and drove to the ready room. We saw the four first-stage aircraft burning and an Ouragan jet strafing these aircraft. We ran to get our helmets and flight gear, but we found the ready room smashed by a bomb. We carried our gear out from the rubble and ran out to try and reach the second-stage alert aircraft that were out near the runway. Before we could drive 200 meters these aircraft blew up and were on fire from Israeli attacks, so we turned back. We ran to a MiG-15UTI that had a cannon, but it would not start. When we came back with a starter cart, it had been hit and damaged.

At this point an Il-14 landed which had the vice president and other VIPs on board. All those on the plane ran out when it stopped. After that it was strafed and caught fire. We gathered at the mess. Our leaders told us to go to Fāʾid village which was about 2 km from the airfield, and so we got a Russian Volga car and drove off. From there we watched further Israeli attacks against our base. It was a disaster.

I saw a Super Mystère, which had a nose like a shark, make an attack on our air base. Then it came by our location at a very low altitude. So I pulled out my 9 mm Beretta pistol, took a deflection, and fired at him several times. I think I might have hit him because I remember very clearly that the pilot looked at me, and I saw his eyes. My companions yelled to me that the airplane was coming back. My back was to him, and when I looked back I didn't know what to do. I thought that if I ran to the right or left he could correct and easily shoot me so I ran toward him where there was a wall that would give me some cover. As I ran toward the wall, I saw him fire, and the shells struck several feet to my right. I jumped and hit the wall and hurt my hand,

> but I was still alive. I wanted to know the name of this pilot. He came from Israel to Fāʾid to kill me—why?[7]

Abū Ṣuwayr in the Canal Zone, 9:00 A.M. A Mirage strafed an Il-28 which burst into flames. Maj. ʿAwaḍ Ḥamdī, (who later retired with the rank of major general) worked feverishly with several mechanics to prepare a MiG-21 for action:

> We scrambled after the first [8:45 A.M.] attacks without orders. The Israelis bombed the runway intersections with anti-runway bombs. It was our first experience with these weapons. I figured out that by flying an aircraft without any drop tanks, I could get into the air. Our main runway was 27, and I scrambled from 22 which crossed it. Our base commander marked where I left the ground. I cleared the holes by only 6 to 8 meters.
>
> After I took off, I patrolled and then joined up with my wingman. A short time later I spotted the Israelis and made a rendezvous with a group of four aircraft. They made a break. I picked one and was certain that I would get him. My wingman fired an Atoll at an Israeli jet, but we were at low altitude, and it just flew into the ground. The MiG-21FL I was flying had no cannon, just two Atolls. I aimed at the centerpoint of the exhaust of a Mystère and launched my missile. It hit close to the Mystère, and then I engaged again. However, they did not stay and fight. They had a mission to attack and go, and this is what they did.[8]

Israeli jets pounded Egyptian air bases for more than an hour (8:45 A.M. to about 9:45 A.M. Cairo time). Nine squadrons of Israeli fighter-bombers flew some 170 sorties during this first hour. Only a dozen fighters were held back to defend Israeli airspace. Flights of four Israeli jets arrived over the principal UARAF bases about every ten minutes and maintained their attacks until they ran out of ammunition. Israeli jets blasted the home bases of Egyptian bombers (Tu-16s and Il-28s), first-line fighters (MiG-21s and MiG-19s), and airfields in the Sinai. They also struck bases housing fighter-bomber units that could immediately retaliate against Israel. The Egyptian bases that were attacked during the first series of raids included Abū Ṣawayr, al-ʿArīsh, al-Mulayz (Biʿr al-Thamādah), Biʿr Jifjāfah, Banī Suwayf, Cairo West, Fāʾid, Inshāṣ, Kibrīt, and Jabal Libnī.

Because of the air-raid warning sirens and radio broadcasts, Egyptian citizens realized that war had begun. Many Egyptian airfields were located near urban areas or villages, and citizens could plainly see jets fly over and hear the explosions and anti-aircraft fire.

## 10:15 A.M., 5 June: The Second Israeli Assault

After a half-hour break to service aircraft and assess results, Israeli jets flew a second series of raids against Egyptian airfields and other targets. About 115 attack sorties were flown during this series of attacks. Many airfields attacked earlier were struck again, and Egyptian bases at al-Manṣūrah, Ḥulwān, and al-Minyā were hit for the first time.

Kibrīt, 10:45 A.M. Brig. Gen. Muṣṭafá Ḥāfiẓ (ret.) was a lieutenant and an instructor at the Flying College at Bilbays before being posted to a squadron of MiG-17PF night-fighters at Kibrīt air base in May 1967. Muṣṭafá Ḥāfiẓ was able to take off despite Israeli attacks:

> I was patrolling over our airfield and had only 400 liters of fuel left. At about 10:30 A.M. we had some SA-2s [Egyptian surface-to-air missiles] fired at us, so we decided to land. We were on final when I saw the enemy. I retracted my undercarriage and engaged them. One of the Israelis shot at me from head-on and it made me furious. I made a steep turn to the left to get behind him. While reversing I was hit by fire from another Israeli aircraft: a Mystère, I think. There were four of them. My rudder was jammed, and my fuel tank caught fire, but I was able to land. I still have pictures of the damaged aircraft.[9]

Egyptian radio announced the beginning of war but claimed that dozens of Israeli jets had been shot down and that the Egyptian Army was fighting a victorious battle in the Sinai. Acting on the basis of prewar treaties and confident pronouncements from Egypt, the countries of Jordan, Syria, Iraq, and Lebanon struck at Israel with aircraft and committed their armies to the ground war. Reality was different. By 11:00 A.M. more than 150 Egyptian aircraft had been damaged or destroyed. Egyptian fighters and anti-aircraft units claimed some fifteen Israeli aircraft, and Israel later admitted the loss of nine (four to Egyptian fighters) during the first series of strikes.[10]

## Dealing with Disaster

The surprise Israeli air assault had succeeded. Most of Egyptian air power was out of action by noon on the first day of the war. Brig. Gen. Qādrī al-Ḥamīd (ret.) said: "We had twenty MiG-21s and one MiG-15UTI at the base [Biʾr al-Jifjāfah], and in no time they were all destroyed. I was up watching all this from the tower. I cried—all we had trained for was gone in just three minutes."[11]

Brig. Gen. Samīr ʿAzīz Mīkhāʾīl remarked: "In a maximum of five minutes all of our MiG-21 and Su-7 aircraft had been destroyed. No one was able to take off because they had destroyed all the alert aircraft and had hit the runways with bombs."[12]

Many Egyptian bases already lay in ruins when, at 10:45 A.M. the Il-14 transport carrying the chief of staff and EAF commander finally landed at Cairo International Airport. During the two hours since the first attack, most of Egypt's senior commanders had been out of contact. The Egyptian chief of staff, Field Marshal ʿAbd al-Ḥakīm ʿAmr, the Air Force commander general, Maḥmūd Ṣidqī Maḥmūd, and many other senior Egyptian military officers had been trapped in this transport aircraft. Most of the senior Egyptian officers who directed the ground and air defense of the Sinai were at Biʾr al-Thamādah waiting to greet the Egyptian delegation.

Egyptian military leaders rushed to the military headquarters where they found confusion and disarray. Nothing but bad news was flooding into the office. Only gradually did the full extent of the disaster become apparent. Soldiers and aircrew were forced to fight against an enemy that now enjoyed control of the air and held the tactical initiative.

At Fāʾid, Inshāṣ, Cairo West, and other airfields UARAF personnel and soldiers worked feverishly to clear away wreckage, extinguish fires, repair runways, and evacuate casualties. More than a hundred Egyptian planes had survived the Israeli attacks, while others had suffered only minor damage and could be made ready to fly. But the surviving aircraft were spread out at many airfields, and many local commanders were not sure how to react.

## 5 June: After the Initial Strikes

Throughout the morning, surviving Egyptian fighters scrambled. Lt. Col. Zuhayr Shalabī led a flight of three MiG-17s dispersing from Cairo West to stand alert at Ḥulwān at about 9:30 A.M. Shalabī was Egypt's leading test pilot, and had been flying prototypes of the Ha-300 fighter before being sent back to a fighter squadron just before the June War. Ḥulwān was the test center for the Ha-300 and other Egyptian aircraft development programs. At around 11:00 A.M. another hour-long series of

Israeli air strikes began. This time in addition to air bases, air defense stations and radar sites were targeted.

While Shalabī's MiGs were being serviced for another mission, four Israeli Mirages and two Vautours struck Ḥulwān. They damaged the runways and destroyed the three MiG-17s along with a MiG-19 and several Ha-200 trainers. An An-12 engine test-bed for the Ha-300 program was also damaged, although the Ha-300 prototype escaped unscathed. (Lt. Col. Shalabī was killed two days later while flying a ground attack mission over the Sinai.)

Inshās, 11:00 A.M. Maj. Nabīl Shuwakrī (who retired with the rank of maj. gen.) was then with a MiG-21FL squadron at Inshāṣ. His actions in trying to recover from the initial Israeli strikes were typical of many in the EAF:

> Most of our aircraft were destroyed on the ground by Israeli attacks, and they bombed the runways. However, we had some MiG-21FLs and two or three MiG-21F-13s stored at a depot nearby. We went to the depot and prepared these aircraft to fly. We found a place on subrunway 22 that was 900 meters long. I know because I measured it in a car. I was the first to take off. Then the wing commander took off, in a MiG-21F-13. After that, two captains took off. I flew a patrol over Bilbays, which is very close to Inshāṣ.
>
> A short time later Inshāwi, one of the captains, joined me over Bilbays. We were at 15,000 feet. The ground controllers told me that the Israelis were now attacking the base. I dove down but didn't see any. Then I climbed again and did some hard maneuvers. My wingman got lost and flew away. Then the controllers called and said that there were Mirages over Inshāṣ. So I flew there and saw two Mirages flying to the left. I came behind them until I saw the crash helmet of the wingman. I was just looking at him and couldn't shoot because of the limitations of the Atoll missile. If they kept in formation, I would be behind them. They tried to get behind me using horizontal maneuvers, but they didn't succeed. All the time I was hoping that they would turn back toward home.
>
> Then they dived down and ran, and I got my chance. I followed them and fired an Atoll at the leader. The first missile hit the Mirage, and there was heavy smoke. I looked for the wingman and didn't see him, so I decided to launch the second missile at the smoking aircraft. The Mirage exploded and pitched up in a stall turn, and the nose then went straight down. The MiG-21FL was very limited. If I had had a cannon I could have shot down both Mirages, but I had to wait and work to get a missile shot.[13]

Despite IAF air strikes against airfields in Jordan and Syria, Israeli jets continued to hit targets in Egypt throughout the day. The Bilbays Air Academy was bombed around 1:30 P.M. At about the same time the air base at Hurghada (al-Ghardagah) on the Red Sea was struck by a flight of Mirages. The Israeli jets bombed the base and claimed to have shot down three MiG-19s. At about 1:45 P.M. long-range Vautour fighter-bombers attacked Luxor. They met heavy anti-aircraft fire—a Vautour was hit and crashed on the airfield—but they destroyed most of the MiG-17s and Il-28s at the base. Late in the afternoon even the remote ex-RAF airfield at Raʾs Banās, far to the south on the Red Sea coast, was hit by Vautour fighter-bombers.

## Reacting to Reality

The officers at UARAF headquarters could hardly believe the magnitude of the damage caused by the Israeli raids. Morale dipped further as news of destroyed aircraft, wrecked squadrons, damaged airfields, and blasted radar stations continued to pour in. Humiliated by the serious losses, senior air force leaders were slow to report the true extent of the defeat to Army HQ and President Nasser.

Many air force commanders tried to organize a fighting force from what remained. Maj. Gen. Muḥammad Nabīl al-Masīrī (ret.) said: "After two or three attacks we still had some aircraft. I told the headquarters this and then asked, what shall we do? We had some MiG-17s and some Su-7s. These jets were assembled at Cairo West at that time and were brand new. The Tu-16 bombers were all destroyed in the first series of attacks. After a while we took all the aircraft we could to Quwaysina, an airfield in the Delta. [The answer, as in the 1956 Suez War, was dispersal.] This was just a runway with a few support facilities. But we kept flying and fighting."[14]

Senior Egyptian military leaders initially believed that the Israelis had suffered heavy losses during their air assault against Egypt. When Syria, Jordan, and Iraq entered the war, Egyptian military leaders hoped that the Arab-Israeli military balance could be restored, but this hope was quickly overturned as the Israeli Air Force battered the air forces of these countries. Israeli air losses were small compared to Arab casualties and had minimal impact on the course of the conflict.

Late in the afternoon of 5 June, President Nasser was informed of Egypt's true losses and that Israeli tanks were already pushing through Egypt's defensive positions. For a time, and at the cost of appalling casualties, Egyptian tanks and infantry held off the advancing Israelis.

However, when Israeli ground forces ran into trouble, Israeli jets were summoned to crush Egyptian resistance with bombs and cannon fire.

The destruction of the UARAF totally undercut the army's plan of operations in the Sinai. Without air cover, Egypt's ground forces had to endure constant bombardment. Since the threat of interception was low, the IAF could effectively employ every one of its aircraft. In fact, Israeli Fouga Magister armed trainers flown by reservists proved to be effective ground attack aircraft, though they suffered many losses to anti-aircraft fire.

Yet the Egyptian Air Force had not been totally wiped out. Many aircraft and facilities had been damaged or destroyed, but aircrew losses had been relatively light. During the first day of the conflict Israeli attacks destroyed or seriously damaged more than 300 aircraft.[15] About thirty Egyptian pilots were able to take off and engage the attackers. Many of these brave pilots were shot down, forced to eject due to fuel exhaustion, or were killed while attempting to land on damaged runways.

Egypt's radar surveillance network had also been seriously hurt, and nearly half of the air bases had been heavily damaged. Nevertheless, runways could be rapidly repaired, and most air bases were quickly made operational. Egypt still possessed nearly a hundred serviceable front-line combat aircraft, and several hundred support planes remained operational. Reacting to the difficult military situation, on the night of 5 June President Nasser had finally agreed with Field Marshal ʿAbd al-Ḥakīm ʿAmr's request for a general withdrawal from the Sinai to save the army from destruction.

During the night of 5–6 June 1967 all available air force and army personnel, supplemented by civilian workers and construction equipment, were called out to clear runways, fill craters, and prepare surviving aircraft for action. The lack of proper runway repair and bomb disposal capability meant that these efforts had to be hastily organized. Air force leaders took stock of what remained and prepared for action at first light. The success of these efforts came as a surprise to the Israelis.

## 6 June: Fighting Back

On the morning of 6 June, Egyptian Army units were ordered to pull back from the Sinai, but relentless Israeli attacks caused a breakdown of Egyptian command and control resulting in a disorganized withdrawal.

The Egyptian Army was shattered by Israeli ground and air attacks. The UARAF now threw into action all the aircraft that had survived the Israeli air assault to assist the Egyptian Army. Egyptian fighters flew air patrols because the Egyptian radar network had been badly damaged in an effort to drive away Israeli jets. Fighter-bombers struck at the advancing Israelis. These attacks caused Israeli casualties but in turn suffered heavy losses to Mirage fighter patrols.

From one base two MiG-21FL night-fighters took off just after dawn to attack Israeli armored units in the Sinai, the use of such aircraft underscoring the desperate position of the UARAF. Major ʿĀdil Naṣr (who retired with the rank of major general), the pilot of one of these MiG-21s, surprised an IAF Super Mystère over the Sinai and shot it down with a salvo of S-5 unguided 57 mm rockets carried for his ground attack mission.[16]

Col. Taḥsīn Zakī (ret.) described operations on the second day of the 1967 War:

> On the morning of 6 June at 7:00 A.M., two Su-7s took off piloted by the late Midḥat al-Malayjī and Aḥmad al-Simarī to attack al-Khalsa airfield [Haluza in Israel]. They found no aircraft there, and so they kept flying very low and penetrated further inside Israel. They crossed over Beersheba railroad and attacked an airfield there [Nevatim]. On the return journey, however, both had to gain altitude to conserve fuel, and thus they were attacked by Israeli Mirages some 30 km from al-Fāʾid. Aḥmad al-Simarī's Su-7 was hit. Midḥat ordered him to eject at once but got a reply that the Su-7 was still flyable and under control and that he would land at al-Fāʾid. Midḥat headed for Abū Sawayr to give his wingman priority to land first at al-Fāʾid. Aḥmad al-Simarī managed to land his damaged plane on the subsidiary runway at al-Fāʾid where his airplane was repaired and was used on operations again after four days.
>
> On the same day, another formation of Su-7s from Cairo West under the leadership of Lt. Col. Jalāl ʿAbd al-Ālim, Pilot Muḥammad ʿAlī Khamis, and Pilot Muḥammad Shihāta were ordered to create an umbrella [CAP] over the Suez Canal from Port Said to Ismāʿīlīyah. While airborne, they received orders to proceed to al-ʿArīsh airfield and engage the Israeli planes attacking the base. They fought with eight Israeli Mirages. Muḥammad Khamis managed to shoot one down in the ensuing melée. The Egyptian formation lost two of the three Su-7s. Khamis and Shihāta managed to bail out safely and returned to the west bank [of the Suez Canal] after three weeks full of adventures. Jalāl ʿAbd al-Ālim ran out of fuel at the Rafaḥ area and bailed out safely and returned in the evening to Cairo West airfield.[17]

Egyptian air activities over the Sinai interfered with Israeli support aircraft, and in response the IAF launched Mirages on air defense patrols. Early on 6 June two IAF helicopters were surprised by Su-7s near Abū Awayjilah. The helicopters maneuvered to escape, and, after several gun passes, two Mirages arrived and downed one of the Su-7s. Later in the day, two Egyptian MiG-21s jumped an IAF S-58 helicopter. The MiG pilots were unable to get their Atoll missiles to lock onto the low-flying helicopter, which landed among the trees while the angry MiG pilots flew off toward home.[18]

Gen. Yehu Gavish, commander of the Israeli Southern Command strike force in the Sinai, had more than one brush with Egyptian aircraft. Egyptian Army units detected his headquarters and called in a flight of MiG-17 fighter-bombers. Gavish was warned of their approach and managed to escape the area just in time. On another occasion his helicopter was attacked by two Egyptian Su-7 jets, but the pilot of the scout helicopter was able to escape.

Not all IAF aircraft were so lucky. During the late afternoon Maj. Fatḥi Salīm, flying a MiG-19, reportedly intercepted an Israeli Nordatlas transport in the central Sinai and shot it down with cannon fire.[19]

Brig. Gen. Fārūq al-Ghazzāwī (ret.) said: "I flew two air-to-ground missions against Israeli targets near Mitla Pass in a MiG-21 armed with 57 mm rocket pods on 6 June. It was hit and run, because we had only a small number of aircraft and we didn't have time to evaluate what damaged we caused."[20] Egyptian fighters and army anti-aircraft fire also destroyed several Israeli aircraft. By the end of the second day of the conflict, the IAF had admitted to the loss of some twenty-six aircraft on all fronts.

## 7 June: Determined Resistance

While the IAF was preoccupied with providing air support for the heavy fighting in the central Sinai and on the Jordanian front, the EAF struck back. All reserve aircraft and jets damaged by the Israeli first strike which had been repaired were sent into action to support the army in the battle for the Sinai. Many sorties were flown against the Israeli force advancing toward the Suez Canal on the northern coast road.

Brig. Gen. Fārūq al-Ghazzāwī remarked:

> On 7 June I flew three missions. My last was in mid-afternoon. It was a patrol mission. I was at 8,000 meters orbiting north and south over our territory. The

other section was at 6,000 meters. All of a sudden, in a typical Israeli fashion, they came by surprise and ambushed us from below.

The Russians always told us that the MiG-21 was a high-speed, high-altitude fighter. I had this in mind then. We gained speed, and the ground controllers directed the other section against the enemy. The Israeli tactic was to show themselves and then they started losing height, to get you to the best altitude for the maneuverability of the Mirage. This, by the way, was also the worst altitude for the MiG-21: we had less maneuverability and higher fuel consumption. By going down, no one can see you on the radar. Thus we didn't receive any more support from our ground control interception (GCI) controllers.

The other section was hit, and they killed my friend's No. 2, Dibus was his name. No. 1 got scared and yelled on the radio, "I got hit" and asked me to come down and save him. However, I was too far away, and so I radioed for him to pull up and bail out. He did but came back with a spinal compression from the ejection.[21]

## 8 June: Last Defense

On the fourth day of the conflict, the scale of the UARAF's counterattacks caught the Israelis by surprise. Some forty patrol and attack sorties were flown by Egyptian jets during the day.

Col. Taḥsīn Zakī (ret.) flew one of these missions:

Air Marshal al-Qādrī told me on the phone that all our forces were now west of Mitlā Pass and all forces east of it were Israeli and should be attacked at will. By early light, I took off, accompanied by Zakaryā Abū-Sidā, and we headed to Mitlā Pass. We found no trace of enemy troops. I kept flying over the road, reaching the area of Nakhl, and saw Egyptian forces still retreating on the road. Suddenly, I saw twelve Israeli Centurion tanks and the rest of the force some 30 km from Nakhl.

I flew low and saw that the Centurions were firing at our retreating vehicles and gave the order to my No. 2 to attack. We made three passes and fired, in each pass, from two to four, 57 mm rockets on each tank we targeted. I noticed that not one of the tanks were knocked out, proving that the Soviet 57 mm rockets were ineffective.[22]

In the afternoon a single Il-28s escorted by four MiG-19s swept in from the Mediterranean to attack al-ʿArīsh airfield which had been captured by the Israelis. Intercepted by Mirages just off the coast, most of the Israeli jets tangled with the escorting MiGs while one attacked the Il-28. Its rear gunner forced the Mirage pilot to pull back from his first pass. However, in the end the firepower of the Mirage's 30 mm cannon won out. The gunner's position exploded from a hail of cannon shells, and the bomber crashed on the beach. Two of the escorting MiG-19s were also shot down.

The UARAF continued to attack Israeli units in the Sinai and sent aircraft to strike at Israeli command posts. While traveling along the road opposite Ismā'īlīyah on the afternoon of 8 June, Deputy Chief of Staff Gen. Haim Bar-Lev's convoy was strafed by a pair of MiG-17s. This attack was documented by news photographers.

Col. Taḥsīn Zakī (ret.) commented about continuing Sukhoi operations: "The sorties of the remaining Su-7s continued on 8 June. They engaged Israeli Mirages that were attacking our tanks, and we lost one of our Su-7s piloted by Muḥammad Najīb. The fighting stopped on the 8th. The number of sorties of the Sukhois reached thirty-two."[23]

In the late afternoon several MiG-17s made strafing passes that wreaked havoc among another Israeli convoy. When a flight of Super Mystères appeared, the MiGs defeated them, downing one and forcing the others to withdraw.[24] However, nothing could alter the balance in the air, nor could it change the outcome of the battle being played out on the ground. At this point the Israeli Army had overrun much of the Sinai and the West Bank and were assaulting the Golan Heights. Most of the Egyptian Army was fleeing back to the Suez Canal, and many units had been surrounded and forced to surrender.

Maj. Gen Nabīl Shuwakrī (ret.) said:

> My leader and I flew a sortie on 8 June. We went to attack some Israeli tanks in the northern Sinai. We were at low level, and all the armament we had was two pods of S-5K hollow charge [57 mm] anti-tank rockets. After we passed the canal we saw two Mirages coming in from the left. I told my leader, "two Mirages attacking from the left," but he didn't see them. So I put the afterburner on, dropped the belly fuel tank, and saw that the Mirages were coming in to attack. I put my seat back fully and turned to see where my leader and the Mirages were. Then, while I was turning to the left, I saw a Mirage behind me. I turned very hard. The Mirage left the combat and disappeared. I told my leader, "there is a Mirage behind you," and he reversed. At this point his MiG-21 exploded from cannon fire. I didn't see if my leader got out.
>
> After he shot the MiG-21, the Mirage pilot did a right turn and headed back toward al-ʿArīsh. I put the nose down with maximum afterburner, but I

> had these two rocket pods which made a lot of drag. I came to the same level as the Mirage within a mile but had no way to close with him because he was accelerating. I started firing one rocket at a time from each pod at him, but I didn't succeed. If I had had a cannon, I could have ejected the pods and run after him, but in this case I could not do anything.[25]

## Outcome

The destruction of the UARAF shaped the outcome of the 1967 War. Jordan had counted on Egyptian air cover to augment its two-squadron force of Hawker Hunter fighters. After flying a single strike against Israel (which destroyed a transport and damaged a village), the Royal Jordanian Air Force was totally destroyed on the ground by Israeli attacks.

The Syrian Air Force also managed to strike Israel on the fifth of June. However, less than two hours later, Israeli air strikes destroyed many Syrian Air Force aircraft on the ground, heavily damaged several air bases, and shot down more than a dozen Syrian jets. Iraq sent several Tu-16 bombers over Israel, and one was shot down. Israeli jets then raided the Iraqi base at H-3 on three consecutive days and caused heavy damage. Even so, anti-aircraft fire and Iraqi fighters, some of which were flown by Jordanian pilots whose own Hunters had been destroyed on the ground, downed several Israeli planes.

Israel quickly achieved air superiority and used it to great advantage. Starting on the first day of the war and continuing throughout the conflict, Arab troops were subjected to repeated air strikes that paved the way for a rapid Israeli advance. In six days Israeli ground forces seized the Sinai, the Jordanian West Bank, and the Syrian Golan Heights, routing the armed forces of the three counties. The June War of 1967 was, in fact, a total disaster for the Arabs.

Despite the surprise attack, the Arab armed forces fought back with courage and determination. Maj. Gen. Muḥammad Nabīl al-Masīrī (ret.) remarked: "Our men fought hard and flew many missions. The maximum we could send out was four aircraft, and so we almost always fought outnumbered. Some of our pilots got victories against the Israelis."[26]

During the five days when Egypt was involved in battle, UARAF pilots and aircrew flew more than 150 attack sorties into the Sinai and dozens of patrol missions to defend against Israeli raids. This was a respectable effort considering the conditions under which the air force operated. The cost to Egypt, however, was high: thirty-four pilots and several aircrew were killed in combat, and many others were wounded on the ground. Air force ground personnel and soldiers serving at air bases suffered hundreds of casualties to Israeli air strikes. About forty Egyptian aircraft of all types were lost in air combat during the conflict. Israel listed Egyptian losses in air combat and hit on the ground as 336 UARAF aircraft.[27]

On the Israeli side the losses are harder to determine. Egypt claimed to have destroyed seventy-two Israeli aircraft in air combat and due to anti-aircraft fire. The IAF admitted to the loss of forty-six aircraft, while a further twenty-three were said to have been seriously damaged in combat on all fronts. The IAF also acknowledged that twenty-four air-crew were killed, eighteen wounded, and seven taken captive.[28] Losses on the Egyptian front were not broken out, but probably the bulk of these occurred over Egypt because that is where a majority of the sorties were flown.

The 1967 War was fought with the same tactics as earlier conflicts such as World War II and Korea. It was an old-style war fought with jet fighters, but all the elements of modern warfare were there: electronic warfare, surface-to-air missiles, and advanced weapons. It would not be long before the Middle East would see all elements of modern warfare come together.

Destruction at the hands of the IAF not only influenced the outcome of the war but also changed the map of the Middle East for more than a decade. President Nasser's goal of regional leadership was destroyed by Israeli bombs. Instead, following the war, Nasser was forced to concentrate on regaining the Sinai and restoring Egyptian self-respect.

Egypt's first three military pilots, Fu'ād Ḥajjāj (right), ʿAbd al-Munʿim Mīqātī (center), and Aḥmad ʿAbd al-Rāziq (left), being trained by the British RAF at Abū Ṣuwayr in the Suez Canal Zone in 1929. (photo Air Vice Marshal A. M. Mīqātī)

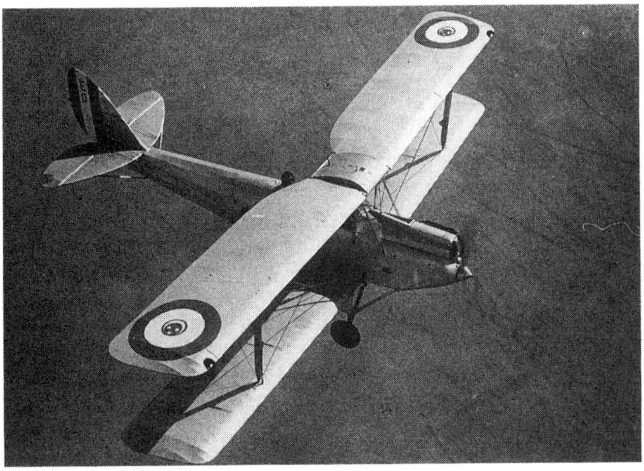

One of the Egyptian Army Air Force's (EAAF) first aircraft, a De Havilland DH 60M, over the desert near Cairo around 1933. The serial number, E106, shows it to have been among the first sold to Egypt. The national markings are those carried by Egyptian military aircraft until 1958, though they differ from the markings carried when these aircraft were first delivered in 1932. (photo Air Vice Marshal V. H. Tait)

Qāʾim-maqām Victor Hubert Tait, the Canadian first director of the EAAF, introduces the Egyptian ambassador in Great Britain to one of the pilots who will fly Egypt's first squadron of Avro 626s from England to Cairo in 1933. Behind the ambassador stands Fuʾād Ḥajjāj who would be killed during this long delivery flight. Both Tait and Ḥajjāj wear the old Egyptian Army uniform, still worn by the EAAF. (photo *Flight International*)

One of the EAAF's two Westland Wessex trimotor transports at Marsá Maṭrūh, in the Western Desert, in 1938. The aircraft appears to be guarded by a bedouin tribesman of this area. (photo Sq. Ldr. I. Blair)

Egyptian fighter pilots of No. 2 Squadron walking past their Gloster Gladiators during an inspection by Air Chief Marshal Longmore of the British RAF on 14 June 1940. Their commander, acting Squadron Leader Ibāhīm Abū Rabīʿah, is at the center. Behind the Gladiators are two REAF Avro Ansons of No. 3 Squadron. (photo Imperial War Museum, London)

Men of No. 6 Squadron REAF with one of their newly acquired, but ultimately unsatisfactory, Curtiss P-40 Tomahawk IIAs in 1943. (photo EAF)

Egyptian groundcrew bombing up an Avro Anson Mk. I, probably of No. 3 Squadron, during the Second World War. (EAF photo)

Supermarine Spitfire LF9s of the REAF over Tel Aviv on 17 May 1948. This was the third day of raids on military installations in the Israeli city. (photo Associated Press)

King Fārūq and his staff at al-ʿArīsh air base during an inspection visit on 7 July 1948. In the foreground small bombs are suspended from the undercarriage winglet of an obsolete Westland Lysander communications aircraft. To the left of the king is Wing Commander Abu Rabīʿah, while second to his right is ʿAbd al-Munʿim Mīqātī, now an air commodore. Two days later Mīqātī, while flying this same aircraft, forced an Israeli Avia fighter to crash into the sea. (photo Associated Press)

A newly delivered Macchi MC205 fighter of the REAF's No. 2 Squadron, serial number 1214, which landed short of fuel at the RAF base of Fā'id in the Suez Canal Zone during the final week of the Palestine War, December 1948–January 1949. (photo E. Thomason)

The first jets delivered to the REAF were a Gloster Meteor F4 jet fighter, serial number 1401, and a Meteor T7 dual-control trainer, flown to Egypt on 27 October 1949 where they formed the nucleus of a new No. 20 Squadron. They are seen here over England before their delivery flight. (photo *Flight International*)

The first jet fighter-bomber sold to Egypt was this De Havilland Vampire FB5. The rest of the order was made up of tropicalized FB52s which then formed No. 30 Squadron. Once in REAF service, the serial identifications of these aircraft were changed from European to Arabic numerals, though the numbers remained the same. (photo Hawker Siddeley Aviation)

The Egyptian Revolutionary Command Council in June 1953. It included two future presidents: Col. Jamāl ʿAbd al-Naṣr (Nasser) (3d from left, front row) and Lt. Col. Anwār Sādāt (Sadat) (far left, back row). Egypt's new revolutionary leadership also included three air force officers: Sq. Ldr. Hasan Ibrāhīm (2d right, back row), Wing Comdr. Jamāl Salīm (far left, front row), and Wing Comdr. ʿAbd al-Laṭīf Baghdādī (3d right, front row). (photo UPI)

A formation of Soviet-supplied Ilyushin Il-28 jet bombers flying over the Bilbays Air Academy during graduation ceremonies on 17 September 1956. During this event President Nasser delivered a fiery speech threatening to retaliate against any attacks from outside Egypt. Israel, Britain, and France took this threat seriously and made great efforts to destroy the Il-28s of Nos. 8 and 9 Squadrons during the Anglo-French air assault of October 1956. (photo UPI)

A reconnaissance photograph of Dākhilah airfield near Alexandria, taken by a British Fleet Air Arm strike aircraft during the tripartite invasion of Egypt in 1956. In fact, only a handful of light aircraft were based here, but other British and French squadrons operating from Cyprus, Malta, Israel, and aircraft carriers in the Mediterranean destroyed most of Egypt's air power during an intense five-day assault in late October and early November 1956. (photo Royal Navy)

A line of Russian-built MiG fighters at Almāẓah outside Cairo in March 1957. The three MiG-17s nearest the camera are probably newly delivered aircraft as they have no serial numbers. Beyond these aircraft are older MiG-15s which probably fought in the Suez War of 1956. The MiG-17 flew with the EAF for almost thirty years, serving in the fighter, ground attack, and training roles. (photo Associated Press)

A MiG-17C of the EAF takes off from Almāẓah with full afterburner in 1957. This aircraft is seen in its early "clean" version, as a fighter. EAF MiG-17s would later appear with an assortment of underwing rocket and bomb racks, particularly during the late 1960s and 1970s. (photo EAF)

A Mi-4 helicopter in the new markings of the United Arab Republic Air Force (UARAF) taking off in front of an audience of military cadets in 1959. The two-tone color scheme, divided by a narrow "lightning flash" line, was applied to various headquarters communications helicopters and fixed-wing aircraft. (photo EAF)

Wing Commander Taḥsīn Zakī, commander of the UARAF's 2d Air Regiment, climbs into the cockpit of his MiG-17F at al-ʿArīsh, probably in 1965. The serial number of his aircraft is 2038. Taḥsīn Zakī was also a member of the UARAF's aerobatic team, flying other MiG-15s with special markings. He continued to fly aerobatic displays after retiring from the Air Force in the 1980s. (photo Taḥsīn Zakī)

The two-stage al-Ẓafir ballistic missile developed in Egypt during the mid-1960s on display in Cairo during an Independence Day Parade. Note the new tricolor flag with two stars adopted by the UAR in 1958, and which also formed the basis of new national markings for the UARAF. (photo *Aviation Week*)

The Ha-300 prototype lightweight jet fighter developed and built in Egypt, seen shortly before the June 1967 War. (photo EAF)

President Nasser gets a snappy salute from a G-suited MiG-21 pilot of the UARAF during the president's visit to a front-line air base in April 1964. The name of the senior officer in the center is unknown, but his uniform includes the stiff shoulder boards copied from Soviet Air Force dress. (photo UPI)

One of the UARAF's huge turbojet Mi-6 transport helicopters, serial number 824, photographed from another Mi-6 over the Cairo Citadel in January 1966. Since these were operational troop transport aircraft, they have desert camouflage and were based at Biʾr al-Jifjāfah in Sinai in June 1967. (photo UPI)

This photograph, taken by an IAF Mirage, documents the severe blow inflicted upon the UARAF by Israeli jets on 5 June 1967. These three Egyptian MiG-21 fighters were destroyed on the ground by cannon fire at an air base said to be Inshāṣ before they could take off. (photo IAF)

Few pictures illustrate the Egyptian Air Force's determination to hit back after the disaster of 5 June better than this photograph of UARAF MiG-17s strafing an Israeli Army column in northern Sinai on 8 June 1967. (photo UPI)

An EAF Su-7 fighter-bomber lands at Bilbays air base following a training mission. This large, fast Russian-built aircraft served with the EAF for nearly twenty years in the attack and reconnaissance roles. It saw considerable action over Sinai during the 1967, Attrition, and 1973 Wars. (photo *Flight International*)

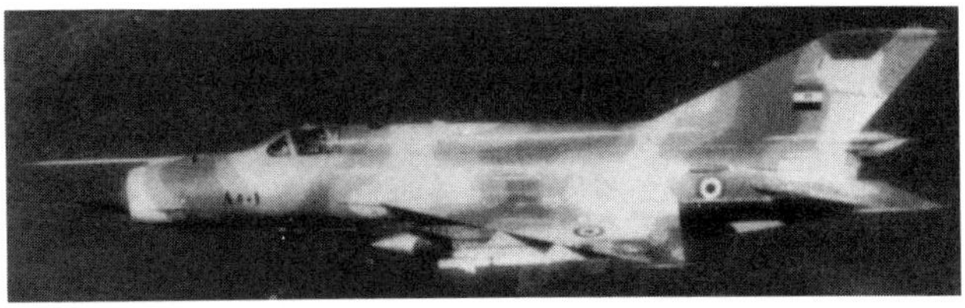

A MiG-21MF of the EAF in the new national markings carried by most, but not all, Egyptian aircraft during the October 1973 War, serial number 8501. Egypt's MiGs were given a variety of different camouflage schemes, reflecting their assorted combat roles. (photo EAF)

A long-range Tu-16 jet bomber of the EAF with the new national marking adopted in 1973. Beneath its wings the Tu-16 carries two Kelt stand-off missiles as used in the October 1973 War. (photo *Flight International*)

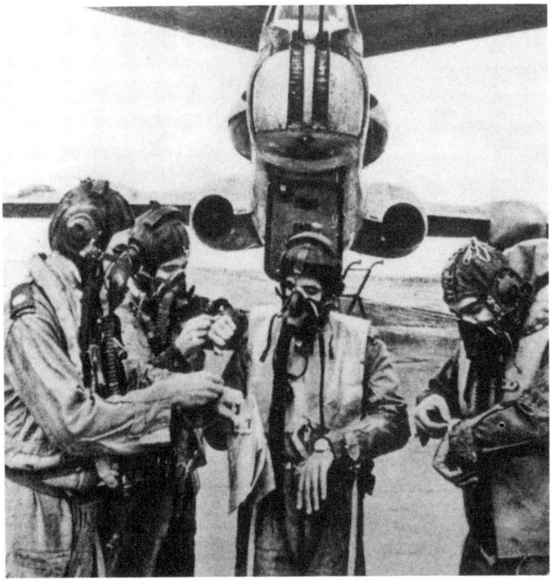

The crew of an Egyptian Tu-16 heavy bomber coordinate their watches before a mission, taken around 1973. These large, slow aircraft were highly vulnerable to Israeli interception and almost always operated with strong fighter protection. (photo EAF)

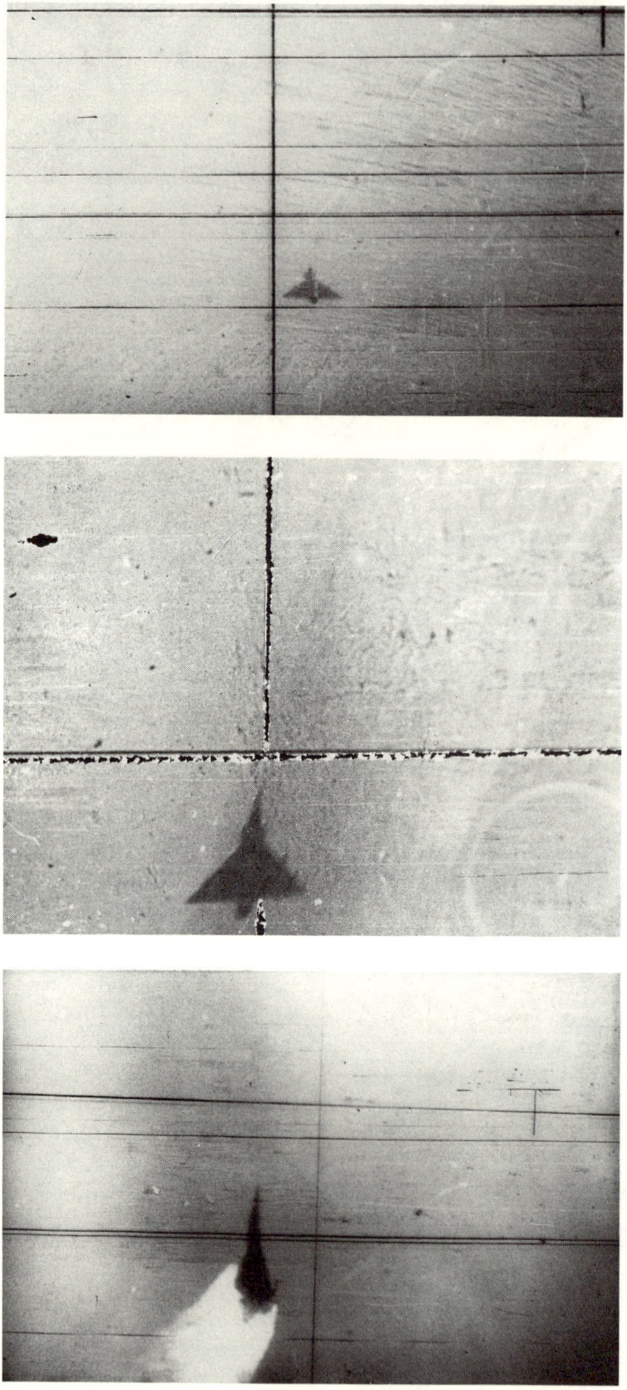

This sequence of gun-camera footage confirms the destruction of an Israeli Mirage fighter by an Egyptian MiG-21 pilot during the October 1973 War. The Egyptian has clearly closed to very close range before opening fire, perhaps reflecting the EAF's lack of really modern targeting and air-to-air missiles, having to rely—as Egyptian fighter pilots put it—on "Eyeball Mark 1." (photo EAF)

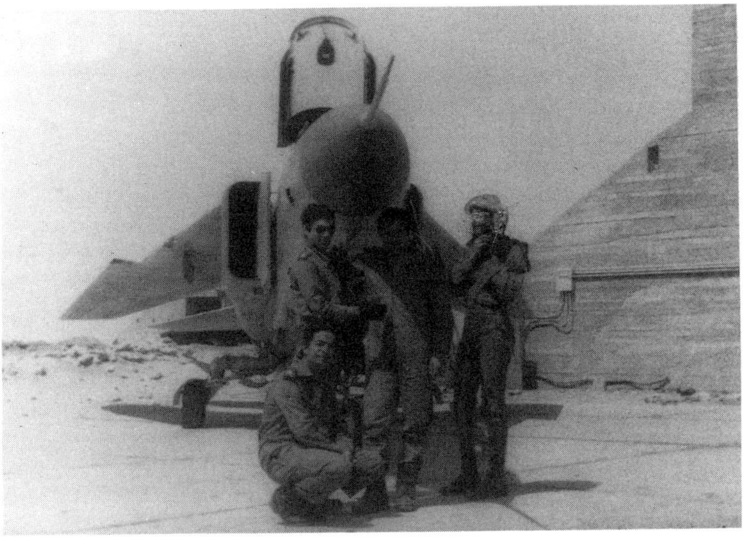

Husayn S—, one of the most successful Egyptian fighter pilots who shot down several Israeli aircraft during the October 1973 War. The Egyptian preoccupation with security mean that the full name of only one Egyptian fighter "ace" has been released. Husayn S— is seen here with his ground crew in front of a swing-wing MiG-23 a couple of years after the 1973 conflict. Note the concrete hardened hanger on the right. (photo confidential source)

After the break with the Soviet Union, Egypt turned to China for some of its aircraft. This photograph, taken at Fāʾid air base in 1989, shows a Chinese-built copy of the Russian MiG-21F known as the F-7. The EAF bought more than one hundred of these aircraft at very low prices in the late 1970s and early 1980s. They were then upgraded with British avionics, French and American air-to-air missiles, and Italian ECM systems. (photo L. Nordeen)

A Chinese F-6, license-built version of the Soviet MiG-19, of the Egyptian Air Force photographed at Almāẓah a few years after the October 1973 War. This particular aircraft, serial number 3860, has an overall two-tone grey camouflage scheme, suggesting that it was used as a high altitude interceptor. Also note the black-edged yellow identification panels on the tail fin and upper fuselage spine. In addition, the machine has the Muslim Declaration of Faith—*Lā illāha ill'Allāh wā Muḥammad rasūl Allāh,* "There is no god other than God, and Muhammad is the Prophet of God"—written next to the serial number on the front fuselage. (photo L. Nordeen)

An EAF Alpha Jet moving out for takeoff. The Franco-German jet was assembled in Egypt at the Hulwān factory and serves in both the training and ground attack roles. In common with most other current EAF aircraft, its Arabic serial number 3501 is repeated in European numerals on the tail fin. (photo EAF)

An F-16A of the EAF at Inshāṣ air base in 1989. This aircraft represents the new generation of advanced American fighters now operated by the EAF. The serial number 9317 is again repeated in western numerals on the fin, which also carries a black-edged yellow identification panel. (photo L. Nordeen)

Today's EAF now uses the EMB Tucano as its primary training aircraft. This photograph taken in the Hulwān factory in 1989 shows several at the stage of final assembly. The Tucano's red and white training color scheme is very like that of British RAF trainers and may reflect a revival of British training influence in the EAF. Dozens of these aircraft have been constructed in Egypt, many being exported to other Arab countries. (photo L. Nordeen)

An EAF Mirage V fighter-bomber photographed at Ṭanṭá air base in 1989. This French-built fighter was the first western tactical aircraft to enter service with the EAF after the end of the "Russian" era in the mid-1970s. Egyptian pilots had earlier flown Libyan Mirages in action against the Israelis during the 1973 War. (photo L. Nordeen)

An F-4 Phantom prepares for takeoff at Cairo West air base. Egypt received forty Phantoms as a reward for signing the Camp David Peace Treaty, but they never won the affection of Egyptian pilots as they did in many other air forces. (photo L. Nordeen)

The EAF has invested heavily in equipment to deter war and monitor events along its huge borders. This truck has a Skyeye remotely piloted vehicle ready to launch. The Skyeye carries a TV camera and radio systems to collect reconnaissance information up to 160 kilometers away. (photo EAF)

# 17

# Egyptian Phoenix (1967)

Israel emerged from the June 1967 War totally victorious. Egypt's chances of regaining the Sinai and military self-respect appeared to be negligible. Yet the following few months were to prove a turning point. President Nasser retained his position and created a long-term plan to confront Israel as well as rebuild the Egyptian military. A massive Soviet resupply operation plus Arab commitments of support set the scene for continued confrontation. On 9 June 1967, President Nasser accepted responsibility for the defeat and resigned during a speech on Egyptian television. But when thousands demonstrated their support for Nasser, the National Assembly passed a resolution calling on him to remain in office. Two days later Nasser fired most of the military leaders who held senior positions during the war. Gen. Muḥammad Fawzī was nominated the new Egyptian commander in chief. Gen. Maḥmūd Ṣidqī Maḥmūd and dozens of other senior military officers were arrested, while Gen. Madhkūr Abū al-ʿIzz became the new air force commander.

## Examining the Lessons and Rebuilding the EAF

The immediate task for the Egyptian authorities after the cease-fire was the rescue and rehabilitation of troops involved in the war. Many of these men needed medical attention, but those in good health were promptly returned to duty. Teams were organized to quantify losses, reorganize combat units, and develop a plan to rebuild the army and air force. Within days of the cease-fire, a high-ranking delegation of Soviet officers traveled to Egypt to assess the situation. This group was led by Marshal Matvei Zakharov, the Russian chief of staff. He was a blunt-speaking and direct man who answered Egyptian pleas for replacement equipment with the crushing reply: "Arms? What do you need arms for? To deliver these to the Israelis too? What you need is training, training. Then we shall see about arms."[1]

Soviet President Nikolai Podgorny also visited Cairo on 21–24 June 1967. During their meetings President Nasser was very critical of the performance of the Soviet-supplied weapons. Podgorny naturally defended Russian systems, whereupon the Egyptian leader retorted: "Very well then, in this first defensive phase I am prepared to leave the entire air defense of Egypt to the Soviet Union—but no base, no Red Flag!"[2] The Russians refused to take any public responsibility for Egypt's air defenses, but massive support and hundreds of advisors were immediately sent to Egypt.

The painful task of examining Egypt's defeat in the 1967 War was conducted by a team led by Gen. Ḥasan al-Badrī. The report, the detailed text of which remains classified to this day, blamed senior Egyptian military commanders for hiding the truth from President Nasser for several hours and faulted Field Marshal ʿAmr for his lack of leadership and mistaken decision to withdraw from the Sinai.[3] In a series of military show trials, Egyptian military leaders were blamed for the nation's humiliation. Field Marshal ʿAbd al-Ḥakīm ʿAmr escaped court-martial because he reportedly committed suicide. But on 29 February 1968 a military court did sentence the former UARAF commander, Maḥmūd Ṣidqī Maḥmūd, to fifty years in jail for dereliction of duty. Ismāʿīl Labīb, the former director of Air Defense, received a ten-year sentence. Many senior army officers were also jailed, and hundreds of other military officers were discharged from service.

The public attitude toward the military in Egypt was so bad at this time that students rioted protesting the "leniency" of these jail terms. Only following the War of Attrition and the Ramadan War of 1973 was public confidence in the Egyptian military restored, and several of the imprisoned senior officers were quietly released.

Immediately after the 1967 War, Yemeni Royalists launched an offensive to take advantage of Egypt's weakened situation. President Nasser sent ground forces and a composite force of MiG-17s and Il-28s that had survived the Israeli attacks to Yemen to stabilize the situation. A bombing campaign and aggressive ground action drove the Royalists back into the mountains. The Royalist offensive was also seen as an act of treachery to the Arab cause throughout much of the world. Fighting continued for many months, but in September 1967 Nasser struck a deal with Saudi Arabia which brought an end to Egyptian participation in the

Yemeni civil war. By November most Egyptian military personnel had been withdrawn.

After the cease-fire the EAF was rapidly resupplied and reorganized. Brig. Gen. Qādrī al-Ḥamīd (ret.) described his experiences during this period:

> After the war we were divided into two MiG-21 pilot groups. I was selected to be a night-fighter pilot. The Russians sent us many jets in a short time. We got MiG-21FLs and MiG-21BFMs from Soviet and East German squadrons. These weren't new aircraft. They were shipped in Antonov cargo planes and ships and quickly reassembled. In a very short time we had a new air force because we didn't lose many of our pilots or ground support troops. I remember Moshe Dayan said that one of their greatest mistakes was that they did not destroy the Egyptian pilots and the ground troops. They destroyed the material but not the people. Maybe we lost thirty pilots in the 1967 War.
>
> After the war we started to fly at very low altitudes. We knew how to fight, and I flew fifty or more hours in the July heat. Flying and flying and flying, I stayed at the air base fifty days without a day off. We flew umbrella missions for the Russian resupply and many training sorties. These efforts improved the standards of our pilots because we were under incredible pressure from the enemy and our commanders.[4]

## Striking Back to Build Morale

To demonstrate their defiance, Egyptian forces started hitting back at Israeli units in the Sinai shortly after the cease-fire. The first major clash, which became known as the Battle of Raʾs al-ʿUshsh, took place on 1 July 1967 when Egyptian Army commandos ambushed an Israeli armored column near Port Fuʾād. This minor victory contributed greatly to morale. In fact Port Fuʾād, at the very northwestern tip of the Sinai, remained in Egyptian hands from 1967 to 1973 and was never captured by the Israelis.

To keep up the pressure and secure information, EAF jets also flew many reconnaissance sorties along the Suez Canal and over the Israeli-held Sinai. On 4 July Israeli anti-aircraft fire downed a MiG near the Gulf of Suez. Four days later a patrol of MiG-21s was ambushed by Israeli Mirages over al-Qanṭarah, and one of the Egyptian jets was shot down.

Col. Taḥsīn Zakī (ret.) described the operations of his unit just after the June War:

> The Sukhoi wing made many reconnaissance sorties deep in the Sinai after the cease-fire. At dawn on the ninth of July, General Barakah phoned, informing me that the Israeli forces will try to cross to the west bank, and asked me to attack with the most jets available. . . . Just after dawn I took off accompanied by my No. 2, Pilot Muḥammad Jabāra, to proceed with the mission. We found there was a very heavy fog covering the Delta and Canal Zone and called back to the rest of the formation to return to base since there is not much time for the Su-7 to loiter until the fog lifts . . . its damn range again!
>
> I continued on my own after my No. 2 returned and managed to reach the al-Firdān area in the Sinai and saw some Israeli tanks through the fog that was lifting. As I started to line up for a pass, the base came through, and ordered me to scrub the mission. As I turned back my No. 2 was trying to land at Inshāṣ but failed because of the fog and so I headed to Bilbays. I tried to land three times without success. Therefore, I returned to Inshāṣ and tried to land twice but failed. The plane was almost out of fuel. When I tried to land for the last time and succeeded, the engine stopped as I taxied. Lucky!
>
> General Madhkūr Abū al-ᶜIzz was on the phone, and I informed him about what had happened, and he asked about the concentrations of Israeli forces, and I said that I saw nothing of the sort when I made my reconnaissance over the Sinai. He mentioned that President Nasser was on the line, and he is very happy with your reconnaissance. . . .
>
> During the afternoon, military intelligence asked for another reconnaissance in the same area, and thus two Su-7s under the leadership of Muḥammad ᶜAbd-al Raḥmān with Pilot Aḥmad al-Simarī flew the mission. The reconnaissance was a success, but the leader's Su-7 was shot down, and he managed to bail out successfully and crossed the west bank at night. Ahmad al-Simary's Su-7 returned back with wing damage and was repaired.[5]

Gen. Madhkūr Abū al-ᶜIzz pushed his men to rapidly reform squadrons with Russian-supplied aircraft, repair damaged facilities, and train for action. Meanwhile Egypt detected Israeli efforts to move up to the Suez Canal in the northern Sinai. The air force was directed to bomb these forces. Attack plans were developed under the direction of Col. Muḥammad Husni Mubarak.

Maj. Gen. Farīd Ḥarfūsh recalled this period of action:

> After the 1967 War the Russians flew an air bridge for some forty days, bringing in aircraft, missiles, and supplies. The air force commander at the time was General al-Izz; he was a super guy. Our morale was very low at this time, so he planned some operations to hit back. We launched many fighter-bombers to hit the Israelis. I don't think they were well fortified at that time with bunkers and defensive positions, and they were vulnerable. We hit them

hard with air strikes and artillery in July, and we believe that they suffered many losses. We had confirmed information that they asked for helicopters and trucks to evacuate wounded back to Israel. We launched tens of fighter-bombers and tried to keep the Israeli fighters busy.

I was a 1st lieutenant stationed at Cairo West during this time. We were flying caps at all hours of the day, and we sat alert a lot. On 15 July 1967 I was on alert, and they launched us. I was No. 4 of the group. The fighter controllers told us to go east. . . . We went east as fast as we could, and we had instructions to join an air engagement between Israeli and Egyptian airplanes. I saw two Mirages and reported them to my leader. We were flying supersonic at this time at an altitude of 7,000 meters. The flight leader ordered us to drop our fuel tanks. We were flying fast, and my fuel tank hit the skin at the rear of my plane. I didn't know that it caused a lot of damage until I landed later. We did a split "S" and got behind the Mirages. I saw another pair of Mirages coming from the east, so my No. 3 and I went after them and left the flight leader and his wingman to fight with the first two. We engaged with the Mirages, and it was really a severe combat. After some hard turns I saw my No. 1 fighting with a single Mirage. I didn't know what happened to the other Mirage or his No. 2. So I came in and tried to protect him. My No. 1 launched a missile, and it went out and hit the Israeli jet. The Mirage exploded, and I didn't see the pilot bail out. Then I saw another two Mirages come in at me. I told the leader about them and made a hard break. I then looked at my fuel and found that it was very low. I told my leader on the radio, and he said "I will come too." I made a steep dive and accelerated to very high speed—1,100 mph or close to it. I landed at Inshāṣ since I could not make it back to Cairo West because of my low fuel state. Then I refueled and flew back to Cairo West.

In my flight the leader shot down a Mirage; I saw it and confirmed the victory. . . . This was my first combat. In my section we didn't have any losses, but some were shot down from the other flights.[6]

Israel admitted the loss of a Mirage on this day.

Col. Taḥsīn Zakī (ret.) described the role his fighter-bombers played in this same attack: "On the fifteenth of July, Su-7s took off accompanied by twelve MiG-17s, the sweetest plane to fly and the best at that time for dogfighting to attack the Israeli forces opposite Suez and Ismāᶜīlīyah town. The attack was a success since it was the first major raid by our air force, except that we lost one Su-7, piloted by Murtaḍá al-Marāghī. During this attack, some Su-7s joined with MiG-21s as an air Combat Air Patrol over the area between Inshāṣ and the capital."[7]

## The Arabs Take a United Stand against Israel

In September 1967 President Nasser and other Arab leaders held a summit conference in Khartoum, Sudan. At the meeting, Arab leaders agreed to cooperate in the fight against Israel. Following the conference a communiqué was issued stating that there would be no recognition of Israel, no negotiations, and no peace—the famous "three Nos." Oil-rich states pledged to assist Egypt, Jordan, and Syria in rebuilding their military forces and repair war damage.

A week after the conference President Nasser stated "all that had been taken by force, would be returned by force." He detailed an Egyptian military strategy to avenge the 1967 defeat which had three phases: "defensive rehabilitation, active defense and liberation." This speech made it clear that as soon as Egypt had rebuilt its military capabilities, hostilities against Israel would intensify.[8]

By the fall of 1967 about 80 percent of the weaponry and military supplies lost in the war with Israel had been replaced. The EAF received some two hundred combat aircraft and dozens of support planes and helicopters from the Soviet Union, east European nations, and Algeria.[9] This massive Soviet resupply operation and pledges of support from Arab allies helped restore Egyptian confidence. Russia's prompt rearmament of Egypt has been described as "one of the most decisive Great Power acts since the Second World War," and it snatched political victory from Israel following its military success.[10]

In the fall of 1967 Egyptian forces regularly shelled Israeli forces in the Sinai and maintained the pressure with ground and air patrols. Following days of Egyptian shelling, in late September 1967 Israeli guns hit back. Thousands of Israeli shells struck Suez, al-Qanṭarah, Ismāʿīlīyah, and other towns along the border, killing and wounding hundreds of civilians. On 21 October 1967 Egyptian missile boats sank the Israeli destroyer *Elat.* Israel retaliated with a massive artillery barrage that set fire to the Egyptian refineries and chemical plants around the town of Suez. Continued fighting then prompted President Nasser to order the evacuation of 400,000 civilians from the towns of Suez, Ismāʿīlīyah, and Port Said.

President Nasser committed his country to a massive investment in money, manpower, and effort to prepare for war with Israel. The resupply and retraining of the EAF and development of a new air defense shield were major elements of this program. It did not take Russia long to reequip Egypt's shattered forces. However, it took much longer to create an effective Egyptian military force capable of conducting sustained offensive operations against Israel.

## Soviet Influence: The Cost of Support

The Soviet Union had sent more than 15,000 advisors and instructors to Egypt by late 1967. Soviet officers and advisors were assigned to all elements of the Egyptian military. In the air force, Russian personnel were assigned to every EAF regiment, in many cases down to the squadron level. Russian officers and enlisted advisors also directed the Egyptian Military Academy and training units.[11] Hundreds of Egyptian aircrews and ground support personnel were sent to the Soviet Union during this period to attend training courses.

Soviet advisors took on a much more influential role after the Egyptian defeat in 1967 than they had before the June conflict. Gradually, with the tacit approval of the Egyptian high command, Soviet advisors were given a voice not only in training and support but also tactics, mission planning, officer assignments, and promotions.

Soviet officers penetrated the Egyptian military and exerted influence over operations in a way never before achieved in a non-Communist country. Yet many Egyptian officers and soldiers remained uncomfortable with this new arrangement, although it was career limiting to speak out or work against the Soviets.[12] The massive resupply put Egypt even deeper in debt to the Soviets in both economic and political terms. Even so, Egyptian nationalism grew to a high level, and this conflicted at times with Soviet concepts. Army and air force personnel were anxious to strike back at Israel immediately and were unwilling to follow the gradual Soviet training plans.

## Retraining and Preparing for Battle

On 2 November 1967 Gen. Shalabī al-Ḥinnāwī took over command of the UARAF. During this period, except for reconnaissance missions over the Sinai and fighter patrols along the border, the air force was held back from battle so that the pilots and support personnel could hone their skills. The Egyptian pilot training program was reduced in length from three years to eighteen months to rapidly generate additional aircrew for the air force. Pilot trainees were sent to the Air Academy at Bilbays or Imbābah air base for primary training where they attended ground school and flew the Jumhūrīyah and Yak-11 aircraft.

Those who made it to the second phase of the course went on to al-Minyā or Marsá Maṭrūḥ to fly additional hours in Yak-11s and finally L-29 Delfin jet trainers. There were two training sessions per year, and each class held from sixty to seventy students. Students that passed this portion of the program (about a third failed) were then sent to the five-month MiG-15/17 operational conversion unit at Darāw north of Aswan. New pilots showing high ability were then transferred to Jabal al-Bāsūr for MiG-21 conversion or Jiyānklīs for Su-7 training. Pilots assigned to MiG-17s, transports, or bombers were sent to their operational units for type conversion.

Helicopter pilots followed their own course and trained on Mi-1 and Mi-4 helicopters before proceeding to operational squadrons. Navigators were trained on Il-14 flying classrooms for duty in transport or bomber squadrons. Officers specializing in technical areas attended facilities of the Armed Forces Technical Training Institute at Ḥulwān.[13] All air force officers were full-time career soldiers.

Meanwhile Israeli Air Force crews were at the peak of their skills following the 1967 War. They maintained their expertise by flying five times as many sorties per month than most NATO aircrew. Prior to the 1967 War most Egyptian aircrew flew only about 120 hours per year. Safety restrictions and Soviet doctrine also limited realistic training. On the other hand, many EAF pilots had experienced combat in Yemen. In 1968 the average flying time for EAF pilots was increased to twenty-four hours per month.

Maj. Gen. Muḥammad Nabīl al-Masīrī (ret.) discussed training efforts during this period:

> At the time Israeli equipment was much better than ours—the radars, the aircraft, and the weapons. You see, their training was also much more advanced, while we followed Russian rules which restricted low flying and aggressive tactics. After 1967 we changed, but it took a while to learn. After the war I was commander of the operational training unit. My instructors taught, and at the same time they flew operational missions. In between training they fought the Israelis. This was the best learning, and it developed our spirit. I can tell you we lost more pilots in aggressive training than we lost in war. I know because later I was responsible for the training of all the fighter-bomber regiments in Egypt. Some of the pilots told me that during combat they were more relaxed than in training. This is what we had to do to prepare to fight the Israelis.[14]

## Air Base Hardening to Increase EAF Survivability

Egypt committed itself to an ambitious program to upgrade existing air bases and build several new airfields. An American Defense Intelligence Agency Report commented: "One of the post-war priority projects has been provision of adequate bomb protection for tactical aircraft. By August 1968 hangarettes had been constructed for 280 aircraft, and it is anticipated that plans include provision of bomb proof shelters for all fighters and bombers in the inventory. In addition six new military airfields have been constructed since June 1967; at least three, possibly five, highway airstrips have been made available for dispersal and emergency use."[15] The Soviet Union had already begun a similar air base hardening program in eastern Europe, but Egyptian engineers modified Soviet designs for reinforced concrete aircraft shelters to fit the hot desert environment.

These shelters, which accommodated one or two aircraft each depending upon the design and aircraft size, provided protection from all but a direct hit by a large bomb. Most of these shelters were built by Egyptian labor. This massive engineering effort consumed millions of kilograms of concrete and involved thousands of workers. Egypt's air defense force was also considerably strengthened. Many more SA-2 missile batteries and anti-aircraft guns were emplaced around airfields, and additional runways and taxiways were built.

Nasser's commitment to confront Israel saw the end of Egyptian investment into its domestic aviation industry, the Ha-300 fighter and guided missile development projects being abandoned. Since the Soviet Union was supplying large numbers of advanced aircraft and the country was involved in almost daily conflict with Israel, this was a luxury the country could no longer afford. Egyptian industry concentrated on the repair and upgrade of Soviet-built aircraft. Government factories and private firms produced high usage spares for Soviet aircraft and other equipment such as external fuel tanks, bombs, rockets, and ammunition. Successful modifications included the addition of racks to allow for the carriage of bombs, rockets, and external fuel tanks by the MiG-17 and similar upgrades to the MiG-21. Egyptian technicians also developed training systems for Soviet aircraft and air defense equipment. This was necessary since the Russians refused to export their latest training systems and techniques.[16]

Egypt's strategic situation was much worse after June 1967 than it had been before the war. The Israeli occupation of the Sinai gave them a buffer, while the cease-fire line was close to the Egyptian heartland. Most Canal Zone airfields and many cities were within Israeli artillery range.

President Nasser was, however, determined to win back the Sinai and avenge the defeat of 1967. To fight Israel successfully Egypt had to reestablish a new network of airfields and air defenses as well as an effective command and control network to defend against Israeli troops and jets that were based only minutes away. Nasser made a deal with the Russians to acquire the support necessary to accomplish this goal. The cost of this massive supply of arms and assistance was Soviet influence and control. Several Egyptian air bases were taken over by the Russians, and ships of the Soviet Navy regularly docked at Alexandria. Russian Tu-16 bombers and Be-12 anti-submarine planes, marked with Egyptian insignia but flown by Soviet crews, kept tabs on American and NATO naval forces in the Mediterranean.

18

# Fighting Back (1967–1969)

At the Arab summit conference at Khartoum in November 1967 President Nasser signaled that he would lead a coalition against Israel and eventually regain the Sinai no matter what the cost. Few observers anticipated the speed with which Egypt would act, the determination of the military to succeed, and the sacrifices the country was willing to make to achieve these goals. Less than a year after the 1967 War, both Egypt and Israel were engaged in another round of fighting. This was the prolonged, savage, and costly "War of Attrition." An uneasy calm prevailed along the Sinai during the first year after the end of the 1967 War. Both sides used jets, artillery attacks, and commando raids to test the reaction of their adversary. It was relatively calm because Israel hoped that the status quo would harden into a fait accompli, while Egypt was busy rebuilding its military strength.

## African Involvement

In May 1967 a savage civil war broke out in Nigeria when a southern province of the country broke away and declared itself the new country of Biafra. Libya, Sudan, Great Britain, the Soviet Union, and Egypt supported the Federal Nigerian government in its effort to reunite the country.

Biafran ground forces supported by mercenary pilots initially forced back Nigerian Federal troops. The tiny Biafran Air Force included one B-25, a single B-26, plus several transports and light aircraft, all flown by foreign volunteers. Successful Biafran air raids also caused many casualties. At this time Nigeria had no air raid warning system, few anti-aircraft guns, and no shelters to protect against air attack. Egypt provided pilots, support crews, and military supplies to the Federal government as the two countries shared a long relationship. Egyptian instructors had trained some of the first Nigerian pilots in the early 1960s. Meanwhile

Russia provided most of the aircraft flown by Federal forces and encouraged Egyptian participation in the Nigerian civil war.

Beginning on 13 August 1967 Soviet Aeroflot An-12 airlifters delivered disassembled jets to Kano airport in central Nigeria. Biafran B-26s destroyed several of these aircraft during raids on 19–20 August. Federal Nigerian Air Force L-29s and MiG-17s, probably flown by Egyptian pilots, began operations on 30 August, and repeated strikes soon destroyed the Biafrans' few combat aircraft. By early 1968 the Nigerian Air Force fielded ten Czech-built L-29 Delfin attack aircraft, ten MiG-15s, several MiG-15UTI trainers, sixteen MiG-17s, six Il-28 bombers, and several dozen transports and liaison aircraft. These jets were flown by a mixed group of aircrew including forty Egyptians as well as mercenary pilots from Europe and South Africa.

Around-the-clock raids by Nigerian aircraft caused heavy damage and many Biafran casualties. Jean Zumbach, a French mercenary who performed many effective strikes flying the sole Biafran B-26 bomber remarked: "The Nigerians received some Czech L-29s, with Egyptian pilots, and they settled the outcome of the war. The better armed Lagos troops invaded the rich eastern province. . . . First Onitsha fell, then Enugu. The Biafrans were finished."[1]

Beginning in the spring of 1969 the Biafrans were, however, aided by a small contingent of Swedish volunteer pilots led by Count Carl Gustav von Rosen. Flying five Swedish-built MFI-9B light aircraft armed with machine guns and rockets, these pilots flew dozens of very successful raids against Nigerian airfields which destroyed more than a dozen aircraft.

Maj. Gen. Nabīl Shuwakrī (ret.) flew MiG-17s in support of Federal forces: "I made many reconnaissance and ground attack sorties during my tour in Nigeria. On one mission I remember well, I spotted an aircraft camouflaged off the end of the runway. I came back despite heavy ground fire and strafed it with my guns, and it caught fire. I think it was one of the small aircraft that the Swedish count had used very effectively against the Nigerians in hit-and-run attacks."[2] The blockade, heavy application of air support, and a successful ground offensive by Nigerian units brought this savage civil war to an end in early 1970.

## Soviet-Style Air Defense Shield

During the 1968Ð70 time period Egypt invested more than 40 percent of its gross domestic product in the military or supporting infrastructure.

The air force and air defense force received a significant share of this investment as Egypt concentrated on training and preparing its armed forces to confront Israel. Initially President Nasser and his senior military leaders counted on a renewed fighter-interceptor force, supported by a radar network, to defend against Israeli air attack. Yet it took years to create experienced fighter pilots. Combat losses and attrition due to accidents, medical problems, and transfers took their toll of Egypt's experienced aircrew.

While intensified training paid off, the Israeli Air Force always seemed to be one step ahead. Finally, President Nasser and Chief of Staff Marshal Fawzī decided that Egypt would follow the Russian air defense approach.[3] A massive network of anti-aircraft missiles and guns would be Egypt's first line of defense supported by EAF fighters.

On 1 July 1968 the Egyptian Air Defense Force was created as a separate, independent service. This new military force was organized along the lines of the Soviet PVO Strany air defense force and commanded by Maj. Gen. Muḥammad ʿAlī Fahmī. The new Egyptian Air Defense Force immediately set out to follow the harsh lessons learned in 1967. These included the need for continuous radar coverage of all Egyptian territory, improved command control and communications, and better weaponry. The air defense force also assumed control of Egypt's surveillance network, missiles, and anti-aircraft guns. New command and control procedures were established so that once an unknown aircraft was detected, this information was passed to a sector command post that housed both air defense officers and air force fighter controllers. In this way both services could work together in a coordinated fashion to defend Egyptian airspace.

Egypt was divided into five air defense zones, each having a control center and assigned radar battalion with ten to fifteen individual radar stations. In early 1968 radar coverage of the Nile Delta and the areas around Alexandria, Port Said, and Suez was very good. However, there were still many low-level gaps in coverage. Observation posts were established along the coastlines and at known radar blind zones to correct this problem.[4] The Egyptian Air Defense Force fielded nearly fifty SA-2 sites and more than one thousand 23 mm, 37 mm, 57 mm, 85 mm, and 100 mm anti-aircraft guns, most of which had radar fire control, while the air force committed more than half of its fighter force to the air defense mission. Sections of MiG-21 fighters at air bases throughout the country sat on alert, ready to intercept any intruder.

According to a U.S. Defense Intelligence Agency report: "The air defense system has been active in detecting and tracking intruding aircraft; tracking and navigation assistance is being given to EAF aircraft, and integrated service air defense exercises are performed periodically. Fighters are on a quick reaction alert status; 3½ minute or less scrambles have been noted. The system has not been noted training with electronic countermeasures, but simultaneous engagement by weapons systems is being exercised."[5]

Even so, it took considerable time to develop an integrated air defense network. Time was one thing the Egyptians did not have as the Israelis kept increasing the pressure. The Israelis regularly flew into Egyptian airspace to gather reconnaissance information and evaluate the effectiveness of the defenses. Unfortunately as a result of this pressure, mistakes were made and a few Egyptian Air Force and civilian aircraft were shot down by accident.[6]

Brig. Gen. Qādrī al-Ḥamīd (ret.) commented:

> The Israelis by 1968 started to intensify the air war against us. They would ambush us all the time. We now had shelters, so Mordechai Hod [the Israeli Air Force commander] said that he would start another war against the Egyptians. He said that they have shelters and are staying under pyramids, so I will bring them up into the air and shoot them down. So this was the nature of the new air war in the area. They started it against one of our squadrons, the MiG-17s, stationed at Hurghadah. The first day I landed there in my MiG-21, my senior officer took me to lunch. Then I heard cannon fire, so we ran out to see what was happening. We saw one of our MiG-17s spiraling down, and it crashed into the Red Sea. Our boats went out and found the pilot, but he was dead. Another of our MiG-17s was hit, but the pilot landed safely.[7]

Maj. Gen. Muḥammad Nabīl al-Masīrī (ret.) remarked on air operations during the War of Attrition:

> We could not say that the Israelis were not good, but they gave the impression that they were *all* really good. However, they planned their actions and pushed forward the best pilots to do the mission. When it came down to it, Egyptian pilots were also good. Many of the experienced Egyptian pilots were better then the average Israelis. But the Israelis chose the rules of the game. They always did it as an ambush. They sent aircraft out—our fighter controllers were aggressive and had good experience, and they pushed out our fighters—and we got caught. After a while we understood the rules—we spoiled this game of theirs. We could not act the same every time or you lose, so we acted in our own way.[8]

## Battles along the Canal

In the summer of 1968 Egypt initiated a new phase of fighting along the Suez Canal with a series of commando raids and heavy artillery fire. Egyptian fighter-bombers and reconnaissance jets were also sent into the Sinai. During one of these missions in September 1968 Capt. Ḥusayn ʿIzzat won the Medal of Excellence, Egypt's highest military award. Captain ʿIzzat was flying his Su-7 on a photo reconnaissance mission deep into the Sinai when he was intercepted by four Israeli Mirages. The Israeli pilots tried to shoot Captain ʿIzzat down, but his sharp turns spoiled their cannon attacks, and all the infrared-homing missiles fired locked onto the hot desert. Israeli Shafrir and Sidewinder air-to-air missiles, while better than the Atoll, were not very effective against low-flying targets. So the captain escaped from the four Mirages and brought back his reconnaissance photographs.

On 26 October 1968 an Egyptian artillery bombardment hit targets all along the canal, inflicting heavy casualties. Israeli counterbattery fire was followed by a new Israeli tactic. On the night of 31 October Israeli commandos destroyed two bridges spanning the Nile and an electrical substation at Naj ʿḤammādī. Such Israeli helicopter raids prompted the Egyptians to strengthen their defenses around potential targets.

Brig. Gen. Qādrī al-Ḥamīd (ret.) recalls his involvement in these efforts: "The Israelis attacked our electric station at Naj ʿḤammādī with troops from helicopters. So our defense minister ordered our night-fighter squadron to Hurghada to intercept Israeli helicopters. It was a difficult mission because we didn't have guns, and it was nearly impossible to intercept a helicopter with an infrared seeking missile. Also our radar had a difficult time finding a helicopter at low level. Hurghada was on the Red Sea, and the range to the nearest airfield, Luxor, was very great. Hurghada was a very windy place, and if there was a sandstorm, we didn't have an alternate airfield nearby."[9]

## Nasser Declares a War of Attrition, March 1969

Between October 1968 and March 1969 there was only limited action along the border between Egypt and the Israeli-occupied Sinai. Then, on 8 March 1969, following a noon air battle that saw the loss of a MiG-21, Egypt struck back with a massive nine-hour artillery assault. Israeli targets all along the front from al-Qanṭarah to Suez were plastered with shells.

In an evening speech President Nasser announced that "a War of Attrition" had begun against Israel. The next day Egyptian anti-aircraft fire downed an Israeli Piper Cub near the canal, and the Egyptian chief of staff, Gen. ʿAbd al-Munʿim Riyāḍ, was killed by Israeli shell fire while visiting-front line positions near Suez.[10] General Riyāḍ had learned his trade as an anti-aircraft officer under British guidance during the Second World War.

Egyptian strategy was to inflict continuous casualties on the Israelis. More than 100,000 Egyptian troops and 1,000 guns and mortars were dug in along the border. Here Israel was outnumbered more than ten to one in manpower and fifty to one in artillery.[11] Nasser was determined to use Egyptian superiority in numbers and artillery to force the Israelis to pull back from the waterway or change the status quo.

The EAF also played an important part in Nasser's new strategy. Fighter-bombers hit targets in the Sinai beyond the range of artillery, and interceptors escorted the strike aircraft and defended against Israeli jets. Transports and helicopters supported the army along the Suez Canal and performed special missions.

According to a U.S. intelligence assessment, by early 1969: "The EAF has developed a good capability in close air support, aerial reconnaissance, troop lift, paradrop and air supply. The Su-7 and MiG-15/17 fighters are used mainly for ground attack and in close army support roles; the MiG-21, a point defense interceptor, also has a ground attack capability."[12]

Using hit and run tactics, the Egyptian Air Force struck frequently at Sinai targets. Maj. Gen. Muḥammad al-Masīrī commented:

> You see, during the War of Attrition the most successful aircraft [in the ground attack role] was the MiG-17. It was a light aircraft, easy to maneuver. We had done some modifications to carry more weapons on the MiG-17 and still have the auxiliary fuel tanks. We modified them to have eight rockets fitted to launchers under the wing—80 mm rockets with armor-piercing or high explosive warheads—besides two 100 kilo bombs, two fuel tanks, and three cannon. One of them was a 37 mm cannon—you can't imagine—when it hits it makes a hole a foot in diameter. Also the MiG-17 had two 23 mm cannon. When they hit they made a smaller hole but many splinters.[13]

Air raids were often coordinated with artillery and commando raids in order to inflict heavy casualties. Israeli counterattacks against well-entrenched Egyptian forces had limited impact even though they did cause casualties. The EAF suffered losses in air battles with Israeli fighters and from ground fire. Nevertheless, Egyptian leaders were willing to suffer the losses necessary to sustain this offensive. In fact Egyptian commandos, artillery, and air units struck at the Israelis more than four hundred times in April and May 1969.

Brig. Gen. Samīr ʿAzīz Mīkhāʾīl (ret.) discussed one of these operations:

> A reconnaissance in force was planned for 14 April 1969. We had two groups, one was to escort two Su-7s over Mitlā Pass and a second to fly down from the north on patrol. We took off, escorted the Su-7s flying at zero feet to the Suez Canal, and then we climbed up to 5 km. The Su-7s were slower than our MiG-21s, so we started to spread out. I stayed about 3 km behind to cover the Sukhois. We made a turn to the right for the Su-7s to take photos, and then I heard the ground controllers tell us that Israeli aircraft are coming from Biʾr Jifjāfah, 40 km behind us.
>
> The Su-7s dove down, and our flight of two stayed in position. When the ground control interception [GCI] controller said that they were 3 km behind me, my knees began to shake—this was the first time I was afraid. I thought, if I get shot down over the Sinai, I will be captured. At this point I got mad and made a very steep turn to the right, a hard break. At that moment a missile exploded near us, between me and my wingman. I started making very hard turns to avoid their attack. The two of us fought eight Mirages, and I turned toward the Suez Canal. My wingman got separated and radioed that he was going west and would land at Biʾr Arāda.
>
> So I am alone against eight Mirages. More Israeli jets were waiting near the Suez Canal to cut us off. I looked out and saw a circle—Mirage-Mirage-MiG-21. I was all alone. I thought, I survived until now, try and hit one of their aircraft. I crossed the circle and picked out a Mirage. He thought that I had a cannon, which I didn't have because I was flying a MiG-21FL, so the Israeli broke to avoid my attack. Then the circle of jets broke up, and everyone split.
>
> One of the Mirages turned away, and I put my gunsight on him and fired an Atoll. It hit him in the tail and made a big flash. I couldn't believe it. Then I fired my second missile at him. They taught us when you start to shoot, look behind and when I did, I saw a Mirage coming for me, and he was within cannon range. I made a hard turn and dove straight down to escape. As I turned I saw heavy smoke coming from the Mirage I had shot. He was flying east into the Sinai. I returned to my home base and told my friends that I had shot down a Mirage.[14]

Throughout the spring of 1969 Egypt maintained the pressure on Israel. An intense series of raids flown against targets in the Sinai during the last week of May caused heavy damage. On the other hand, Israeli fighters downed four MiG-21s, and one jet was shot down by a Hawk surface-to-air missile. This was the first EAF loss to this American-built weapon.[15]

The EAF did not have access to defensive systems such as electronic jamming pods or decoy chaff and flares—like the United Stares was then using in Vietnam—to disrupt Israeli missiles or fire control radars. The Soviets were just beginning to field these systems with their premier units and refused to supply them to the Egyptians for fear that these items would fall into Israeli and American hands. As a result Egyptian pilots could only use the element of surprise and low-level tactics to avoid Israeli defenses.

Israeli leaders threatened serious reprisals if the Egyptian attacks did not cease. Heavy Israeli shelling destroyed the Suez oil refinery complex and forced the final evacuation of Port Said, Port Fuʾād, and other towns near the battle zone. Commando teams flown by helicopters raided several Egyptian military and industrial centers.

On 17 June 1969 a pair of Mirages flew over President Nasser's home at supersonic speed which broke windows for miles around. The president used this "attack" as an opportunity to change military leadership. The EAF commander, Major General Shalabī al-Ḥinnāwī, retired and was replaced by Major General ʿAlī Baghdādī. Changes were also made at the senior levels of the air defense force.[16]

Israeli jets also continued to fly over Egypt to send a warning. Brig. Gen. (ret.) Tamīm Fahmī ʿAbd Allāh described one of the missions he flew during this time:

> I had a good combat in a MiG-21FL on 26 June 1969. We were transitioning into the MiG-21M at Cairo West. At the same time we still had to stand alert with the MiG-21FLs. We were on a third-stage alert to support the fighters from Inshāṣ. If they needed help, we would go. The horn went off, and we were in the classroom going over our studies on the MiG-21M. We ran to the jeep, drove to the aircraft. After a quick takeoff, my wingman and I turned to the operations channel. We were told to intercept the enemy and support the other Inshāṣ fighters. We were directed toward a low-flying target south and east of Cairo at a range of about 40 miles. I got near the area and I started looking, and I saw a DC-3 Dakota.
>
> The controller said no, combat turn, and we flew about 35 miles to the east, about 15 miles from the Gulf of Suez. Then I saw a Mirage all by himself going back toward the Sinai. I dove and tried to catch him. At about one mile away he turned hard into me. He turned left, and I came to within a half mile of him but I didn't have a gun, so I maneuvered with him to try and get a missile shot. He was turning hard, and I did a high speed "yo-yo" to keep

> with him, and we did this about three times. Then he reversed and dove toward the deck.
>
> I went below him to get a missile tone, and then I heard the buzz of the Atoll [indicating that the missile seeker was locked onto a heat source], and I thought it was good. I was about one kilometer back, and I pulled the trigger. The missile went out and hit near him, and the Mirage started to trail light smoke when my wingman said "turn left and I will shoot now as I only have some 600 litres of fuel left." I made a combat turn to the left. But the Mirage pulled away. We climbed to 20,000 feet to save fuel and glided to Ḥulwān and landed on fumes.[17]

During a week of heavy air action the Israelis claimed to have downed seven EAF aircraft in air combat and hit several other Egyptian jets with ground fire. While Egyptian pilots and air defense troops fought bravely and scored some victories, these actions again demonstrated Israel's superiority in the air. Egyptian leaders responded by moving dozens of SA-2 missile sites and anti-aircraft batteries into positions parallel to the Suez Canal.

## Israeli Air Power Advantages

Despite the massive Egyptian investment in its air force and air defenses, Israel still held the lead in air power. Soviet aircraft were inferior in range, armament, and electronics compared to the French-built Mirages, and the new U.S.-supplied A-4 Skyhawks and F-4 Phantoms had even greater superiority. In the fighter-interceptor role, Egypt relied on the MiG-21 Fishbed, but sometimes the older MiG-17 Fresco was also pressed into service. The highly maneuverable MiG-17 was respected as a capable dogfighter by Israeli pilots. But compared to the Mirage and F-4, it was slow (the MiG-17 was limited to subsonic speeds), had short range, and carried no missiles. The MiG-21 could climb fast and fly at supersonic speed, but it had limited range and ineffective weaponry. In 1969 only half of the MiG-21s in service were the Fishbed F model armed with a 30 mm cannon.

The Israeli Mirage matched the MiG-21 in speed, had longer range, and was armed with two 30 mm cannon and a pair of air-to-air missiles. The Mirage could carry three different types of missiles: the American-built AIM-9B Sidewinder, Israeli-made Shafrir, or French Matra 530. All of these heat-seeking missiles had superior performance to the Soviet Atoll.

In late 1969 the IAF received its first F-4E Phantom fighter-bombers from the United States. While the F-4 is not as maneuverable as the MiG-21 or Mirage, it was fast, had twice the range, and carried a wide variety of weapons. It could perform both ground attack and air combat missions. The Phantom was armed with a 20 mm cannon and up to eight air-to-air missiles. One Israeli F-4 carried the same number of missiles as a flight of four Egyptian MiG-21s. The F-4 also had a better radar and AIM-7 Sparrow missiles that could be used at night and in poor weather conditions.

The EAF's principal ground attack aircraft were the Su-7 and the MiG-17. In the ground attack role the MiG-17 could carry two 220 lb. (100 kg) bombs plus eight 80 mm rockets and two external fuel tanks that were necessary to reach a useful tactical range. The MiG-17's powerful cannon armament was effective against ground targets. The Su-7 was faster and had greater range, however, on a typical mission only two 250 kilogram or 450 kilogram bombs could be loaded since external fuel tanks were required to reach targets in the central Sinai. EAF Su-7s could carry about the same load as an Israeli Mirage but could not fly as fast or as far.

By now Israel had two jets with even better attack performance, as an Israeli official commented: "The Skyhawk carries twenty bombs where the Mirage took two. The Phantom is fantastic! Two of them can do what a squadron of Mirages did."[18]

Israeli fighter pilots were among the most experienced and well-trained aircrew in the world . The IAF pilot training program was very demanding, and standards remained high. An IAF squadron commander once remarked: "We have the world's greatest rate of washouts from flying schools. Generally speaking, Arab pilots have not improved since the Six Day War. It's not more training they need, it's mentality, willpower, integrity. Some of them are damn brave knowing they don't have a chance. Judging from our experience, it will take more than a generation before they can change. You have to motivate people to be good fighting pilots. Flying itself is not so difficult."[19]

A U.S. Defense Intelligence report of 1969 stated: "Intense training since the June 1967 hostilities has improved [the EAF's] overall effectiveness; its combat capability is considered good but primarily defensive in nature. . . . Standards of skill and discipline in the staff, flying and technical departments of the air force have improved measurably.

The EAF has gained considerable experience in night and interservice operations."[20]

Many Egyptian aircrew had flown in Yemen, Nigeria, the 1967 War, and air battles over the Sinai. Through skill and determination, EAF pilots pressed on, hit Israeli ground targets, and fought off Mirages. Many Egyptian pilots got into firing position against Israeli aircraft, and some achieved victories. Egyptian pilots and fighter controllers honed their skills in combat, the toughest learning environment of all.

## Israeli Air Power Unleashed, July 1969

By mid-1969 Egypt had gained the initiative along the Suez Canal as a result of its attrition tactics. As Israeli casualties mounted, the IAF was called in to strike back and make up for Egypt's superiority in firepower and numbers.

During the night of 19–20 July 1969 Israeli commandos stormed an Egyptian fort at Jazīrat al-Akhdah (Green Island) in the Suez Gulf and eliminated its defenses. This created an opening in the Egyptian air defense network. The next day the IAF flew more than a hundred attack sorties against Egyptian fortifications along the Suez Canal. This was the first time the IAF flew a large number of strikes against Egyptian targets since the 1967 War and signaled a new phase in the fighting. Israeli raids continued, and by early August dozens of air defense sites and artillery positions had been blasted with bombs.

Egypt retaliated with intense shelling and air strikes against targets in the Sinai. Air and ground battles occurred daily, and on 20 July the IAF claimed to have downed five Egyptian planes in two days of heavy air fighting but also admitted the loss of two aircraft, including a Mirage that fell to an EAF MiG-21 during a dogfight. Egypt said that its only loss was a single Su-7 piloted by Major Nabīl Saʿīd, who ejected and was captured by the Israelis after his jet was hit by anti-aircraft fire.[21]

The Israeli air offensive increased Egyptian losses, but the EAF and air defense force fought back aggressively. On 19 August the IAF lost two A-4s over Egypt.[22] EAF interceptor and fighter-bomber squadrons flew hundreds of sorties during this period, while EAF transport and helicopter squadrons received little publicity, performing vital support missions for Egyptian units near the Suez Canal and throughout the country.

On 9–10 September 1969 the Israelis conducted a daring assault into Egypt using T-54 tanks and other vehicles captured during the 1967 War. In retaliation on 11 September more than a hundred EAF aircraft penetrated into the Sinai and blasted the Israelis.

Brig. Gen. (ret.) Samīr ʿAzīz Mīkhāʾīl participated in these attacks:

> I flew from al-Manṣūrah [in a MiG-21]. We were to escort MiG-17s that were to hit a Hawk site in the northern Sinai. I was the flight leader of a finger four, and we escorted the MiG-17s at zero feet until they popped up to hit the site. The MiG-17s didn't find the site. It was very well camouflaged until they fired at us. Then they plastered the Hawk site with bombs and rockets and knocked it out. When we turned back we found that Mirages had cut us off at the Suez Canal. So the MiG-17s continued at zero feet and we climbed up to 5 km and fought with the Mirages. One of our pilots was jumped by Mirages which fired missiles at him, but somehow none of them hit him. I told him that there was an aircraft behind him, but he didn't hear. After some maneuvers I came to the Mirages and shot an Atoll at one. It hit, and I saw a piece of his tail separate and fall off. Now I had only about 300 liters of fuel, enough just to land. After I hit the Mirage the others started to turn into me, but I turned to make them think that I would attack with a cannon, so they turned away.
>
> I think that if I had had a cannon I could have shot down many of the Mirages. This was the second Mirage I had hit with a missile, and it was very easy. The Mirage trailed smoke but turned away into the Sinai. I made a break and headed back to al-Manṣūrah where I landed on fumes.[23]

The IAF admitted the loss of a Mirage lost in air combat and the pilot was captured, but Israel claimed to have destroyed seven Egyptian MiG-21s, three Su-7s, and a MiG-17.[24]

Egyptian air, ground, and naval assaults into the Sinai continued throughout the fall. The ferocity of these attacks caught the Israelis by surprise. In addition to attacks by fighters, EAF helicopter pilots flew many daring missions into the Sinai. In one of these raids on 28 September an army commando unit attacked an Israeli barracks 52 miles east of al-Qanṭarah, causing serious damage. Israeli public relations sought to minimize the impact of Egyptian air attacks.

During a November speech, Israeli Chief of Staff Lt. Gen. Haim Bar-Lev stated: "What the Egyptians are doing is not serious. Their planes attack, but they do not hit anything. They don't take time to get a proper angle of attack. They get rid of their bombs as soon as possible. They do everything to be over our territory as short a time as possible."[25]

In November the Egyptian Navy and Air Force coordinated their efforts to strike against Israeli targets in the Sinai. Brig. Gen. (ret.) Samīr ʿAzīz Mīkhāʾīl flew in support of one of these assaults:

> On 18 November one of our destroyers went to attack Baluza and Romana, Israeli bases on the northern Sinai coast. I was on night alert duty. We saw the red flare, I scrambled and flew to Dumyat, west of Port Said on the coast. I saw flares and was told by the fighter controller that there were aircraft making air-to-sea attacks. I caught one aircraft on my radar and moved it to attack. At this moment I heard noise on the radio. They were jamming my radio frequency with the fighter controller very loudly. The aircraft on my radar dove, and I lost the target in the clutter. My wingman called me and said "come back—there are Mirages in the area." So I broke away and flew back. During this period the destroyer turned north and made a successful escape.[26]

Despite intense efforts by the EAF and air defense force, the punishing Israeli raids continued. The constant rain of bombs battered Egyptian positions, caused thousands of casualties, and damaged Egyptian morale. When EAF interceptors challenged the Israeli formations they were often harshly dealt with. In the last four months of 1969 the EAF lost twenty-five aircraft. Air defense units fared no better: dozens of SA-2 sites, radars, and gun positions were demolished with bombs. But while the IAF seemed supreme, Egyptian leaders resolved not to give up.

## Air Order of Battle, February 1969[27]

| Egypt | Israel |
|---|---|
| **Fighters** | |
| MiG-21 × 160 | Mirage III × 60 |
| MiG-15/17 × 140 | Super Mystère × 20 |
| Su-7 × 50 | Mystère IV × 30 |
| | A-4 Skyhawk × 48 |
| Total 350 aircraft | 158 aircraft |
| **Bombers** | |
| Il-28 × 30 | Vautour × 20 |
| Tu-16 × 15 | |
| Total 45 aircraft | 20 aircraft |

19

# Attrition War (1969–1970)

As 1969 drew to a close the United States and the Soviet Union become more involved in the ongoing conflict between Egypt and Israel. President Nixon and Secretary of State Henry Kissinger saw the Middle East as a battleground for East-West confrontation. Arming Israel with advanced aircraft was seen as a way to frustrate Soviet designs in Egypt and the region. Soviet leaders reacted as they saw the military position of their ally, Egypt, deteriorate. The air war during the later part of the War of Attrition reflected this superpower intervention. After U.S-supplied A-4s and F-4s turned the tide of the air war, Soviet forces became directly involved in the air defense of Egypt. A cease-fire was arranged only after months of intense fighting and several direct confrontations between the Israeli Air Force and Soviet forces. By the end of 1969 the initiative in the ongoing conflict along the Suez Canal had shifted to Israel. The massive application of Israeli air power had turned the tide against Egypt. Egyptian Air Defense forces and Army units along the Suez Canal had suffered four months of intense bombardment. Israel often integrated commando raids, artillery, and air strikes. While Egypt struck back, inflicting heavy Israeli losses, Egyptian casualties were mounting. With every Egyptian success the Israeli bombing seemed to intensify.

## The EAF and Air Defense Force under Assault

The EAF was active in defense and attack, but the cost was high. Israel's surveillance network was good, Mirages patrolled aggressively, and Sinai targets were ringed with Hawk missile sites and anti-aircraft guns. During the dozens of air battles fought between May and November 1969, Israel claimed to have downed forty-eight EAF aircraft including thirty-one in air combat, nine from anti-aircraft fire, and eight as a result of Hawk missiles. Egypt reported the destruction of more than two dozen Israeli jets during this period, with most falling to anti-aircraft defenses, while Israel admitted the loss of only seven aircraft.[1]

Despite continued losses, Egypt continued to fly reconnaissance sorties and fighter-bomber strikes. While the military impact of these missions might have been low, it was politically necessary to continue to carry the fight to the Israelis in the occupied Sinai. Egyptian pilots did everything possible to ensure their success; they used low-level tactics, surprise, and decoys to minimize their vulnerability. Many EAF attack and reconnaissance missions reached their targets and returned home successfully.

The MiG-21 escorts were often less lucky; many fell to the frequent Israeli ambushes and suffered heavy losses. Since Israeli Mirage pilots could not usually catch EAF strike aircraft because of their hit-and-run tactics, Israeli frustration was taken out on the MiG-21s protecting the withdrawal of the attackers. Even when intercepted, low-flying Egyptian MiG-17s and Su-7s were difficult to shoot down. Israeli pilots did not want to fight over the heavy Egyptian air defenses unless they had to. Once EAF aircraft flew back through the Egyptian Surface to Air Missile (SAM) belt, they were relatively safe.

Without the cover of EAF MiG-21s, more fighter-bombers would have been lost and Israeli jets could have followed back the returning EAF planes. EAF MiG-21 interceptors also fought many air battles around the Great Bitter Lake and the Gulf of Suez to stop the Israelis from penetrating into Egyptian airspace where air defenses were thin. Even so, despite heavy losses, Egyptian air and ground forces continued to resist with determination.

Col. Aḥmad ʿĀṭif flew many air defense missions:

> I was on first-stage alert at Inshāṣ on 9 December 1969. We had scrambled, and we joined four other MiG-21s from Ṭāmīya (Kawm Awshim). The Israelis were flying a reconnaissance with Mirages against Banī Suwayf. The first four MiG-21s went to intercept the two Mirages who were flying at low level. We scrambled, and the ground controllers told us about the situation. The first flight of MiG-21s could not reach the Mirages because they were flying too fast. So the MiGs went back to their base. We flew an umbrella to protect their landing. After that the GCI told us that there is a target over the Suez Gulf approaching at about 6,000 meters. We had another four MiG-21s scramble from Inshāṣ to support us. The Israeli object was to intercept our flight of four MiG-21s—there were eight F-4s coming to get us. We started heading north to meet them at a height of 6,000 meters. Behind us at a distance of about 15 km was the second group of MiG-21s.

The GCI fed us information about the situation—our targets' height and strength. At this time we knew that we were going into combat. We were told, "go to maximum speed, release belly tank, go to afterburner on, your target is at 35 km at eleven o'clock." I have the GCI tape of this whole combat. They told us to make a combat turn and climb to 10,000 meters. The F-4s were flying in two sections of two, and the second flight was further away. After I made my combat turn I saw the number three and four Phantoms. My leader said, "I have two aircraft in sight." Then we saw a missile in flight—an Israeli air-to-air missile. It passed through our formation but did not hit anything. The leader of the Israelis made a very smooth turn, and his No. 2 delayed a little. My leader went after the No. 2. When this Israeli saw the two MiGs approaching, he made a quick turn to the right and forgot about his leader and escaped. He made a very good egress and was gone to the coast.

I looked and saw an aircraft at about 5 km, and my leader said that they were the approaching MiGs. I looked and saw their noses were black; all of our aircraft have green noses—the Su-7s, MiGs—all have green noses. So I went after him. Now I know he was a poor pilot, he made some turns but could have gotten away if he had dove down and run. I was flying a MiG-21F-13 and could not have caught him if he did this. We maneuvered, and I reached about 200 meters where I opened fire with my cannon. He made a very gentle turn and put his afterburners off; I was at about 1,000–1,200 meters behind him. I believe he did not see me. We were at low level, and I knew that below 500 meters altitude the Atoll will go into the ground if fired down at a target. I dropped below him to make him the only heat source in the sky.

This was my first time to shoot an Atoll. I was on vacation when the squadron shot Atolls in training. I remember the speed was 1,250 km per hour, which was out of the limitation of the MiG-21F-13, but I didn't care. I had a tone from the missile, and I launched it. The missile went to the F-4, and the explosion was below the aircraft. Nothing happened, just a bang, smoke, and the aircraft was still flying. So I fired the other missile. There was an explosion below the F-4, and I think the left engine started to smoke, but it was still flying. I couldn't imagine what is going on, so I changed to the gun and I approached the jet. He started slowing down, and I didn't want to open fire until I put him in my lap. Then my No. 4 said, "break, break, we have two aircraft behind." So I forgot about the smoking Phantom and turned away. I left him southeast of Suez, and the jet crossed into the Sinai. I pulled 10 "Gs" and returned back to Inshāṣ at very low level. I think the F-4 crashed on the way home.[2]

After this Aḥmad ᶜĀṭif was known as FFK (First Phantom "Killer") among his fellow pilots.

Just under four months after Aḥmad ᶜĀṭif shot down his first Phantom, a Syrian MiG 21 pilot named Capt. Bassām Hamshū downed another Israeli Phantom over the Golan area on 2 April 1970. As a result he became as celebrated in the Syrian Air Force (SAF) as "FFK" ʿĀṭif did in the EAF.[3]

## Constant Pounding Damages Egyptian Morale, December 1969

Soviet advisors worked closely with the Egyptian Air Defense Force and EAF to develop new tactics and techniques to counter the Israeli air offensive. Egyptian jets intercepted IAF strikes, and a dozen SA-2 missile batteries with seventy-two ready to launch missiles plus hundreds of anti-aircraft guns were moved forward to positions in the Suez Canal area. During the last weeks of 1969 Israeli jets hammered this air defense network with bombs.

The IAF capitalized on the many lessons learned by the United States in its five-year battle against the Soviet-supplied air defense network in North Vietnam. Using the American-built F-4s' and A-4s' battle-proven tactics, the IAF battered the Egyptian air defense network. These attacks inflicted heavy casualties on Egyptian missile and gun crews and killed several Soviet advisors.[4]

Israeli Mirage pilots and F-4 crews hunted down any Egyptian fighters that took to the air to prevent them from interfering with the bombing campaign. Electronic jamming of radio communications and direct attacks against fighter control stations crippled Egypt's interceptor direction organization. The centralized Soviet-style ground control intercept network was, in fact, vulnerable to this type of assault. Lack of success and heavy losses hurt the EAF's morale.

On the night of 26–27 December 1969, Israeli commandos captured an advanced Soviet-built P-12 Barlock radar that was being operated by the Egyptian Air Defense Force (EADF) near the Gulf of Suez. Israeli helicopters lifted out the radar, and technicians rapidly identified the limitations of this new system. This was a major embarrassment for the Egyptians, and several officers were reportedly executed or jailed as a result of this loss.

Despite the Israeli successes, EAF pilots and EADF troops contested IAF strikes, and fighter-bombers continued to blast targets in the Sinai. But these air actions plus Egyptian artillery attacks and commando assaults did not change the strategic situation. Relentless Israeli attacks and air engagements had severely hurt the Egyptian Air Defense network. Some Egyptian Army artillery units dug in along the Suez Canal were afraid to fire due to the rain of bombs that were sure to follow.

President Nasser sent an Egyptian delegation headed by Vice President Anwar Sadat and chief of staff Marshal Muḥammad Fawzī to Moscow to seek new weapons and financial support.

## IAF Deep Strike Raids, January 1970

On 7 January 1970 the IAF initiated a new air offensive. Israeli jets began bombing military bases and industrial sites near Cairo and other targets well behind the battle zone. The goal of this campaign was to destroy President Nasser's credibility with a rain of bombs and demonstrate that Soviet assistance could not limit Israel's actions. This air offensive was made possible by U.S. delivery of a squadron of F-4 fighter-bombers. While A-4s and older French-made jets participated in the air assault, only the F-4s had the range and payload to strike Egyptian rear area targets and the speed to evade EAF interceptors and air defenses.

The following day Egypt retaliated by sending jets into the Sinai. These strikes damaged the Israeli-held oil wells at Raʾs Sudr and several targets near Sharm al-Shaykh. However, two Su-7s were lost to ground fire.[5]

Maj. Gen. Muḥammad Nabīl al-Masīrī commented: "During the War of Attrition we sent in the MiG-17s with no escort. On the way back we sent in MiG-21s to CAP them, and they often fought with the Mirages. The Su-7 was a much bigger and heavier aircraft, maneuverability was not the same. It was faster [than the MiG-17], but it was a much bigger target. We suffered more losses with the Su-7 than with the MiG-17s."[6] As a result the EAF employed MiG-17s in the fighter-bomber role until they were finally phased out of service.

Israeli jets hit three targets near Cairo on 10 January and came back again to bomb the al-Khānkah supply depot 15 miles north of Cairo three days later. In response to this new Israeli air offensive, many radars, missile batteries, and anti-aircraft guns were redeployed to defend the vital military bases around Cairo.

Israeli jets now targeted these repositioned missile sites, and losses among the EADF rose. Though EAF interceptor pilots often tried, they were largely unable to interfere with these damaging attacks. Most of the Israeli deep-strike missions were flown by IAF F-4 crews who flew to their targets at low altitude and high speeds. Mirage fighters jealously protected the attacking planes from interception by MiG-21s. Nevertheless, on 16 January Egyptian air defenses shot down an Israeli A-4 as it was returning from a deep-strike raid.

In an interview with the press, Israeli Prime Minister Golda Meir stated: "Nasser is not happy. The war has been brought home to him. He can't fool his people anymore. You can't lie anymore when people hear planes right over Cairo."[7] But the Israeli deep-strike raids did not crack Egyptian morale. In fact it increased support for the struggle. These attacks also drove President Nasser to give the Soviet Union a larger role in defending Egypt.

As the Israeli attacks continued, Soviet Premier Kosygin sent a personal note to President Nixon calling for the United States to restrain Israel. However, Israeli fighter-bombers continued to pound targets deep in Egypt. Egyptian jets increased their air defense patrols, and on 9 February an EAF MiG-21 shot down a Mirage. Israeli military spokesmen admitted this air combat loss.[8]

Egyptian pilots also saw action against Israeli jets on the Syrian front during this period. An EAF MiG-17 unit was stationed at Mazzah air base near Damascus in late 1969 to bolster Syrian air defenses. By early 1970 Egyptian pilots were flying air defense missions over Syria. On several occasions in the spring of 1970, EAF MiG-17s fought with Israeli jets.[9] The crews sent to Syria included some very experienced fliers, such as Muḥammad Fikrī al-Jandī, who had served in Yemen in the early 1960s and fought over the Sinai in the 1967 War.

During an interview with the American press in early February 1970, President Nasser stated: "The Israelis think they are strong—all right they are strong. They know they have air superiority." He went on, "The problem is not the airplanes really. The problem we feel here in the Arab countries—not only here in Egypt—is pilots. We have more planes than pilots; the Israelis have two pilots for every plane; so the Israeli have air supremacy." Nasser stated that the Israeli attacks demonstrated "the arrogance of power" and that soon dramatic steps would be taken to change the situation.[10]

During a raid against a military depot in the Nile Delta on 12 February 1970, Israeli F-4s "accidentally" bombed an Egyptian factory, killing some seventy workers and wounding nearly a hundred. This attack generated worldwide condemnation of the Israeli air campaign. For the first time in public, U.S. diplomats questioned if Israel had gone too far.

## Soviet Combat Forces Deploy to Egypt, February 1970

President Nasser flew to Moscow for secret meetings with Soviet leaders 22–25 January 1970. There Nasser asked for immediate delivery of offensive weapons (including Tu-16 bombers with stand-off missiles, MiG-23 fighter-bombers, and surface-to-surface missiles) to strike back against Israel. The Russians would not supply these weapons but did agree to bolster Egypt's defenses with Soviet interceptor squadrons and more advanced air defense weaponry.

Egypt and the Soviet Union repeatedly sent diplomatic signals to Israel and the United States to halt the deep-strike raids. When the Israeli raids intensified in early February, Soviet leaders responded with a massive show of support for Egypt. In February and March 1970 a Soviet Air Defense Division arrived in Egypt. For the first time, Soviet pilots and missile crews were assigned combat roles in the Middle East.

Two decades later Soviet newspapers published articles admitting the open secret of Russian combat operations in Egypt:

> Soviet pilots appeared in Egypt in March 1970. They arrived there after President Nasser requested that the Soviet government beef up Egyptian air defenses. This was a time of despair for Egypt. The Israeli command, having rejected attempts to break through Egyptian ground defenses on the Suez Canal, had selected the "air terror" tactic—inflicting massive air strikes. An avalanche of Israeli bombs crashed down on Egypt. Units of the Egyptian forces deployed along the canal suffered heavy losses.
>
> An air defense unit commanded by Alexei Smirnov was rapidly formed and sent to Egypt. A Soviet Fighter Regiment and an independent fighter squadron from the Soviet Union were rushed to Egypt soon thereafter. Their personnel were very carefully selected—pilots 1st and 2nd class, and each had several thousand flying hours. They took only volunteers. The equipment was the latest, for that time, a version of the MiG-21 fighter aircraft. [The Soviets sent MiG-21 Fishbed J interceptors to Egypt. These were the newest version of the MiG-21 with an internal 23 mm cannon, upgraded engine, and improved avionics.] Soviet pilots were tasked to prevent Israeli air strikes against Egypt's rear area targets including industrial enterprises, military bases, Nile River crossings, and population centers.[11]

Soviet Col. Konstantin Andreyevich Korotyuk (ret.) discussed his experiences in Egypt at this time: "My regiment's primary airfield was at Banī Suwayf, and the second was at Kawm Awshim. . . . They [the Israelis] immediately located us from the air as soon as we landed at Egyptian airfields. And several days later Golda Meir's statement [was transmitted] via Israeli radio: 'We know the Russian pilots are in Egypt. We are prepared to fight. When we are required to do so, we will.'"[12]

The Soviets assisted the Egyptians in rebuilding their air defense network. Many of the new missile sites had all Russian crews, and Soviet troops took over responsibility for several air defense control centers. Until the spring of 1970 only SA-2 missile batteries with the older Fan Song B model radar were provided to the Egyptians. In February 1970 Egyptian Air Defense received the improved SA-2D and SA-2F versions of the air defense system which featured radar and SA-2 missile improvements resulting in higher performance and greater resistance to U.S.-developed electronic countermeasures.

## Soviet SA-3 Goa Air Defense Missile System

In March 1970 SA-3 air defense systems, manned by Soviet crews, were deployed at air bases manned by the Soviets and around Cairo, Alexandria, and the Aswan Dam. The SA-3 missile was smaller and faster than the SA-2 but had a shorter range of about 20 kilometers and an effective altitude of about 11,000 meters. The SA-3 Low Blow tracking radar and processing equipment could guide the missile to hit low-flying aircraft. This plugged a critical weak spot in Egypt's air defense shield.[13]

A dense network of surveillance radars, SA-2 and SA-3 missile sites, light (12.7 mm–57 mm) anti-aircraft guns, plus EAF and Soviet interceptors was being established. This shield threatened Israeli aircraft flying at from zero to 60,000 feet. Thousands of Egyptian workers and army engineering troops worked around the clock creating with cement and sand hundreds of decoy and real SAM sites and anti-aircraft gun positions. This network slowly moved forward toward the Suez Canal.

The Soviet-piloted MiG-21MF Fishbed J interceptors patrolled over Cairo and the Nile Delta but refrained from flying near the battle zone. As the new Soviet-manned air defense missile sites and fighter patrols expanded their protective shield over Egypt, the area in which the Israelis could operate without risking a confrontation with Soviet forces shrank. By April 1970 the Soviets had taken over five major EAF air bases, operated from more than a dozen others, and had established dozens of Russian-manned SA-3 sites throughout Egypt.[14]

## Struggle for Air Superiority, January–April 1970

Israel initially ignored the Soviets and continued their deep-strike raids. Air defense sites and artillery batteries that moved to within 160 kilometers of the Suez Canal were bombed to rubble. During the first quarter of 1970 Israeli jets flew some 3,300 sorties over Egypt, dropped 8,000 tons of bombs, and admitted only six losses.

An IAF base commander remarked:

> We have a very peculiar enemy which the west can hardly begin to understand. Their way of thinking is so different. We have signs that they speak of attacking us in the same way as we are attacking them from the air, so they may have learned something from us. The question is whether they will be able to attack us, and whether they are anywhere near ready. Except for the Sukhoi and the Tu-16 I don't think they have suitable aircraft. They could hammer Tel Aviv but could not do much harm to a small and well-defended target. . . .
>
> The quality of the pilot is becoming far more important than the sophistication of the aircraft. After all, a Phantom can always disengage from an action. It has longer endurance and better acceleration than the MiG-21. As far as I know, only one of our aircraft has been shot down. [In air combat] It just goes to show that even an Egyptian pilot can get into a missile firing position occasionally.[15]

Such remarks revealed the arrogant attitude Israeli pilots had at the time for their Arab adversaries.

EAF interceptors continually scrambled to challenge the Israelis, while fighter-bombers flew regular attacks into the Sinai. Israeli fighters and air defenses inflicted steady losses on the EAF, with more than twenty jets being lost during the first quarter of 1970. Already short of trained aircrew, these casualties further bled the EAF, but despite such heavy odds, Egyptian pilots still took off to perform their assigned missions.

For several months the IAF refrained from bombing SA-3 missile sites known to be manned by Soviet crews. Even though Israeli strikes knocked out many of the new SAM and anti-aircraft sites, each time one was destroyed another air defense unit took its place. Egyptian forces suffered serious losses, and several Soviet air defense advisors were killed or wounded, but they would not back down.

The Israelis also withdrew from several air engagements to avoid conflict with Russian-flown aircraft. Following a confrontation between Israeli and Soviet aircraft on 18 April Israeli leaders decided to abandon their deep-penetration raids over Egypt. Israeli leaders, however, announced that they did not want to confront the Soviets but would continue to strike targets in the battle zone near the border.

Soviet Air Force Lt. Gen. Yuri Nastenko commanded one of the Soviet interceptor squadrons flying in Egypt during this time:

> After this, the Israeli Air Force ceased raids deep into Egypt. Internal areas were no longer bombarded. Only reconnaissance aircraft appeared which attempted to avoid contact with our pilots. Our presence alone had already cleared the situation in the Egyptian skies. The Israeli Air Force transferred its primary activities to the Canal Zone. A war of nerves and character and a struggle to wear [us] down and [create] exhaustion began for us. . . .
>
> There was constant tension, and everyone's nerves were on edge. As a rule, only one group of Israeli aircraft flew at high altitude as a show of force. But another group, invisible to our radars, flew just over the ground somewhere to the left, to the right, or just behind [the group at high altitude]. We also had to manage to intercept and neutralize it. Therefore, as soon as one flight took off, all others were placed on combat readiness. Pilots had to sit in aircraft cockpits parked in shelter in total readiness—just press the engine ignition button, taxi out onto the runway, and take off. The pilots sat in their high altitude suits in shelters with neither a draft nor a breeze, and it was over 40 degrees [centigrade]. And the takeoff could wait for hours. . . .
>
> The enemy was serious. They had a wonderful pilot school and some—mainly Jews of American descent—had Vietnam combat experience behind them. They fought intelligently and prudently, although at times somewhat stereotypically. They avoided combat with equals, but they were cautious.[16]

## The EAF Returns to the Offensive, April–June 1970

The arrival of the Russians boosted Egyptian morale and allowed the EAF to concentrate on striking targets in the Sinai. Egyptian jets flew a massive series of raids against targets all along the front 21–23 April 1970, at the cost of three aircraft lost. The Israelis admitted that these were the heaviest and most successful air strikes Egypt had flown since the 1967 War.

President Nasser voiced Egyptian optimism in a 2 May 1970 speech: "A change has taken place. Our armed forces have regained the initiative with bold military operations in the air and on land."[17]

Between 25 April and 5 June 1970 EAF fighter bombers flew eleven major assaults into the Sinai. Maj. Gen. Muḥammad Nabīl al-Masīrī (ret.) discussed these operations:

> I discovered during the War of Attrition that some squadrons would go hit the Hawk sites near Biʾr al-Jifjāfah and keep on attacking—starting with the bombs, launching the rockets, and then attacks with the guns. Some of my pilots would come home without a bullet. I had to interfere and give the order for them not to make more than two attacks because of the probability that they would be intercepted by Mirages that were circling nearby.
>
> We gave strict instructions not to make more than two attacks. If you accurately put your bombs and rockets on the target, keep your guns to defend yourself. Best to hit them and return home early before they can intercept you. Why take the risk? Your aircraft is inferior relative to the Mirage. To fulfill the mission and return safely is much better than being a hero and we lose some of our aircraft and pilots.[18]

The IAF and USAF/USN both had considerable experience defending against, and attacking, anti-aircraft missile sites (SA-2 in Vietnam and SA-2/3 in the Middle East). As a result they had perfected tactics to evade Soviet missiles and attack well-defended missile sites. In addition, American forces fielded a full range of defensive electronic countermeasures and Shrike anti-radar missiles to disrupt or destroy the SA-2s' tracking radar. The EAF did not have the benefit of these tactics nor ECM systems, and its anti-Hawk tactics were rather basic. Nevertheless, through experience and training, EAF pilots improved their skills.

Israel was now losing the initiative in the air as a result of Soviet intervention and aggressive Egyptian attacks. The expanded air defense network and Russian fighter cover limited Israeli options. The IAF also suffered increasing losses. Throughout May and June of 1970 Egyptian and Soviet air defense units fought to establish a dense air defense shield parallel to the Suez Canal to protect front-line positions from the constant Israeli bombing. A belt of Egyptian SA-2 sites, anti-aircraft guns, and Soviet SA-3 batteries were emplaced about 25 kilometers back from the Suez Canal. Israeli jets plastered these air defense sites with bombs and intercepted any aircraft that flew near the combat zone. The missile and gun sites of the air defense belt steadily moved forward despite the heavy casualties suffered.[19]

On the night of 30 June 1970 Egyptian and Soviet forces used their air defense shield offensively. During the hours of darkness they boldly moved fifteen SA-2 and SA-3 missile sites and anti-aircraft batteries forward and camouflaged them in the trees near the Suez Canal. When a patrol of Israeli F-4s flew over the sites at dawn, they were ambushed and two were shot down.[20]

## Desperate Struggle, June–August 1970

During the summer of 1970 Israeli jets and Egyptian/Soviet fighter and air defense units were locked in a desperate struggle. While IAF jets battered air defense positions and shot down both Egyptian and Soviet fighters, Israel was unable to knock out the air/ground air defense shield. During a month of intense air action Israel lost five F-4s, and Lt. Col. Shmuel Hetz, the commander of the IAF's first F-4 squadron, was killed.[21]

Senior U.S. and Soviet leaders became increasingly concerned that the situation in the Middle East could get out of control. U.S. Secretary of State William Rogers visited Israel and Egypt in June 1970 to broker an end to the fighting, while Soviet diplomats met with American officials and pressed Egypt to consider a UN-sponsored cease-fire.[22]

The EAF continued its heavy assault into the Sinai, and Russian pilots were also becoming more aggressive. On 25 July 1970, two Russian-flown MiG-21s ambushed and damaged an Israeli A-4 during raids along the Suez Canal. Five days later the Israelis set a trap for the Russians. Three flights of MiG-21s were scrambled to intercept several Mirages flying a decoy mission. The Russian pilots were surprised by more than twenty Mirages and Phantoms. In the brief battle five Soviet planes were shot down, and only one Israeli Mirage was damaged.[23]

The poor showing of the Soviet pilots prompted Air Marshal Pavel S. Kutakhov, commander of the Soviet PVO Strany Air Defense Force, to fly to Cairo with a team to investigate the air battle. In addition, a Soviet regiment of MiG-21s and a squadron of Su-15 Flagon all-weather interceptors were sent to Egypt to bolster defenses.

The Egyptians reacted with ill-concealed delight when they heard of the Israeli ambush. For years Soviet advisors had lectured them about their failings and boasted of their own superior skills. Egyptian pilots and air defense crews were quick to point out that the Soviets had fallen prey to the usual Israeli ambush tactic, and they wondered if the Soviet pilots were really as good as they thought they were.[24]

Heavy air action continued into August. Both Egyptian and Israeli jets flew daily patrol and attack sorties. Many vicious air battles were fought.

The IAF lost a Phantom and a Mirage while bombing air defense sites near the Suez Canal on 3 August 1970, and three EAF jets were shot down by Israeli ground fire during the first week of August.

As the result of U.S. and Soviet diplomatic intervention and the inability of either side to secure a clear advantage, Egypt and Israel finally agreed to a UN call to halt the fighting. The cease-fire came into effect on 7 August 1970.

## Conclusion

During the War of Attrition the EAF flew thousands of operational sorties, but the force was pushed nearly to breaking point by constant combat, heavy losses, and the rapidly changing military situation. When EAF interceptors proved incapable of halting Israeli attacks, prime responsibility for the air defense of Egypt was given to a new military branch, the EADF. This force also proved unable to succeed until the Soviet Union sent in combat units including more potent SA-3 anti-aircraft missiles, plus PVO Strany MiG-21 Fishbed J and Su-15 Flagon interceptor squadrons. Close integration of the EAF with the new air defense force and Russian units again required a major change in air force organization and doctrine. The War of Attrition was also the first conflict in the Middle East in which electronics, including electronic warfare, radar surveillance, communications, and missiles, shaped the course of battle. Israel initially demonstrated better flexibility in these areas, though the EAF and EADF rapidly learned the new rules.

The EAF contributed to President Nasser's political goals and the defense of Egypt through sustained combat against Israel from mid-1967 until the cease-fire of August 1970. During this period EAF units also operated in Syria to enhance air defense against Israeli attacks and for political reasons. Egyptian fighter pilots flew thousands of air defense missions over Egypt and escort sorties over the battle zone. Even so, the IAF continued to penetrate Egyptian airspace and inflicted heavy losses. When the EAF and the EADF were unable to stop Israeli raids, President Nasser was forced to turn to the Soviets. This was the first time the Soviets deployed combat forces in the Middle East and fought to support a client state. This expanded Soviet-U.S. confrontation which had already been stressed by the cold war and Vietnam conflict.

EAF fighter-bomber pilots and bomber crews also flew hundreds of attacks into the Sinai. These raids damaged dozens of Israeli targets and caused numerous casualties. On the other hand, Israeli fighters, Hawk missiles, and anti-aircraft fire downed many EAF aircraft. Despite these casualties, the EAF kept coming back. While they received little publicity, EAF transport, communications, helicopters, training units, and maintenance and support crews worked tirelessly to support the Egyptian war effort.

The cease-fire came at a considerable cost. Thousands of Egyptian Army and EADF personnel and many Russian troops were killed or wounded by Israeli bombs. Clearly Egyptian and Soviet forces deserve considerable credit for their determination and persistence under such circumstances.

According to U.S. sources, between July 1967 and August 1970 Egypt lost 109 aircraft and ninety aircrew in combat with Israel and to operational causes. In addition, another forty-five to fifty aircraft were lost in training accidents. These sources credit Egyptian pilots with shooting down only two Israeli Mirages, although SAM units downed seven IAF jets, and anti-aircraft guns claimed eight additional Israeli aircraft.[25] Claims by Egyptian sources were considerably higher, but no official information has been released by the Egyptian government.

During this long period of fighting the EAF paid a heavy price, but air and ground crews gained considerable experience. Many lessons were learned during these operations, information that was to help the Egyptians plan for the liberation of the Sinai.

20

# New Directions (1970–1973)

The end of the War of Attrition saw a new situation along the Suez Canal. Egypt, with massive support from the Soviet Union, had fought the Israelis to a draw. During three years of fighting Egypt had suffered higher casualties and equipment losses than the Israelis, but the Egyptians had learned many valuable lessons. Israel had once again demonstrated its military expertise, especially in the air, during the three years of fighting. Even so, Israeli superiority had been challenged considerably since the 1967 War. Soviet support and the Egyptian decision to keep on fighting, no matter what the cost, had started a resurgence. The August 1970 cease-fire came about as a result of the changing political and military situation between Egypt and Israel and superpower pressure. Israeli leaders did not want to accept a cease-fire when things seemed to be moving in Egypt's favor. On the other hand, Egypt and Israel were under intense pressure from their respective superpower sponsors to end the fighting. Also both countries and the Soviet Union were losing men and material at an exhausting rate. Following the cease-fire, the air defense shield along the border was strengthened by the relocation of Egyptian and Soviet missile sites. The EAF and EADF examined the lessons of the War of Attrition and restructured to prepare for renewed confrontation with Israel. A split with the Soviet Union and new political leadership led to a change in direction. Subsequently President Sadat went to war with Israel to shatter the status quo and restore Egyptian pride.

## Death of President Nasser, September 1970

President Nasser died of a heart attack on 28 September 1970, and on 15 October 1970 Vice President Anwar Sadat was selected by the National

Assembly to lead the country. Sadat had been a close associate of Nasser's dating back to their years together at the Egyptian Military Academy. The two were also leaders of the 1952 Revolution.

President Sadat was determined to win back the Sinai and restore Egypt's place as a the leader of Arab nations. To achieve these goals he had to consolidate his power and prepare Egypt's armed forces and the country for renewed conflict with Israel. The EAF played an important role in Sadat's plans.

## Lessons of the War of Attrition

Despite serious losses both in the air and on the ground, Egyptian morale had been greatly enhanced by the War of Attrition. For the first time Israel had suffered a military setback. The Egyptians had also learned a great deal about Israeli military doctrine and Soviet combat tactics. Egyptian personnel demonstrated that they could stand up to the Israelis after the disaster of 1967. As a result of this experience, air force personnel were more confident in their skills, if not entirely in their equipment. The lessons of more than three years of fighting were intensely examined following the cease-fire. These studies showed that when the EAF and EADF worked together in an integrated manner, they were effective. Carefully planned and practiced EAF attack and support operations also were successful.

However, in general Egyptian and Soviet doctrine during the War of Attrition remained defensive. Egypt had the largest army in the Middle East, and massive Soviet aid provided the country with a sophisticated air defense network. But Egypt was lacking in offensive firepower. Israel, in contrast, possessed a smaller military that was trained to strike first rather than absorb an Arab attack. This doctrine was reflected in the composition and employment of the EAF and IAF during the Attrition War. Israeli A-4s and F-4s repeatedly bombed targets throughout Egypt, despite the air defense network. Egyptian aircraft could only hit targets in the western Sinai because of their limited range, payload, and survivability.

The 100,000-strong air defense force remained the backbone of Egypt's defenses against the IAF. Air defense units were located throughout the country, but most were concentrated around Cairo and along the Suez Canal. The EADF included dozens of surveillance radars, some sixty SA-2 batteries (each with six missile launchers), and hundreds of anti-aircraft units, all tied together by an integrated communications system. This system was supported by thirty Soviet-manned SA-3 batteries, special P-15 Flat Face surveillance radars and control sites, plus Egyptian and Soviet interceptors.[1]

The EAF's first priority was providing interceptors that were integrated with the air defense force. In late 1970 the EAF fielded four such regiments with nine squadrons of MiG-21s. This included both day-fighter F-13 versions and all-weather FLs and Ms. At this time there were also four Soviet squadrons with MiG-21MFs and at least one Su-15 all-weather unit flying in Egypt.[2] An important lesson of the War of Attrition was the importance of maintaining surveillance and communications over the battle area. The ability of the Israelis to monitor and jam Egyptian and Soviet communications had always been a major factor in the IAF's success in air combat. Many Egyptian and Russian fighter pilots were shot down because they were not kept informed of the rapidly changing tactical situation and were caught by surprise.

Following the cease-fire the Soviets deployed to Egypt advanced command and control systems with the NATO code name Swamp and Markham.[3] New communications systems better tied together the surveillance network, missile sites, and interceptors. The EAF and EADF also expanded their electronic listening (ELINT) and intelligence capabilities. Egyptians who spoke Hebrew were sought out and integrated into intelligence units to monitor Israeli radio communications, assess intent, and develop deception techniques. Egyptian and Soviet forces in Egypt fielded upgraded data links and used more secure wire communications and radios that were more difficult to monitor and jam.

Attacking enemy targets and battlefield support were major EAF priorities. By 1971 Soviet resupply following the end of the War of Attrition enabled the EAF to field fifteen fighter-bomber and bomber squadrons. These units had a total inventory of some 220 aircraft, but the EAF still struggled to maintain a one-to-one pilot/aircrew to aircraft ratio. Intensified training was nevertheless slowly expanding the cadre of experienced pilots and aircrew.

This sizable force included two fighter-bomber regiments (five squadrons) with one hundred MiG-17s, three fighter-bomber regiments (six squadrons) flying about one hundred Su-7s, and two bomber regiments. Each bomber regiment had two squadrons; one had twenty Il-28s and the other thirty-five Tu-16s. A small number of Tu-16s were equipped to fire the AS-5 Kelt stand-off missile.[4]

The real offensive power of the EAF's attack force remained limited. During the Attrition War, EAF MiG-17s, and Su-7s flew hundreds of strikes into the Sinai and Il-28s made a few night attacks. However, fighter-bombers and their escorting interceptors suffered heavy losses to Israeli defenses. The EAF bomber force was mostly held in reserve during the War of Attrition. The EAF's tactical strike aircraft were limited by their range and payload performance. To increase their survivability against Israeli Hawk missiles, anti-aircraft guns, and fighters, Egyptian fighter bombers flew to and from their targets at low altitude and top speed.

While rugged and fast, the Su-7 could only carry a load of two 500 kilogram bombs to a range of 320 kilometers flying the low-altitude, high-speed profile necessary to survive. The range from Bilbays, a front-line Su-7 airfield, to Israeli targets in the central Sinai was 240 kilometers, while it was 290 kilometers to the major Israeli base at al-ᶜArīsh. The MiG-17 had only a 240-kilometer range with two 250 kilogram bombs and rockets.[5]

Israel's attack force was slightly smaller than the EAF's, but the U.S.-built A-4 and F-4 could fly farther, faster, and carry more bombs than Egypt's Soviet fighter-bombers. In early 1971 the IAF fielded three squadrons of F-4 Phantoms (70 aircraft) and four A-4 Skyhawk squadrons (100 planes). The A-4 Skyhawk could each deliver 2,750 kilogram and the F-4 Phantom 6,350 kilogram of weaponry in a single strike and carry these bombs to a range of more than 480 kilometers.

## EAF Upgrades, 1970–1972

The Egyptian Air Force examined combat lessons and made changes to prepare for confrontation with Israel. Egyptian air bases were upgraded with steel reinforced concrete aircraft shelters that could protect planes from all but a direct hit by a large bomb. All major EAF air bases were upgraded with additional runways, taxiways, and a rapid bomb damage repair capability. Airfields were surrounded by SA-2 and SA-3 missile sites and anti-aircraft batteries. The EAF even went back to World War II techniques to deter Israeli air attacks. Hydrogen-filled balloons with strong steel cables were installed at several airfields to deter low-level attacks. These barrage balloons were kept in storage containers but could be rapidly inflated and sent aloft. Shelters, air defenses, and other techniques also denied Israel the opportunity to easily destroy the EAF on the ground. Such a high level of airfield defenses and hardening started a worldwide trend.

Egypt was in fact far ahead of most of the world in protecting aircraft and air bases from attack. NATO and Warsaw Pact countries followed this lead, and soon most airfields throughout Europe, the Far East, and the Soviet Union sported protective aircraft shelters, improved air defenses, and other hardening efforts.

New tactics were tested, and the EAF developed its own training curriculum which combined Soviet and Egyptian doctrine. Despite extensive training efforts, the EAF was still short of experienced unit commanders and skilled pilots as a result of combat, training losses, medical washouts, and transfers. Skilled technical personnel were also in short supply. These factors were addressed, and the situation improved during the next several years as aircrew and technical personnel became more experienced.

Egypt also increased multiservice coordination and training with other Arab nations. Many EAF officers were assigned to the EADF to coordinate fighters and anti-aircraft weapons. EAF crews flew on exercises and were detached on exchange duty with the air arms of Syria, Jordan, Iraq, and other Arab countries to train and exchange ideas.

## Expanding the Air Defense Screen

Soon after the August 1970 cease-fire, Israel accused Egypt of increasing the number of SAM sites near the border and moving batteries closer to the Suez Canal, thus extending the Egyptian missile screen into the Israeli-occupied Sinai. Israel complained to the United Nations that this was a clear violation of the terms of the cease-fire.

An IAF squadron commander commented about Egyptian actions after the cease fire:

> Sure, we feel frustrated. But we felt that way when the cease fire was first agreed to. We knew what would happen. We were sure of it. And, as it turned out, we were right. . . . Before the cease fire, they were starting to get pretty good. The missiles were coming a little bit closer from one mission to the next. Seems like less were being thrown up just for fun. It wouldn't be easy to go back there today. As I said before, I've encountered as many as a dozen at one time. Two dozen would be a different matter altogether. And now they have

> the SAM-3s in there. They are more maneuverable, better at low altitudes than the SAM-2s, and we all know by whom they are being operated.

The squadron commander commented about his adversary in a MiG-21: "He's getting better too. But he's still afraid to fight. He makes one run at you and flees for home [reflecting the lack of emphasis by the EAF on air combat maneuvering—emphasis on Soviet-style intercept techniques] . . . I couldn't care less whether an Egyptian or a Russian were piloting those planes today. As long as a MiG-21 is trying to attack me, there could be any foreigner inside and I'll go after him."[6]

In fall 1970 Israel released to the press reconnaissance photographs of many missile sites that had been moved into positions after the cease-fire. Israel and the United States demanded that the United Nations take action to condemn Egypt's cease-fire violations. Egypt and the Soviet Union ignored the UN protests and stated that they had honored the cease-fire agreement.[7] In response to Egyptian and Soviet actions the United States provided Israel with additional F-4s, A-4s, and weaponry to maintain the balance of power. For the first time Israel was given AGM-45 Shrike anti-radar missiles that proved effective in destroying missile radars in Vietnam combat. Advanced electronic jamming pods including the ALQ-101 were also sent to Israel.

The Israeli government also upgraded its defenses in the Sinai. This $250 million dollar program expanded Israeli air bases and fortifications and added many new supply depots and roads. A new air base, Eẓyon, was built in the southeastern Sinai and facilities at Refidim (Biʾr al-Jifjāfah) and al-ʿArīsh were improved to allow additional IAF squadrons to operate or deploy to the region.

## Sadat's Goals: Return the Sinai and Independence

President Sadat initially followed the same general course set by Nasser. He moved slowly to consolidate his power amid an almost nonstop series of economic, political, and military challenges. Egyptian military leaders were obsessed with the idea of striking back at Israel and recapturing the Sinai. Sadat supported this goal, but not at the expense of becoming a client state of Russia. A deeply patriotic man, Sadat had struggled against Britain during the 1940s, and he planned to reduce the Russians' influence in Egypt just as his predecessors had forced out the

British. Sadat was also determined to reduce Egypt's political and military dependence on the Soviet Union.

In 1971 some 3,000 Soviet Air Force personnel and more than 12,000 air defense crews were serving in Egypt. In addition, there were thousands of Russians working as advisors and technicians attached to the Egyptian Air Force, Army, Air Defense Force, and Navy. The Soviets not only controlled six Egyptian air bases but also managed the country's early warning network and thirty SA-3 batteries. Soviet Navy aircrew flew a squadron of Tu-16 jet bombers with Egyptian markings on regular patrols over the Mediterranean to keep tabs on the U.S. Sixth Fleet. A number of Soviet-flown Be-6 amphibious anti-submarine aircraft with EAF markings also flew from Alexandria harbor.[8] Other Soviet-flown aircraft based in Egypt during this period were An-12 "Cub C" electronic countermeasure machines and Il-38 "May" reconnaissance aircraft.

During the early 1970s relations between Egypt and the large force of Soviets stationed in the country became increasingly strained. Most Soviets treated the Egyptians as second-class citizens and blamed them for all the problems that cropped up with Soviet equipment. While some EAF officers and enlisted men developed close personal ties with individual Soviets, most Egyptians maintained only a working relationship with these advisors. Many EAF personnel resented the overbearing attitudes exhibited by some Soviet personnel, men who were less experienced in combat than the Egyptians themselves. The Egyptians also felt that the Russians generally underestimated their Israeli foe. Meanwhile the Soviets tried to hide the losses suffered by their missile crews to Israeli bombs during the War of Attrition and the poor showing of Soviet Air Force pilots in the air battle with the IAF.

The Soviets tended to teach Egyptian technicians only the basic operation of electronic equipment. Whenever an item malfunctioned, they instructed their students to replace the defective part rather than repair it. This ensured continued Egyptian reliance on Soviet-supplied spare parts. There was also a widespread feeling that the quality of Soviet advisors and operational units had steadily declined since the cease-fire in 1970. It seemed that Russia was now using Egypt as a fine-weather practice area for its junior pilots and missile crews.

President Sadat and his military leaders were upset that the Soviets still had not provided long-promised offensive weapons. In March 1971 President Sadat sent ʿAlī Ṣabrī, a retired senior air force officer, to Moscow with a personal letter requesting Tu-22 Blinder jet bombers, more Tu-16s armed with air-to-surface missiles, SCUD surface-to-surface missiles, and MiG-23 Flogger fighters. Soviet leaders agreed to supply SCUD missiles but no Blinders or Floggers. These would be delivered only on the condition that they be supervised by Soviet advisors. Moreover, Soviet leaders would have to give permission before these systems could be employed against Israel. Sadat naturally refused this request.

## Egyptian Operations in Sudan, 1970–1973

Egypt's primary focus during the years after the 1970 cease-fire was preparing for renewed confrontation with Israel. At the same time the military had to maintain full alert in the face of Israeli threats and support important foreign operations. Both branches of the vital Nile River flow through Sudan which has consequently been watched closely by Egyptian leaders. The country is divided ethnically and religiously between largely Arabic-speaking Muslims who live in the north and center, and other Muslims, Christians, and animists who make up the majority of the people in the south.

Beginning in 1958 a rebel movement known as the Anya Nya created unrest in the south largely as a result of resentment of the greater political power and economic position of the northerners. This long-simmering conflict resulted in hundreds of thousands of deaths and forced more than a quarter million people to flee the country. Israel provided arms and training assistance to the Sudanese rebels. Consequently, in late 1969 the government of Sudan entered an informal alliance with Egypt and Libya to coordinate anti-Israeli strategy. Sudan also sent a battalion of troops to Egypt.

When an increase in rebel activity threatened the Sudanese government in 1970, a large contingent of Egyptian Army and Air Force personnel were sent to aid the government in restoring order. EAF Mi-8 helicopters and MiG-17 fighter-bombers supported the Sudanese Army fight Anya Nya rebels in the southern provinces. Egyptian MiGs were stationed at Wādī Sayyidnā, north of Khartoum. The EAF also provided aircraft as well as training to the Sudanese Air Force, and flew in direct support of anti-rebel operations. EAF and Sudanese pilots and aircrew flew transport and armed reconnaissance missions and regularly attacked convoys and rebel villages. A small number of Il-28s, MiG-17s,

Mi-8 helicopters, and Il-14 transports remained in Sudan until just before the 1973 conflict.[9]

## The Foxbat Affair, 1971–1972

In mid-1971 Soviet pilots flew a number of high-speed missions over the Egyptian desert in a new jet, the MiG-25, which was given the NATO code name Foxbat. These sorties received worldwide press attention when in October 1971 two Soviet-flown MiG-25s flew parallel to the occupied Sinai coast and took pictures of Israeli defenses without harm.

The MiG-25 outperformed any fighter the West had in service at this time. Israel's Mirages, Phantoms, and Hawk anti-aircraft missiles simply could not catch the Mach-3, high-flying Foxbats. In November 1971, pairs of MiG-25s made three high-speed runs over the occupied Sinai, but no flights penetrated the airspace of Israel proper.[10] Nevertheless, these jets shattered windows with their sonic booms and left Israeli F-4s lagging far behind. Israel replied to these reconnaissance sorties with its own supersonic window-smashing flights by F-4s over Cairo. In addition, Israel sent unmanned BQM-34 Firebee drone aircraft over Egyptian territory near the Suez Canal to take pictures and tease the air defense network.

While Egypt gained a morale benefit from these MiG-25 sorties, these operations were solely a Soviet affair. The primary task of these aircraft was to monitor the U.S. Sixth Fleet, which was little help to the Egyptians. Some excellent photographs of Sinai defenses were provided, but when the Egyptians asked the Russians to take more photographs, the answer was no. This again underscored the worsening of relations between the two nations.

Foxbat flights disturbed the Israelis and scared western military planners, but they did little to change the situation along the Suez Canal. In fact, Israeli jets increased their flights along the border to stimulate a reaction and stepped up their penetrations of Egyptian airspace. These sorties were contested, both sides suffering losses. Egypt soon became fed up with this Israeli campaign to test the coverage and reaction time of its strengthened air defense shield along the Suez Canal. Then, on 17 September 1971, a lumbering IAF C-97 intelligence gathering aircraft flew along the Suez Canal within range of an EADF SA-2 missile battery and was promptly shot down by an Egyptian missile.

## Intensified Training

Despite these challenges, Egypt concentrated on building up its military strength and training. Soviet training techniques were notably slow and thorough. To correct such a situation the EAF instituted new training programs that combined Soviet and Egyptian training techniques and better fitted the Middle East environment.

Maj. Gen. Muḥammad Nabīl al-Masīrī (ret.) commented on EAF training efforts during this period:

> The real contribution I made was when I was inspector of fighter-bombers. I was in charge of training those pilots. We created realistic ranges with runways, shelters, anti-aircraft guns, Hawks, tanks, and columns of trucks (in the Western Desert near the Libyan border). We concentrated on operational flying at very low altitudes—to approach the targets and attack with minimum time and return back as fast as possible using good maneuvers. The pilots made great sacrifices in training and developed high standards. I would stay at the range with a tape recorder and then go to the regiments one by one and brief the pilots on what they did right and their mistakes. . . . "You didn't maneuver—you're dead," until they reached a very high standard. After a time they stopped improving, so we then started training with real ammunition, rockets, bombs, and cannon. The greatest danger to our pilots was the Israeli Mirages. If you approached low and maneuvered hard, you could escape the Hawk. Anti-aircraft fire was tough but not very accurate.
>
> At that time we had Russian experts here. We used to train our pilots to attack our missile sites for both sides to get training. They tried to defend themselves against our aircraft according to their plans, and we did the same. They confided in me that aggressive maneuvers over the target destroyed the capability of their missiles. So we learned.
>
> I remember one of the maneuvers we learned in practice with the air defense force. Later we used this on the Hawks. Our jets would come in at low altitude, pop up, put their nose down, fire, and come out. We had a very high standard at that time, especially with the MiG-17s. And these were the younger pilots. This was the path to the MiG-21s, Sukhois, and Mirages.[11]

Egyptian preparations for renewed hostilities with Israel were now running at a high level. The EADF had established a training center in the desert south of Marsá Maṭrūḥ. EAF aircraft flew against this range and practiced against replicas of Israeli Sinai targets. Colonel Qaddafi's government in Libya initially developed a close working relationship with Egypt, and some EAF aircrew trained at the ex-RAF base at al-

ʿAdam in Libya. A small number of EAF personnel still attended classes in the Soviet Union, but most instruction was conducted in Egypt.

The Egyptian armed forces, and especially the EAF, worked hard to train young soldiers to operate and maintain the new generation of Soviet-supplied military equipment. A good portion of the EAF was short-term conscript soldiers, but intensified training gave them the necessary skills to support Soviet systems. The high level of unemployment in Egypt during this period also encouraged many enlisted men and officers to remain with the EAF. This improved the proficiency level of the force as experienced personnel were kept on.

Motivation and commitment among Egyptian troops were high. The defeat of 1967 and loss of territory had given everyone a common national purpose—the ending of Israel's occupation of the Sinai and restoration of national pride.

## The Expulsion of the Soviets, 1972

President Sadat repeatedly asked Soviet leaders for Tu-22 bombers and MiG-23 fighters. He wanted these to threaten Israel. When he saw there was little chance of receiving these systems, Sadat changed his strategy. Once again, as seen in 1955, the inability of Egyptian leaders to secure advanced military equipment forced a high-level foreign policy decision. Egypt under Nasser turned from the British and the West in 1955 to buy Soviet military hardware and secure financial support for the Aswan Dam. In 1972 it was the refusal of the Soviets to provide offensive systems, and their involvement in anti-Sadat activities, that resulted in their being ejected from Egypt.

The simmering Egyptian-Soviet relationship came to a boil in the summer of 1972. In April, for the third time, a high-level Egyptian delegation went to Moscow seeking weapons. The EAF was especially keen to get MiG-23s, which had not yet been exported outside the Soviet Union, or a similar advanced aircraft to counter Israel's Phantoms. President Nasser and Egyptian military leaders had been informed that it would take several years to train the EAF to fly and maintain the MiG-23, whereas senior EAF officers maintained that conversion from the MiG-21 could be done in less than a year.

Soviet leaders again demanded that advanced aircraft or surface-to-surface missile systems provided to Egypt must remain under Soviet control. Sadat now regarded this as an unacceptable demand. The Soviets had frequently used advisors and the flow of weaponry, spare parts, and training as a key tool of their foreign policy. Egypt was tired of being used in this way. The relationship between the two countries was also damaged by Russian involvement in Egyptian political activities. Sadat had earlier been obliged to act against his vice president, ʿAlī Ṣabrī, and other senior officials to prevent a coup d'état in May 1971. He was not now going to allow Russian actions to challenge his authority.

In July 1972 President Sadat issued an ultimatum to the Soviets: either supply the requested arms and stay out of Egyptian politics or your troops and advisors will be expelled. When he received no satisfactory reply to his demands, he gave the Soviet advisors, technicians, and pilots ten days to leave the country. Soviet forces began their withdrawal on 17 July 1972. The Soviet Union also lost the right to fly maritime reconnaissance missions with Tu-16 bombers and Be-6 ASW planes and to dock naval vessels at Alexandria. Only a few hundred Soviet advisors were allowed to remain in the country. In an attempt to retain some influence, the Russians left behind aircraft and missile systems and offered Tu-22 bombers. The MiG-21MF interceptors of Russian squadrons were also transferred to the EAF. However, Russian electronic warfare detachments and most radar batteries were removed.

Before the Russian expulsion several new systems were given to the Egyptians. The EADF received the SA-6 mobile SAM which had the range, firepower, and low altitude capabilities of the SA-3 system but was mounted on four-tracked vehicles. For close-in defense, the tank-mounted ZSU-23–4 anti-aircraft gun was unmatched. This system combined four liquid-cooled 23 mm cannon and a tracking radar in the turret of an armored, tracked vehicle. Small numbers of SCUD surface-to-surface missiles were delivered, and a squadron of Su-20 fighter-bombers were provided to the EAF.

## Su-20: The Swing-Wing Fitter

Soviet engineers fitted the Su-7 airframe with variable geometry wings and an improved engine to upgrade the aircraft's performance. These new jets formed the backbone of the Soviet Frontal Aviation tactical strike force and were widely exported to Warsaw Pact and allied air forces.

Col. Rifʿat Yusrī discussed the good points of the swing-wing Fitter: "The Su-20 was a very good plane. The Su-7 needed almost 2 km to take off on a hot day with a minimum load. With the Su-20 it needed just 1,200 meters with a maximum payload. Landing run in the Su-7 was long, and we always used a braking parachute. We had no radar in the Su-20, but we had a good gyro gunsight. The Su-20 was faster and could go much farther than the Su-7. It was also more maneuverable, very stable at low altitude, and could carry a much greater weapons load."[12] A number of EAF aircrews traveled to the Soviet Union to qualify on the Su-20, and conversion from the Su-7 to the Su-20 was an easy one for most EAF pilots. In addition, the Su-20 was equipped to carry up to four Atoll air-to-air missiles for self-defense as well as air-to-ground ordnance.

Though the equipment gap was narrowing somewhat, the Israelis still possessed more capable aircraft and weapons and maintained a clear edge in skilled aircrew and support personnel. In bomb load, speed, range, and armament, Israeli F-4s and A-4s still had a significant advantage over the MiG-21MF and Su-20, the newest Soviet aircraft delivered to Egypt. The departure of Soviet units and most advisors did impact the readiness of the EAF, the EADF, and the army, but these challenges were eventually overcome. The Israelis were closely monitoring these events and began testing Egypt's defenses.

On 13 June 1972 two IAF Phantoms penetrated Egyptian airspace and were intercepted by MiG-21s scrambled from al-Manṣūrah. When the MiGs chased the Phantoms out over the Mediterranean, they were ambushed and shot down by other Israeli fighters lurking below radar coverage. Israeli jets continued to probe Egyptian defenses, and on 24 July and again on 10 October 1972 IAF aircraft were downed by SAM missiles near the Suez Canal.

In 1972 Air Marshal ʿAlī Muṣṭafá Baghdādī retired and was replaced by Air Marshal Muhammad Husni Mubarak. An experienced officer who had flown Tu-16 bombers and held a series of senior positions including EAF chief of staff, Mubarak's assignment was to assist in planning an attack and then to lead the air arm into action.

Following the expulsion of most of the Soviets, Egypt quietly turned to several European countries seeking aircraft, missiles, and army weapons. President Sadat was determined not to be dependent on a single source of supply. These sales did not immediately materialize because of political and financial complications. However, in October 1972 the British foreign secretary did approve limited sales of Westland helicopters. Quiet discussions were also being held with France concerning the potential purchase of Mirage fighter-bombers. Egypt was committed to the long-term goal of diversifying its source of weaponry.

## Preparing for a Coalition War

Sadat's "Year of Decision," 1971, had passed and another was drawing to a close, with no action. At this time there were reportedly rumblings of discontent within the Egyptian military about the continued Israeli occupation of the Sinai. Although most of the Egyptian armed forces were unaware of it, President Sadat and his closest advisors had already begun planning for war with Israel. President Hafez Assad of Syria agreed to participate in a coordinated assault at a meeting with President Sadat in Cairo on 12 September 1972.

The growing cooperation with oil-wealthy Arab nations improved Egypt's economic, political, and military position. Syria, Egypt, and Libya joined into a loose confederation known as the Federation of Arab Republics in 1971. The three nations formed an Arab Defense Council to coordinate military activities, and an EAF MiG-17 squadron was again sent to al-Mazzah air base near Damascus in 1972. Saudi Arabia invited Egyptian pilots to train on Royal Saudi Air Force BAC Lightnings as part of the regional military cooperation. Saudi and Egyptian fighter squadrons also flew together during several exercises. Egyptian pilots went to France, carrying Libyan passports, to train on the Mirage. The EAF was keen to secure the Mirage for it had long been an adversary and one of the few western fighters available. By gaining experience flying and supporting Libyan Mirages, the EAF could make a rapid conversion when the political climate allowed the sale of French jets. The EAF benefited from these relationships through cross-training on new aircraft and the sharing of tactics and weaponry.

Col. Rifʿat Fatḥī remarked:

> I made my conversion to the Mirage III at Dijon [a French Air Force base] in 1973. A mix of pilots from MiG-17s, MiG-21s, and Su-7s converted to the Mirage. This was good because it was a multi-role jet. We flew Mirage Vs in Libya for training and also some Mirage IIIEs, which were ground attack jets. . . .
>
> There was a big difference between the Su-7 and the Mirage. The Su-7 had a good engine, reasonable payload, but poor maneuverability. The Mirage has

> good maneuverability, better acceleration, good visibility, and much better avionics. It has a good radar and fire control system.[13]

In March 1973 Kuwaiti Lightnings flew to Egypt, and a number of Libyan Mirages deployed to Egypt, several weeks later. Two Libyan Mirage squadrons with a total of at least twenty-five planes were sent to Egypt, including Mirage IIIEs optimized for long-range ground attack and multi-role Mirage Vs. These jets were flown by Egyptians, Libyans, and Pakistani Air Force instructors. Fighter squadrons from Iraq, Abu Dhabi, and Saudi Arabia also trained with the EAF. The EAF was particularly impressed with the skills and experience of Iraqi Hawker Hunter pilots. Such inter-Arab cooperation was helpful in cementing political relationships and provided valuable military training.

## Planning for War

In a speech on 1 May 1973 President Sadat signaled Egyptian resolve and declared: "Egypt wanted peace based on justice, but it could not accept an Israeli fait accompli and had therefore embarked on the state of all out confrontation." He added: "Egypt will break the stalemate."[14]

In their intelligence assessments, Israeli defense planners made a number of incorrect assumptions about future Arab actions. One of these was the belief that Egypt would not go to war until its air force was substantially upgraded and equipped with offensive weaponry. President Sadat was trying to do this, but when he was blocked by the Soviets he moved forward with his plans without the new weapons. Events were later to prove that Israel had misjudged both Egyptian and Syrian resolve.

Only slowly and reluctantly did Egyptian military leaders come to accept Sadat's concept of a war with limited objectives designed to open up the chance for peace with Israel. Gen. Aḥhmad Ṣādiq had been quietly drawing up plans for a massive Egyptian assault into the Sinai. Egyptian war plans started with realistic objectives. The assault planned to take advantage of the small force Israel maintained in the Sinai, its extreme sensitivity to losses, and the fact that reserves had to be mobilized before a counterattack could begin.

Using surprise and massive forces, the Egyptians and Syrians expected to overwhelm Israeli defenses quickly. The war plan called for

Egyptian forces to cross the Suez Canal, seize numerous bridgeheads in the Sinai, destroy Israeli Bar-Lev Line defenses, and then dig in. Egyptian forces would then defeat Israeli counterattacks. During a second stage of fighting, Egyptian forces planned to advance into the Sinai in force and seize the Mitlā and Jiddī passes. A simultaneous assault all along the Suez Canal and Golan Heights was intended to dilute the retaliatory striking power of the IAF and Israel's armored forces.

Air defense shields along the Suez Canal and on the Golan Heights would protect Egyptian and Syrian ground forces from air attack. Missile and gun defenses along the Sinai and Golan now were strengthened, and new equipment, such as the ZSU-23–4 anti-aircraft gun and SA-6 mobile SAM, were fielded. The EAF, SAF, and allied air contingents played major role in these war plans. Arab fighter-bombers would open the war with raids against targets in the Sinai and the Golan to disrupt Israeli command and control and slow the movement of reserves.

General Ismāʿīl, the Egyptian chief of staff, took particular interest in the air force's role in attacking Israeli command and control centers in the Sinai. A second major EAF mission was the flying of fighter patrols over Egyptian territory to defend against Israeli air assaults. The EAF and EADF had worked long and hard to develop effective coordination procedures. These included safe corridors for returning strike aircraft and identification friend or foe (IFF) confirmation before firing SAMs.

Another important assignment was the lifting of commandos into the Sinai by EAF helicopters. Egyptian special forces planned to ambush Israeli reinforcements and create confusion. By design the EAF was assigned a limited role during the initial phase of the conflict. This was done because interceptors were needed to defend Egyptian airspace. The Egyptian bomber and fighter-bomber force would also be available for use in later phases of the fighting.[15] Both Egypt and Syria now embarked on a deception campaign to confuse the Israelis. Rumors were spread that the Egyptian Air Force and Air Defense Force were in poor technical shape and that the army was short of spare parts.

During a meeting of senior Egyptian and Syrian commanders on 2 May 1973, the date for the assault was decided upon. Final plans for the attack were made in August 1973. The assault was to take place on the tenth day of Ramadan, 6 October 1973. This was also Yom Kippur, the most important Jewish holiday of the year.

On 12 September 1973 a fierce air battle took place over the eastern Mediterranean when the IAF lured two squadrons of Syrian MiG-21s into an ambush. A dozen Syrian MiGs were downed for the loss of one Israeli Mirage. Israel heightened its state of alert on both fronts and sent an additional brigade of tanks to the Golan Heights. By 1 October Egyptian and Syrian forces were ready for war. War plans were camouflaged by the fact that Egypt annually held a major military exercise in October and November and that a Syrian military buildup was to be expected after the recent major air confrontation.

On the eve of what became known in Egypt as the Ramadan War, the EAF had a personnel strength of 23,000 men, including some 730 pilots and nearly 800 aircraft. Arab allies sent to Egypt two Libyan Mirage units, three Algerian squadrons (one each of MiG-21s, Su-7s, and MiG-17s), a Iraqi Hunter squadron, and a detachment of North Korean MiG-21 pilots.[16]

The EADF had prime responsibility for the air defense of Egyptian forces and airspace. EAF MiG-21 squadrons were assigned to support the massive air defense network. In 1973 Egypt was divided into four air defense zones. Each of these had its own air defense division and a command center that directed all interceptor, missile, and artillery operations. Some thirty air defense brigades and several dozen radar regiments were attached to the four air defense divisions. Radar sites located throughout the country and a ring of observation posts along the borders provided warning of approaching aircraft. MiG-21 interceptors were the first line of defense. They could patrol over the Mediterranean, Red Sea, and Great Bitter Lake where there were no air defenses and rapidly move to augment ground defenses.

Missile batteries provided defense against Israeli aircraft at medium range and all altitudes. Shoulder-fired SA-7 missiles and anti-aircraft guns were used for close-in defense. This air defense screen would be the primary defense against the IAF. In 1973 Egyptian defenses included some seventy-five SA-2 batteries, sixty SA-3 batteries, and fifteen mobile SA-6 units. Supporting these medium-range missiles were more than 2,100 anti-aircraft guns ranging in size from 14.5 mm to 100 mm. Many of these systems had their own radar fire control. Front-line Egyptian Army units fielded dozens of ZSU-23–4 self-propelled guns and more than 5,000 SA-7 missiles.[17]

At the last moment the Israelis realized that the Arab buildup meant war was about to begin. But it was too late for Israel to mobilize its ground forces, and Prime Minister Golda Meir elected not to authorize a first strike by the IAF.

## Air Order of Battle, 1973, Egyptian Front[18]

**Fighters**

| Egypt | Algeria | Libya | Iraq | Israel |
|---|---|---|---|---|
| MiG-21 × 210 | MiG-21 × 20 | Mirage × 30 | Hunter × 20 | F-4 × 120 |
| MiG-17 × 100 | MiG-17 × 15 | | | A-4 × 160 |
| Su-7 × 80 | Su-7 × 20 | | | Mirage × 70 |
| Su-20 × 20 | | | | Super Mystère × 16 |

**Bombers**

| | | | | |
|---|---|---|---|---|
| Tu-16 × 25 | | | | |

**Support**

| | | | | |
|---|---|---|---|---|
| Mi-4 × 40 | | | | Noratlas × 20 |
| Mi-6 × 40 | | | | C-47 × 15 |
| Mi-8 × 80 | | | | Stratocruiser × 10 |
| An-12 × 30 | | | | C-130 × 2 |
| Il-14 × 40 | | | | Super Frelon × 12 |
| L-29 × 100 | | | | CH-53 × 12 |
| | | | | UH-1 × 45 |
| | | | | Alouette × 5 |

21

# Ramadan War (1973)

> It was a local war in which only conventional arms would be used; that it would have decisive strategic aims to upset the balance in the region and shatter Israel's theories and strategic mainstays. . . . I want to stress from the start that it is a purely Egyptian strategy that was neither imported from the East nor the West. It was drawn from the bitter reality through which Egypt and the other Arab nations lived after the 1967 defeat.
>
> —General Muḥammad al-Jamasī, Egyptian Deputy Prime Minister[1]

War started on 6 October for a number of reasons. It was the Day of Atonement, Yom Kippur, a very important Jewish festival. The Arab High Command assumed that the mobilization of Israeli reserves would be disrupted by the restrictions surrounding the religious holiday. It was also the tenth day of Ramadan for Muslims, the day on which the Prophet Muḥammad had begun his preparations for the battle of Badr, the battle that enabled him to return in triumph to the holy city of Mecca. Hence the choice of the name "Operation Badr" for the Egyptian assault into the Sinai.

Achieving surprise was a critical element of Arab strategy. During the first days of October a well-planned disinformation campaign reached its peak. Egyptian newspapers announced that hundreds of army officers were given leave to go on the ᶜUmrah, or Little Pilgrimage. By day Egyptian and Syrian troops casually sat in the sun or played in the Suez Canal. At night SAMs and tanks were moved up to the front. In Israel, press coverage of the Palestinian raid on Soviet emigrants heading for the Schonau transit camp in Austria captured the attention of the public and senior leadership. Israeli intelligence did detect many signs of an Arab buildup for war, but these were misinterpreted.

On 5 October Israel confirmed that the families of Russian advisors were being withdrawn by airlift from both Egypt and Syria. Agents also detected that Syrian Army field bakeries had moved out of Damascus up to the border. Only then did the Israeli leaders realize that war was about to start.

On the morning of 6 October, the Yom Kippur holiday, Israel ordered a partial mobilization and brought the air force to full alert. Lt. Gen. David Elazar, the Israeli chief of staff, spoke out in favor of an immediate preemptive air assault against Egyptian and Syrian forces massed along the border, but Israeli Prime Minister Golda Meir did not approve this request. She had been warned by U.S. political leaders that America would not provide support if Israel was again seen as the aggressor.

Israeli leaders also knew they would have little success against a prepared foe. The Egyptian and Syrian air forces were armed and ready for battle, safe in their hardened shelters, and their ground forces were protected by a dense air defense shield.

## EAF Missions Critical to the Success of War Plans

The EAF was assigned to perform numerous missions during the 1973 War including strike, reconnaissance, logistics support, fighter escort, and air defense. Disruption of Israeli defenses and mobilization plans was the most critical for EAF fighter bombers and Tu-16 bombers at the start of the conflict. Egyptian war plans also called for continued hit-and-run strikes into the Sinai during the first week of the war. EAF and Arab squadrons equipped with multi-role MiG-21MF and Mirage fighters flew a mix of attack, intercept, and fighter escort missions.

Egypt planned to rely on its massive air defense shield for protection against Israeli air attack over the battle zone and vital rear area positions. The 100,000-man air defense force fielded some 150 SA-2, SA-3, and SA-6 missile batteries, more than 5,000 SA-7 shoulder-fired SAMs, 2,000 anti-aircraft guns, and hundreds of radars.[2] Egyptian and allied Arab fighter squadrons worked closely with EADF radar, missile, and gun units to defend the skies of the country. Intercept support for the EADF was the primary mission performed by MiG-21 squadrons. These Egyptian MiG-21 squadrons were assisted by Algerian and North Korean Fishbed squadrons. Egypt also assigned MiG-17, Su-20, and Mirage V squadrons to air defense missions.

Col. Muḥammad al-Ibrāhīm commented: "I flew the Su-20 in the Ramadan War. This jet has a better engine than the Su-7, and we carried Atoll missiles. At the start of the war we flew patrol missions from Banī

Suwayf armed with Atolls. Later in the war we moved to the ground attack role."[3] A portion of the EAF tactical force was held back as a reserve to be employed later in the conflict if needed.

Long years of training and close coordination between the EAF and EADF paid off. Pilots knew from training and war plans where it was safe to fly and what areas to stay away from. All EAF aircraft were equipped with IFF systems to reduce the risk of accidental fratricide. In addition, EADF and EAF officers worked together at district air defense command centers to coordinate the air battle.

## EAF Jets Strike the First Blow, 2:00 P.M., 6 October 1973

> As chief of staff, I helped plan these initial strikes. More than 200 aircraft crossed the Suez Canal at the same time coming from different aerodromes with the aircraft traveling at different speeds and coming from varied distances. They all crossed the canal at the same time. After the jets crossed the canal, within two minutes the army started its fire preparation with artillery. On their way back we had corridors for the jets to come back through, and every squadron went to their own aerodrome.
>
> Believe me, at many aerodromes the first sorties were done before the mechanics and other men knew that the war had started because they used to do it many times for practice. This time when the pilots took off they flew east into battle. When the pilots came back, they were astonished that the mechanics met them as usual—they didn't know it was war.
>
> —Maj. Gen. Muḥammad Nabīl al-Masīrī (ret.)[4]

The 1973 War was very different from previous Middle East conflicts. This time the Arabs struck first. EAF MiG-17s, MiG-21s, Su-7/20s, and Iraqi Hawker Hunter fighter-bombers swept into the Sinai, while Tu-16 bombers fired AS-5 Kelt air-to-surface missiles against Israeli positions. To maintain the element of surprise, aircraft flew from their regular operating bases. Many pilots did not know they were going to war until they were briefed for the mission on the morning of 6 October.

Arab aircraft hit the Israeli air bases at Biʾr al-Jifjāfah, al-ʿArīsh, Raʾs Naṣrānī, Biʾr al-Thamādah, Biʾr Ḥammah, Balūza, and al-Ṭūr. Fighter-bombers accurately delivered special Egyptian-built bombs—known as dibber bombs—which dug large holes in runways. These 250 kilogram weapons, built in Egyptian state factories, employed a parachute to slow them down and ensure that they fell in a vertical position. Then a rocket fired to drive the bomb into the runway. The warhead exploded after a short delay. Airfield facilities were also plastered with conventional bombs, rockets, and cannon fire. In addition, Arab jets also bombed ten Hawk missile sites; two batteries of 175 mm long-range artillery; three radar command and control sites; two ECM bases; and three logistics areas and a strong point south of Port Fuʾād.[5]

Col. ʿĀdil Sharārah described the mission:

> I flew in the opening strike of the 1973 war. It was a raid against a Hawk site near Mitlā Pass. We went in with twelve MiG-17s. I flew with the second group of six because I was only a captain at that time. We flew from Quwaysina air base. At the initial point we broke into two groups, and the lead group struck first. [This was an example of the importance of tactics as practiced at ranges in the desert against "Israeli" targets. EAF pilots still had no ECM, chaff, and were using low-accuracy weapons like bombs and rockets]. They suffered some losses because they were more exposed. Then I made my attack, and it was very easy, no problem. Then we flew back to base.[6]

Brig. Gen. Muṣṭafá Ḥāfiẓ (ret.) remarked:

> In 1972 I converted my squadron from the Su-7 to the MiG-21MF. We fought the war and flew from Inshāṣ. . . . I led my squadron in the initial attack against Biʾr al-Thamādah airfield. We had sixteen aircraft in the first wave. Our MiG-21MFs used anti-runway bombs, and we flew a low-level attack. However, on the way in I had to climb up at least twice flying toward Biʾr al-Thamādah airfield because the MiG-17s were attacking the Hawk sites. I couldn't help it, I had to climb up to 150 or 200 meters so as not to obstruct their attacks. We met with no opposition that I could tell at the base. We hit the airfield, and the runways were damaged, definitely. We had sixteen aircraft each armed with four runway dibber bombs. And with a dibber bomb you really don't have to aim—you just fly over the runway and let them go. They hit the runway and the subrunway. We dropped them along the whole length of the runway. These were Egyptian bombs we used.
>
> Gladly enough we went in, and all sixteen aircraft came back from the mission. My squadron after that was multi-role; we continued to fly both air-to-air and air-to-ground missions.[7]

Col. Aḥmad Wafāʾī, another MiG-21MF pilot, flew in support of a similar strike against a target in the southern Sinai: "We were based at Wādī Qina (Wādī Abū Shīhāt). My first mission was ground attack against Raʾs Naṣrānī air base. The mission included sixteen MiG-21MFs, and I was the leader of two flights that were targeted to attack the airfield. We were armed with anti-runway bombs and rockets. We attacked the airfield, and after that we flew protection for some naval vessels in the Gulf."[8]

Simultaneously, on the Golan Heights more than a hundred Syrian aircraft bombed and strafed Israeli command centers, anti-aircraft batteries, and logistics sites. Syrian MiG-17s, Su-7s, and MiG-21s again caught the Israelis by surprise, and only one of the attackers was lost to Israeli ground fire.[9] Syrian commandos carried by Mi-8 helicopters also quickly captured the important Israeli observation post on Mount Hermon.

The Syrians were also supported by a regiment of Egyptian Air Force MiG-17s based at al-Mazzah. This force, which was led by EAF Brig. Gen. Muḥammad Fakhī al-Jindī, flew air defense and ground attack missions throughout the war.

## Assault into the Sinai, 2:15 P.M., 6 October 1973

Just after the waves of Arab fighter-bombers passed into the Sinai, the front erupted when nearly 2,000 guns opened fire on Israeli targets. In the first minute some 10,000 shells were fired. While the half-hour barrage continued, some 8,000 Egyptian troops crossed into the Sinai. Egyptian forces assaulted the many Israeli strong points of the Israeli Bar-Lev line, and by midnight fourteen Israeli strong points had been captured. Other troops infiltrated around Israeli strong points, dug trenches, and then set up their guns, portable anti-tank and anti-aircraft weapons, to meet Israeli counterattacks.

Army engineers rapidly set up ferries and Soviet-designed PMP pontoon bridges to carry tanks and other heavy vehicles across the waterway. High-pressure hoses sprayed water against the sandbanks on the Israeli side of the canal to cut openings for these bridges and ferries. In many places the large sand banks had been heightened by the Israelis to form a more formidable barrier against amphibious assault.

Egyptian fighter-bomber pilots and Tu-16 crews had flown thousands of practice sorties to prepare for this assault, and their training clearly paid off. The airfields at Biʾr Jifjāfah, Raʾs Naṣrānī, and Biʾr Ḥasanah were hit hard, and several command centers and radar sites were put out of action. Egyptian air attacks also destroyed a number of Hawk sites and several batteries of Israeli Army 175 mm artillery guns.[10]

EAF strikes against air bases targeted runways in order to hinder the response of the IAF and buy time for the Suez Canal crossing. Delayed-action bombs, exploding over a period of several hours, added to the chaos and slowed Israeli efforts to repair damage caused by the raids. Egypt was successful in this effort.

While these attacks caught the Israelis by surprise, F-4s and Mirages on alert at Raʾs Naṣrānī and Refidim did manage to scramble. The IAF fighters claimed seven Egyptian aircraft, while additional losses were suffered to Hawk missiles and anti-aircraft fire.[11] However, it was some time before the IAF went into action against the Egyptian bridges.

In his book *In Search of Identity,* President Sadat said of these air strikes: "The Egyptian Air Force recovered all it had lost in the 1956 War and the 1967 defeat and paved the way for our armed forces subsequently to achieve the victory which restored the self-confidence of our armed forces, our people, and our Arab nation." He added that only five EAF aircraft were lost during the initial series of air raids.[12]

Egyptian and Iraqi jets flew attack missions throughout the remaining hours of daylight. Arab fighter-bombers and their fighter escorts returned from their missions through air corridors and registered special IFF codes in order to pass safely through the EADF's air defense wall along the Suez Canal and Red Sea. More than a dozen AS-5 Kelt air-to-surface missiles were fired by Tu-16s on the first day of fighting. Several of the Kelt missiles were shot down, but at least five hit their targets. Some Kelts also had anti-radiation seekers, as two of the missiles struck Israeli radar sites in the Sinai and destroyed them.[13] The air strikes and massive ground assault inflicted losses and disrupted Israeli forces in the Sinai for the key first day of the Egyptian assault.

Iraqi Hawker Hunters participated in the initial strike and made three attacks on 6 October. Lt. ʿAbd al-Qādir Ḥādīr from Mosul, Iraq, did not expect to be involved in a war. He thought his unit would fly exercises with the EAF for several months before returning home. His first strike against a Hawk battery went well. However, on his second sortie he was hit and had to eject over the Sinai and was captured.[14]

## Lifting Commando Teams into the Sinai, 6 October

Egyptian transports and helicopters played an important role in the fighting. As dusk fell on the first day of the war, dozens of Mi-8 helicopters carrying troops and supplies flew into the Sinai. A formation of eighteen Mi-8s headed for Raʾs Sudr was intercepted by IAF F-4s and Mirages. The Egyptian helicopter pilots flew at very low altitude and performed evasive maneuvers to escape the attacking Israeli fighters and avoid anti-aircraft fire. Like sharks, however, the fighters swept through the Egyptian formation, and many Mi-8s were shot down. The Israelis claimed to have downed twenty Egyptian helicopters during the first day of fighting.[15]

Israeli soldiers reported seeing Il-14 transports drop supplies and commando teams by parachute near several Israeli Bar-Lev positions.[16] The commandos delivered into the Sinai by helicopter and transport disrupted Israeli operations and interfered with the forward movement of reserves. These special units fought until they were overrun or forced to surrender, or returned to Egyptian Army positions.

Egyptian leaders expected to suffer losses during the initial air assault. Only the small number of Tu-16s that could fire their Kelts from within the air defense shield were safe from Israeli defenses. Arab fighter-bombers did not have ECM pods nor chaff/flare systems to decoy or defeat Hawk missiles or AAA tracking radars. Instead pilots reduced their vulnerability by using the element of surprise and by their use of low-level tactics. Calculated risks had to be taken since it was vital to disrupt Israeli command and control and to slow the movement of Israeli reinforcements to the front.

The Egyptians admitted the loss of ten jets during the first day of combat.[17] Among the EAF pilots lost was President Sadat's half-brother, Capt. ᶜĀdil Sādāt. His plane was hit by Israeli anti-aircraft fire and exploded while he was attacking a target in the Sinai.[18] Egyptian helicopters carrying commandos suffered the heaviest casualties. Just before dusk a pair of MiG-21H reconnaissance jets sped over the Suez Canal and then turned south when they reached the Red Sea. Their task was to collect a full photographic coverage of the Sinai battle zone. These photographs were used to devise future battle plans.

## Israeli Jets Swing into Action, 6 October

Israel had a geographical advantage that had certainly proved its defensive worth again and again during the War of Attrition. From the mountains of the Sinai, Israeli radars and ECM detachments tracked nearly every EAF sortie near the border and monitored all Egyptian radio and electronic activity. Israeli controllers could vector Mirages or F-4s to intercept EAF raids, alert Hawk sites, and direct IAF strikes to evade

MiGs. These elements continued to work in Israel's favor also during the 1973 War.

In the late afternoon, 6 October, Israeli F-4s and Skyhawks began striking back. The Suez Canal crossing points were naturally prime targets, and the bridges, ferries, and vehicles gathered to cross the waterway began to suffer. One such IAF strike killed the deputy commander of Egypt's Engineer Corps who was crossing a pontoon bridge. Israeli aircrews flew into a storm of missiles and anti-aircraft fire. The IAF had faced this air defense wall during the War of Attrition, but by 1973 the EADF air defense shield was more dense and had been enhanced through the addition of new weapons like the SA-6 SAM, ZSU-23-4 mobile anti-aircraft gun, and SA-7 shoulder-fired missiles. Egypt claimed that twenty-seven Israeli jets were shot down on the first day of the fighting.[19] According to Israeli records, the IAF flew nearly 200 sorties on the Egyptian front on the first day of the conflict and lost only four aircraft.[20]

## The IAF Strikes Back, 7 October

On the morning of 7 October the IAF struck at Egyptian air bases including Banī Suwayf, Biʾr Arādah, Jiyānklīs, Khutāmīyah, and al-Manṣūrah in a massive effort to destroy the EAF on the ground. The strike force consisted of F-4s and A-4s, escorted by Mirages. More than fifty Israeli jets swept in low from over the Mediterranean in several waves. The Egyptian coastal observers detected the low-flying intruders, giving the Delta air bases warning time to scramble dozens of MiG-21s and prepare missile and gun defenses. Some attackers came in as low as 10 meters in an effort to escape the SAMs and fighters. Most of the Israelis made it to their targets, but fierce defensive fire punished the attackers. Large air battles developed over the Delta as MiGs attacked the IAF aircraft. Many Israeli crews were forced to drop their bombs and withdraw, and several were shot down.

The lethal web of SAMs, radar-directed guns, and MiGs cost the Israelis at least five aircraft, and they inflicted only limited damage on their targets. From this time on, the IAF no longer included A-4 Skyhawks in its strike forces against Egyptian rear area targets. Nevertheless, Israeli pilots fought aggressively, and a number of EAF aircraft were shot down. In the air defense role, the EAF enjoyed several notable successes during the day.

Brig. Gen. Qādrī al-Ḥamīd (ret.), who fought in this large air battle describes the scene:

> In 1973 I was stationed at al-Manṣūrah. I engaged in an air combat on the seventh of October during the first big Israeli raid against us. My wingman shot down an F-4 that was right behind me with his cannon. The crew went to al-Manṣūrah hospital because they were hurt in the ejection.[21]

Maj. Gen. Muḥammad Nabīl al-Masīrī, EAF chief of staff during the conflict, commented:

> In the 1973 War they never succeeded in stopping any of our airfields from operating except for al-Manṣūrah for several hours. Why? They didn't have the initiative. We had many airfields. To be successful, you had to stop all of our airfields at one time by grounding the aircraft. All of our aerodromes had several runways and taxiways that could be used. They couldn't put our airfields out of action. We had hardened shelters and barrage balloons all around the airfields. These stopped them from making low-altitude attacks against our airfields. The balloons were up at 500 meters and were held by heavy wires.[22]

Israeli jets also flew dozens of strike missions against Egyptian military positions near the Suez Canal and bridges across the waterway. While Israeli bombs hit home and several pontoon bridges were damaged, they were quickly repaired and the IAF suffered heavy losses to Egyptian SAMs and anti-aircraft guns. On the second day of action the EAF flew dozens of raids into the Sinai. A number of Egyptian jets were lost in action to Israeli fighters and air defense weapons. Meanwhile Egyptian MiG-21 interceptors fought many battles with attacking IAF planes and claimed several victories over Israeli jets. Following the conflict, Israeli spokesmen admitted that on 7 October the IAF suffered its greatest one-day losses ever, twenty-two aircraft, as it was forced to fight on both fronts to stem the Arab assault.[23] The Israeli Army also suffered heavy losses in the Sinai with more than 200 tanks knocked out during a series of counterattacks against the Egyptian bridgeheads.

## Battering Your Head against the Wall, 8 October

Overnight and continuing on the third day of the conflict, the IAF again attempted to destroy the bridges spanning the Suez Canal. IAF bombs did hit, knocking out several bridges. However, Egyptian engineers rapidly replaced the damaged portions of the bridges, while Israeli air losses were again very heavy during these strikes.

In fact, Egyptian defenses were so successful that late on the third day of the war (8 October) the IAF commander, Lt. Gen. Benjamin Peled, ordered his pilots to try new tactics to minimize losses and not to approach within 24 kilometers of the Suez Canal unless they were on a priority mission. For the first time a modern army had used an integrated air defense screen to survive in the face of heavy air attacks by a first-rate air force. After failing to destroy the bridges or knock the EAF out on the ground, the IAF tried to destroy missile batteries at the ends of the air defense screen. These raids were also designed to bring EAF interceptors out over the water where they could be ambushed. Most of the Israeli attacks were flown against the air defenses in the north near Port Said, but dozens of strikes also hit Egyptian missile positions in the Suez area. Here EAF fighters patrolled the northern and southern flanks of the main battle front to prevent Israeli jets from accomplishing these missions.

Col. Samīr ʿAzīz Mīkhāʾīl (ret.) flew one of these air defense sorties:

> In 1973 I was second in command of a MiG-21 squadron. On 8 October near last light there was a scramble. I took off and headed toward Port Said to fight with eight Mirages. We had a finger four of MiG-21MFs. Before we engaged I told my pilots to eject their external fuel tanks, but mine hung up because of an electrical failure. I was flying with three tanks but, since I was in the lead, I continued on. They hit Port Said a lot then. I saw that four Mirages were coming at us from each side, and we were sandwiched. So I decided to make a head-on pass against the Mirages. We started to make maneuvers, and my No. 2 was dead about thirty seconds after that. They shot him with a cannon. My No. 3 started maneuvering with them, and I was alone with all my fuel tanks like a sitting duck. Behind me I saw two Mirages coming to shoot me with their cannon. So I made a very hard turn to the right, and one of the Mirages came out in front of me. I reversed my turn to shoot him. I tried the gun but it didn't work. I thought, something is very wrong with my aircraft—the tanks won't separate, the gun doesn't work, and the missiles don't fire. I know they will kill me for sure; everywhere I looked I saw a Mirage. I decided to make a crash with one of them. If I am sure to die, I will take one of them with me. So I tried to hit Mirages twice, but they were very clever at escaping. Then I decided to quit and made a steep dive down toward my base. After I pulled out I said to myself, why did you leave your No. 3 and No. 4 alone? I made a combat turn toward them, and at that moment I felt an impact like a bus hitting a bicycle. A missile hit me hard.

> I lost control, and I saw in the mirror that the fuselage and wings were on fire. The aircraft started to rotate and I ejected, but my back was broken. I came down in a lake, and an Egyptian fisherman came up and started to beat me with his oar because he thought I was an Israeli. I shouted and insulted him, and he finally figured out that I was Egyptian. So they pulled me out and took me to the shore and then on to the hospital. I feel that if I didn't have the three tanks hung up and the electrical problems, I could have fought well and shot at least one of them down.[24]

Meanwhile Egypt sent dozens of aircraft into the Sinai and bombed Israeli radar sites, command centers, and armored columns on the third day of the war. Egyptian MiG-21s escorted these strikes and fought with Israeli fighters. Both sides suffered many losses in these air battles.

## Continuing Air Action, 9–10 October

Egyptian jets continued to send strikes into the Sinai. Nevertheless, after several days of intense fighting, the EAF's sortie rate was reduced. Egypt elected to husband its air power as the EAF was still short of pilots, skilled support crews, and spare parts for Soviet-built aircraft. While fighter-bomber sorties were reduced, many effective missions were flown. Maj. Gen. Avraham Adan (ret.), an Israeli division commander in the Sinai during the Ramadan War, remarked about EAF fighter-bomber operations:

> In a show of daring [on 11 October], the Egyptians dispatched planes that carried out short, low altitude sorties over our lines. Two of these planes hit a point on the Ma'adim Road where four tanks from one of Natke's battalions were reloading with ammunition; two platoon leaders were killed and a crewman wounded.
>
> At 1400 hours two other enemy aircraft bombed a point that was some 15 km east of the front line, on the Ma'adim Road. This was a working site for one of our ordnance companies, repairing tanks. Fuel trucks were there also, as well as vehicles loaded with ammunition and nearly 200 men. The dunes made it difficult to disperse the vehicles, so the company was crowded in an area close to the road. The men overconfidently thought that our air force was in control of the skies and that the Egyptians would not dare send their planes into our territory. This was a serious miscalculation, and we paid dearly for it: fuel trucks caught fire, ammunition began exploding all around, and eighty of our men were wounded.[25]

EAF intercept and fighter patrol sorties were increased to counter Israeli raids against air bases and screen the air defenses around Port Said which had taken a pounding. On the other hand, many MiG-21s fell in action with IAF F-4s and Mirages as the Israelis repeatedly used ambush tactics to inflict losses on the EAF. However, not all air engagements resulted in one-sided claims. Egyptian pilots shot down numerous Israeli aircraft, and these claims were supported by displaying captured IAF aircrew.

The EAF commander in chief, Air Marshal Husni Mubarak, visited with one Israeli survivor who had been shot down during a dogfight over al-Manṣūrah. "What has happened to the standards of your air force," the EAF chief asked the pilot in English. "It is not us," the Israeli replied, also in English, "I think you have changed."[26]

Israeli fighter-bombers flew hundreds of raids against targets around Port Said, Suez, and near the Suez Canal on 9–10 October. After that, most IAF sorties were sent to the northern front to assist in a successful counterattack against Syrian forces.

Col. Aḥmad Wafāʾī, who flew MiG-21s during the conflict, remarked: "Israeli performance at the beginning of the war was not good. We had the initiative. They were trying to avoid dogfights, and their way of striking our air bases was bad. They weren't very determined."[27]

## A Week of War, 6–12 October

The coordinated Egyptian and Syrian surprise assault had driven back the Israeli armed forces. Egypt had challenged Israel and captured territory in the Sinai. During the first week of the Ramadan War the Egyptian Air Force and Air Defense Force had fought with honor and inflicted serious losses on the Israelis.

Lt. Gen. Muḥammad ʿAlī Fahmī (ret.), chief of staff, Egyptian armed forces and former commander of the EADF, remarked:

> There is no doubt that Israel had actually sustained a big loss in its planes and a bigger and worse loss of the elite of its pilots on whom it has spent long years in training and in their preparation. We have actually felt the great effect of this loss when the lower level of Israeli pilots became prominent obviously after the first days of the war. Despite these great Israeli losses, we believe that the greatest loss it has sustained from the Egyptian Air Defense Force is

> the psychological shock to its Air Force high command and its pilots and the fact that their self-confidence has been shaken.[28]

Egyptian pilots had fought dozens of engagements, including seven major air battles, and shot down several IAF jets. One of these battles involved some seventy EAF aircraft. EAF helicopters, transports, and support troops also contributed significantly to the course of the battle. Nevertheless, during this week of intense fighting the EAF lost eighty-two jets and helicopters, and many other aircraft were damaged.[29] Despite such casualties, Egyptian morale soared because of the successful assault into the Sinai. The Egyptian armed forces were proud to have avenged the defeat of 1967.

# 22

# Battles for the Bridgeheads (1973)

After suffering severe losses in the air and on the ground, Israeli forces regained the territory seized by the Syrians on the Golan Heights and pushed further into Syria. There were, however, no more lightning Israeli advances, only a series of savage and hard-fought battles. On the Sinai front the initial series of Israeli counterattacks were costly failures. Then Egypt launched an offensive on 14 October to help ease the pressure on the Syrian front. Egyptian attacks against well-entrenched Israeli defenses made little progress at the cost of heavy air and ground losses. Israel now moved to seize the initiative during the night of 15–16 October. At considerable cost, Israeli air and ground forces pierced the Egyptian line and crossed the Suez Canal just north of Great Bitter Lake. During the next week, Israel pushed 30,000 men and hundreds of tanks across, threatening the Egyptian heartland. Israeli jets flew thousands of sorties over the invasion force, providing air cover and paving the way for the advance with bombs. Israeli troops seized considerable Egyptian territory and tore a massive hole in the EADF's missile wall. Arab interceptors tried to plug the holes in the air defense screen and protect fighter-bombers during strikes against the advancing Israelis. But, while Arab air and ground forces inflicted serious casualties, the Israeli advance could not be halted. Egyptian and Israeli forces finally ceased hostilities on 24 October 1973. Although Israel had again fought to a position of advantage, the 1973 War had accomplished the main Egyptian and Syrian goals of destroying Israeli self-confidence and reshaping the regional political equation. In time the new reality would result in a peace treaty between Egypt and Israel.

## Hold in the Sinai—Withdraw on the Golan, 10–13 October

The primary Egyptian war objectives had been accomplished by the fifth day of the war, 10 October. Egypt had captured a sizable strip of the Sinai and defeated dozens of Israeli air and ground counterattacks. However, on the Syrian front, things were not going so well. Israeli armor had forced the Syrian Army back to their starting positions. The IAF was also hammering Syrian airfields, oil refineries, and electrical power stations in retaliation for FROG rockets attacks against positions in northern Israel.

EAF MiG-17 pilots, based at al-Milāzī air base near Damascus, joined Syrian aircraft and flew dozens of damaging attacks against Israeli units on the Golan Heights. As a result, several formations of Phantoms were sent to blast this base, but the field suffered minimal damage and several Israeli jets were lost. One EAF MiG-17 pilot was also credited with shooting down an F-4 over Damascus.[1]

While Israel concentrated its efforts in the north against Syria, things were not quiet in the Sinai. Israeli tanks continued to assault Egyptian positions, while the IAF flew large raids against EAF bases, Port Said defenses, and battlefield targets. The Israeli Army lost more than 200 tanks, and dozens of IAF jets were shot down during these costly counterattacks.

Over the next three days Egypt moved thousands more men, additional armor, and dozens of air defense batteries into the Sinai to prepare for an assault against Israeli forces. Meanwhile, to bolster the Arabs and replace losses, Russian transports airlifted critical supplies to Egypt and Syria. From 4 to 11 October, forty-four resupply missions had been flown to Syria and fourteen to Egypt.[2]

## Continuing Air Assault: Raids into the Sinai

Since the first day of the war, Egyptian, Iraqi, and Algerian pilots and EAF bomber crews had flown dozens of attack sorties each day against Israeli convoys and fortifications. Egyptian and Algerian Su-7s, EAF MiG-17s, and Iraqi Hunters bore the brunt of this assault, while Egyptian Tu-16 Badger crews fired Kelt missiles against Sinai targets and flew reconnaissance sorties over the Mediterranean and Red Sea. Egypt next stepped up its air attacks into the Sinai on 12 and 13 October, while the army prepared for action.

Col. ʿĀdil Sharārah commented:

> On 12 October I was flying a strike across the Suez Canal with six [MiG-17] aircraft flying at 750 km per hour at very low altitude to hit the Israelis. I looked back and saw a missile coming up from the ground, so I turned hard

> and my speed dropped from 750 to 300 km per hour. It was a very tight turn. I finished my turn, popped up, and dove to attack the target which was a collection of Israeli tanks. We had Su-7s with us and MiG-21 escorts. Then many Israeli Mirages attacked. These planes were painted different than usual. I later found out that they could have been South African because they had different paint jobs and the pilots used new tactics unlike the Israelis. They also had yellow tails and wing tips.
>
> I made my attack with rockets against tanks. The two new pilots flying with me got shot down. They were downed because they continued in the same direction and didn't turn. We flew out of Khutāmīyah [Birma]. Our mission was to help the Egyptian Army in the ground battle.[3]

Egyptian and Arab fighter-bomber pilots usually had to contend with heavy defenses during their raids into the Sinai. Lt. Col. Ṣidqī, who commanded the EAF's 52nd Fighter-Bomber Squadron during the 1973 War, spoke with English reporters following the conflict. He himself had flown dozens of reconnaissance and strike sorties over the Sinai during the War of Attrition and was shot down during one of these missions. His Su-7 squadron was located at Bilbays, which was the home of the EAF Air Academy. Two Egyptian and one Algerian Su-7 squadrons flew from this base during the conflict. Ṣidqī was forced to eject during a strike late in the Ramadan War after his Su-7 was hit in the tail by a missile.

Ṣidqī and his pilots rated Israeli fighters as the most difficult adversaries. However, since his pilots approached their targets at low level and were escorted by MiG-21s, they could usually sneak in, attack, and be on the way out before the Israeli fighters appeared. Hawk SAMs were also very dangerous. Ṣidqī recalled seeing a number of his friends fall victim to these missiles. Most Su-7 losses, according to Ṣidqī, were due to Israeli 20 mm and 40 mm anti-aircraft guns.[4]

The Soviet-built MiG-17s, MiG-21s, and Su-7s, which formed the backbone of the Arab strike force, were very reliable and rugged. During the October War many Arab aircraft limped back to their bases after suffering heavy battle damage. Even so, dozens of Arab aircraft and helicopters fell prey to Israeli fighters and air defenses. Many pilots parachuted to safety, and more than a dozen Arab aircrew were captured by the Israelis. Most were treated well, though unfortunately there were some exceptions, as ever. Several EAF and other Arab pilots who successfully ejected from their aircraft were, however, able to evade capture and walk to Egyptian positions in the Sinai. Egyptian efforts to recover downed pilots were, of course, hindered by the lack of a dedicated helicopter rescue force and heavy Israeli air defenses.

## Renewed Offensive, 14 October

On 13 October Egyptian fighters and missiles attempted to intercept a high-flying target, probably a USAF SR-71 reconnaissance aircraft, that overflew the Sinai at a speed of more than Mach-three. Meanwhile EAF fighter bombers flew a series of heavy raids against Israeli positions in the Sinai. At the same time a pair of MiG-21 reconnaissance aircraft made another detailed reconnaissance of the entire Egyptian bridgehead in the Sinai.

At dawn on 14 October, day nine of the conflict, 500 Egyptian guns blasted Israeli defensive positions. Nine brigade-sized Egyptian armored task forces headed toward the Khutmīyah, Jiddi, and Mitlā passes. Egyptian military leaders were aware of the low success potential of a frontal assault against well-entrenched Israeli tanks and artillery. Yet the worsening position on the Golan Heights meant that the attack had to go forward. The armored columns received air support from MiGs and Sukhois. This was also the first time that swing-wing Su-20s and Mirages were sent into action. Col. Rifʿat Fatḥī remarked: "During the war we had one big Mirage squadron at Ṭanṭá with Egyptian and Libyan pilots. They [the Libyans] didn't participate in the war. . . . We flew many air patrol missions and made our first attacks into the Sinai in the middle of the war."[5]

Egyptian armor surged forward and tried to penetrate Israeli defenses. On two axes they made progress but were forced back by Israeli counterattacks and air strikes. Fighter cover for the assault was limited because most of the interceptor force was assigned to home defense. In fact, as the assault moved forward, IAF jets were simultaneously attacking several airfields in the Nile Delta. Air support consisted of preplanned strikes against known Israeli positions as the EAF and Egyptian Army did not have a coordinated close air support capability.

Once Egyptian columns went out beyond the coverage of the bridgehead air defense shield, they were vulnerable to attack from the air. More than 200 Egyptian tanks and other vehicles were destroyed, while the Israelis suffered about fifty tank losses. Egypt claimed to have downed twenty-four Israeli aircraft, while Israeli officials said its forces shot down seven

Egyptian jets, including two Mirages.[6] As fighting flared in the Sinai, Israeli jets struck at EAF airfields at Ṭanṭá, al-Manṣūrah and Ṣāliḥīyah. Brig. Gen. Qādrī al-Ḥamīd (ret.) flew against one of these strike forces:

> On 14 October I engaged in an air combat. We were coming from a combat air patrol, and I was short on fuel. A wave of F-4s was coming to strike our base. They used to come and the first two would pull up and drop cluster bombs on us to keep the ack-ack gunners down. Then these F-4s were clean, and we got into a severe combat right over the base. It was a hell of a fight.
>
> Wherever I turned I saw a Phantom behind a MiG and MiG behind a Phantom. I pulled behind a Phantom and attacked with my gun—at the same time my engine stalled. I tried to restart but it would not, because of a lack of petrol. These pilots were really good—it was not the same standard of performance we saw at the start of the war. These pilots were much better, either they were foreigners or more experienced, higher-rank pilots. They lost the new, inexperienced ones against our forest of missiles along the Suez Canal. My cannon shells hit the Phantom, and it exploded like the sunlight right over the field near the maintenance shops.
>
> I had engaged in this combat for three or four minutes, which is a long time. To be frank, I didn't watch because once I fired and hit, the Phantom exploded, and I had my own problems. I wanted to make a forced landing to save the airplane, but I was crazy. If I had done it, I would have been killed because other Phantoms had hit the runway and it was full of holes. At 50 meters height I ejected. I got a compression fracture and was in the hospital for four or five days. Then I went back to the squadron, but I could not fly for the rest of the war.[7]

## Superpower Intervention

Another important development of 14 October was the arrival of the first USAF cargo planes loaded with weapons in Israel. Secret supply flights carrying ECM equipment, missiles, and other weapons had been going on using El Al civilian airliners and chartered jets since the second day of the war. The United States also began flying in F-4s and A-4s to replace those lost in action. U.S. Air Force units in Europe were stripped of their F-4s and the jets flown to Israel. More than forty A-4s were taken from U.S. Navy and Marine Corp squadrons. The U.S. also supplied Israel with a dozen C-130 transports, ten CH-53A helicopters, and large amounts of advanced weaponry including TOW anti-tank and Maverick air-to-ground missiles plus electro-optical and Laser-guided bombs.[8] The Soviet Union had been flying a similar airlift of critical war material to Egypt and Syria since the start of the conflict. By 15 October the Soviet Air Force flew some forty-two An-12 and sixteen An-22 transport sorties to Egypt.

The Egyptians noticed a distinct change in the combat styles of Israeli pilots after the first week of fighting. It is well known that IAF pilots tried new tactics to improve their survivability after the first phase of the conflict when the air arm suffered serious losses.

"We were fighting a new war," an Egyptian Air Force general commented, "new planes, new weapons and new tactics. As President Anwar Sadat said, 'we are prepared to fight Israel but not the United States.'"

There is a strong feeling among top Egyptian commanders that American pilots also flew many of the Phantoms they delivered to Israel in combat against the Egyptian and Syrian forces. "These Phantom pilots we met after October 17 had a much different style to their combat tactics than we ever encountered with the Israelis," one Egyptian air defense commander noted. "Whether they were volunteers, reservists, or regular U.S. military pilots we don't know, but they were certainly not Israelis."[9]

Egyptian pilots also noticed that Israeli Mirages sported unusual paint jobs including orange or yellow tails and wing tips. According to Israeli sources, the IAF added these recognition panels to their Mirage and Israeli-built Nesher fighters after it was believed that Libyan Mirages were being used in combat. EAF pilots also noted that late in the war some Israeli Phantoms had a green and tan camouflage that was easy to spot. These were the American F-4s that had been stripped from USAF units in Europe. The first of these jets were delivered on 16 October and were immediately pressed into service.

## Israeli Forces Invade Egypt, 15 October

Late on 15 October a bitter battle developed in the central Sinai, opposite al-Tāsā, which would eventually turn the tide of the conflict. During the night an Israeli force pushed through the Egyptian lines and reached the Suez Canal at the northern end of Great Bitter Lake. This force crossed the canal on rafts and assault boats. Egyptian military leaders initially were not aware of this Israeli crossing. They were concentrating on defeating what they thought was an attempt to roll back their bridgeheads. When they did become aware of the crossing, savage Egyptian counterattacks then led to a series of fierce battles fought in what became known as the "Chinese Farm" area of the Sinai, so called because an experimental agricultural center had been established there with Chinese help before the 1967 Israeli occupation.

Israeli units west of the canal now fanned out and attacked the Egyptian Army's supply lines and demolished portions of the air defense shield. The Israeli prime minister remarked in a 4:00 P.M. speech on 16 October, "Right now as we convene in the Knesset, an IDF task force is operating on the west bank of the Suez Canal." This confirmed reports Egyptian commanders had received about the crossing and eliminated any lingering element of surprise. Late on 16 October, Egyptian forces began plastering the Israeli bridgehead with intense artillery fire, and an Egyptian commando unit was hurriedly flown in by helicopter to defend rear area positions.

On Wednesday, 17 October, fierce fighting continued as the Egyptians assaulted Israeli positions on both sides of the Suez Canal with tanks, artillery, and air power. EAF jets bombed and strafed Israeli vehicles and targeted the crossing point. These strikes and artillery fire inflicted hundreds of casualties and disrupted the efforts of Israeli engineers who were trying to install a pontoon bridge across the canal. The Israelis defended the crossing site with dozens of anti-aircraft guns, while Mirages and F-4s patrolled over the area. More than a dozen Egyptian planes were shot down over this area which became known as the "Deversoir Gap."[10] Arab fighters were called on to defend against another series of heavy Israeli raids against Port Said.

Col. Ahmad Wafāʾī remarked:

> Israeli performance in dogfights got better and better starting on the fifteenth and sixteenth of October. Then we started suffering a lot of losses among the MiG-17s and Su-7s attacking in the bridgehead area.
>
> On 17 October I got in a big dogfight over Port Said. My leader took eight aircraft, and we got into combat with many Mirages, twelve to sixteen. I shot one Mirage, and my leader hit two. There was a lot of close combat. We started the combat from the sea, and we saw the Mirages coming from the east. I made a left turn and climbed to flight level nine [9 km]. This height gave us a good maneuvering advantage for my MiG-21 over the Mirage. We split into two pairs when we saw the Mirages. We were above them at 9,000 meters, and

> they were at 8,000 meters. We got into a close combat, and I shot down one Mirage with an Atoll missile. Then I went back and landed at Ṭanṭá again.[11]

## Arab Air Assistance

At a press conference on 17 October, Egyptian military spokesman Maj. Gen. ᶜIzz al-Dīn Mukhtār maintained that Egyptian air losses were only one-fifth of those suffered by the IAF. Egyptian air defenses and interceptors had claimed dozens of Israeli jets and helicopters, but the five-to-one ratio was, in reality, very wide of the mark.

By this point in the war the EAF had flown some 4,000 sorties. More than a hundred EAF aircraft had been lost in action. Combat attrition was biting deep into Egypt's limited corps of experienced aircrews. The EAF had long had to contend with having more aircraft than pilots. As a result of the heavy pilot and aircraft losses, the EAF was forced to send junior and senior staff officer pilots and aircrew into action.[12] Egyptian pilots, ground support personnel, and air defense crews were also exhausted after having fought almost nonstop for ten days.

Egyptian air units were, however, augmented by squadrons from many other countries. Iraqi and Algerian pilots had flown several hundred patrol and attack missions with a loss of about a dozen aircraft.[13] A North Korean MiG-21 squadron was also flying in Egypt. Based at al-Minyā, the unit included thirty pilots who flew Egyptian-supplied aircraft, plus support crews and air controllers. North Korean pilots flew many air patrol sorties during the October War and first clashed with the IAF on 18 October. Neither side reportedly suffered losses during this engagement.[14]

Brig. Gen. Muṣṭafá Ḥāfiẓ (ret.) remarked: "Look at the situation we faced when the Israelis crossed the canal and took the missile line with their infantry and tanks. We were really at a disadvantage—we knew this—we could only respond with aircraft. This made it an air battle. We could only attack with a limited number of aircraft at a time, and they were waiting for us."[15]

On 18 October at dusk a group of Egyptian Mirages swept in from the sea to raid al-ᶜArīsh air base where many American resupply flights were landing. Col. Rifᶜat Fatḥī flew this mission:

> We were intercepted just thirty-five seconds from the target by Israeli Mirages. I think that the Americans reported the strike as we flew over many ships in the Mediterranean, or they might have had their E-2C airborne early

> warning planes up. It would have been hard for the Israelis to detect us as we were flying very low over the sea. The Israeli Mirages intercepted us. They shot down one of our jets, another crashed into the water, and the others turned back. . . . After that, most of our missions were over the gap. We attacked the Israelis with 400 kilo bombs and 68 mm rockets.[16].

For more than a week EAF and Arab squadrons desperately tried to halt the Israelis and fill the expanding hole in the Egyptian air defense shield. Col. Muḥammad Ibrāhīm, an Su-20 pilot, flew several missions during this time:

> On 19 October at around 3:30 P.M. ten of us attacked Israeli tanks south of Ismāʿīlīyah with S-5K rockets. We were struck the moment we reached the target by Mirages and lost four aircraft. We did our best, but we were caught without warning and had no MiG-21 escort.[17]

Brig. Gen. Muṣṭafá Ḥāfiẓ (ret.) had a similar experience:

> On 19 October I was shot down while attacking the bridges over the gap. A surface-to-air missile hit me—I think it was a Redeye [or a captured SA-7]. I was at very low height, maximum speed, and we were attacking the bridges. We had eight MiG-21MFs from our regiment. We went in sections [of four aircraft] with a time gap of one minute.
>
> I saw three bridges, one was hit and partially sunk, the other we damaged, and the hard bridge was then being put in. We were carrying 250 kilo bombs. The area was loaded with Israeli Mirages. I told my Nos. 2, 3, and 4 to make only one attack. It was impossible to make multiple passes. My wingman didn't acknowledge the message, so I went in, released my bombs, and called him again. He said that he had not dropped his bombs, so I decided that we had to do another attack.
>
> At the initial point I saw a MiG-21 followed by three Mirages, so I decided to support this MiG-21. I was at a range of some 1,200 meters across from the MiG-21, and I tried to cut into them. Then I saw two Mirages pull up and the MiG-21 engulfed in flames, and he crashed. Then I was hit in the right wing, and I started to roll to the left. That was when I ejected. I was lucky, really, as I ejected at very low altitude. My chute deployed at a maximum of 50 meters, and I was concentrating on who was running toward me.[18]

Heavy fighting continued, and by 21 October more than 250 Israeli tanks and 10,000 men had crossed the Suez Canal. Egyptian forces on both sides of the Suez Canal fought aggressively, but they could not stop the Israelis. The IAF provided heavy air support for the Israeli advance because the Egyptian air defense screen had been shattered by ground attacks and air strikes.

## Desperate Defense, 16–24 October

Egyptian, Algerian, and Iraqi pilots flew more than 1,500 strike and fighter patrol sorties over the Deversoir Gap during the last week of the 1973 War. The situation was so serious that even helicopters and jet trainers were sent in to attack the Israelis. During a visit to the bridgehead, Israeli Defense Minister Moshe Dayan was nearly hit by a fire bomb dropped from an EAF Mi-8 helicopter.[19]

Maj. Gen. Aḥmad ʿIzzat was a squadron leader at the Air Academy during the 1973 War. He led his unit into battle over the gap:

> We planned to use the L-29 only in critical attack missions against the Israelis who crossed the canal in the gap near the Fāʾid area. We had two regiments of L-29s then, one at Bilbays and the other at Kawm Awshim. All the L-29 pilots were instructors at the Academy. They were very good pilots, especially in flying skills and ground attack because they had been teaching the cadets. At this time our regiment at Bilbays started to fly combat missions against the Israelis. The regiment commander at this time was Col. ʿAlī Ḥāda. During the six days our L-29s fought, we flew about twenty to twenty-four aircraft every day. The flight from Bilbays to Fāʾid was only six to eight minutes. We used the S-5K high explosive and S-5M anti-tank 57 mm rockets. We didn't have cannon or bombs on the L-29. The aircraft carried eight rockets in two pods.
>
> At this time the Su-7s were also stationed at Bilbays. Normally they would take off at the same time. There is a big difference in the speeds between the Sukhoi and the L-29. During the Sukhoi's return they would find the L-29s passing below them on the way to hit the tanks of the Israelis. I remember we lost about four pilots at this time killed, and we had a lot of damage to our aircraft from ack-ack, but normally we flew back to base O.K. As I mentioned, we flew combat missions for six days. Some pilots had three missions everyday, and others had from one to four each.
>
> We had Mirage V, Sukhoi, and MiG-21s with us, and they made a good umbrella to cover the L-29s. Normally we made one pass and released all our rockets in one salvo. Because they are small rockets, one or two would not do anything. We went against Israeli tanks, camps, and targets like that. Our job was to cut the supply line from the bridges. Some of our people had combat with Israeli fighters, and that is how we lost most L-29s.
>
> With this aircraft we had very good results because the pilots were very experienced and their accuracy of shooting was very good. The L-29 is good at low-level flying, very stable, and the gyro gunsight is very good. The critical point was after the attack when we had no armament. We could not fly very fast, about 450 km per hour. This is why we took off from Bilbays at the same time as the Sukhoi. He hit the target and came back after disturbing the

defenses and kept them in panic. We lost three of the four pilots in air combat with the Mirages and one to anti-aircraft fire.[20]

On 22 October 1973 the United Nations passed Resolution 338 calling for a cease-fire at sundown. However, after a short lull, both sides continued to fight. Israeli forces renewed the assault because they hoped to gain control of additional territory and were trying to cut off the supply lines of the Egyptian Third Army. Egyptian air and ground forces fought against these Israeli attacks.

Col. Aḥmad ʿĀṭif discussed his eventful mission of 23 October 1973:

I was on a free hunt over the Third Army. There were seven of us because my No. 2 had to make a ground abort because his afterburner didn't work. I was crazy because I went ahead and flew single ship. We passed over the Suez Canal near Kibrīt, and my No. 4 saw four Mirages, so he took the lead. . . . our Nos. 3 and 4 fought with the four Mirages.

I engaged an F-4 which came close to me. I think this pilot was not Israeli because of the way he used the aircraft. He fought with very good low-speed maneuvers, something I never saw the other Israeli pilots do. He made a vertical scissors, and I was with him until I reached zero airspeed in my MiG-21. Can you imagine an F-4 and a MiG-21F-13 going vertical, and the pilot of the F-4 was still vertical in full afterburner as I was stalling? He was a very good pilot. Finally he stalled, pointed down, and I saw him put his two afterburners on and escape to the east. I forgot about him because I saw my No. 4 approaching the Mirages very closely, and he opened fire with his cannon. Then I saw his shells hit the Mirage in a series of small explosions. At the same time I saw another F-4 making the last fine aiming on my No. 4. I yelled to my No. 4—break! break! There is a Phantom behind you! He didn't hear me because No. 3 was yelling on the radio. Then I saw a missile come from the F-4 which hit my No. 4, and there was a big explosion. After shooting the MiG down the F-4 made a very smooth combat turn to the left approaching me, coming from above. With his eight missiles the F-4 pilot must have decided to stay in the area to continue the turkey shoot.

I said on the radio, I am not leaving this guy because he just shot down and killed my best friend. I was sure that he died. I didn't know that he found himself in the air after the explosion and safely parachuted. It was easy for me to approach the F-4 at close range because he was making an easy turn into me. He must have lost sight of me. I climbed toward him and approached to within 150 meters. I opened fire with my cannon and saw white puffs from his aircraft. After that it caught fire. Then I felt my aircraft rocking, and I said "Oh, no," I thought someone was shooting at me. The rocking stopped, but I looked at the instruments and everything was out, the engine was at zero speed. When I fired the gun at low speed, I ingested the smoke and stalled the

> engine. I was at about 4,000 meters and I nosed down, and the engine started very slowly. Time stopped for me. I wanted the engine to start quickly because I knew someone would see my problem and come after me. That is exactly what happened. After the engine started I put the afterburner on, and then boom! Something hit me hard. I ejected and landed about 3 km from Binhah.[21]

The Egyptian armed forces surprised the Israelis, seized territory in the Sinai, and inflicted serious losses. The EAF (with support from several allied Arab squadrons) began the war with a massive air strike and continued to carry the fight to the Israelis throughout the conflict. Dozens of F-4s, A-4s, and Mirages were shot down by the EAF/EADF team, and the IAF was denied the freedom to operate over Egyptian forces. This defense was effective until Israeli tanks overran portions of the air defense shield.

Col. Ahmad Wafāʾī reflected on his role during the conflict:

> During the Ramadan War I flew about sixty combat missions. Only one was a strike. This was the one I did against Raʾs Naṣrānī at the start of the war. The rest were CAP missions. I had several air combats over the breakthrough area, and it was a mess. On one mission I was going to intercept four F-4s, and I saw them 25 km away because they were smoking. They didn't see me until I was within 7 km, and they turned into me and passed parallel. We made a snap shot as they passed, and I hit one of the Phantoms with cannon fire. After this attack I engaged two Mirages. I went down to 700 meters and attacked. I launched an Atoll which hit one of the jets. I left because I was running short on fuel. I was not a hero. I saw many other pilots do well, and we all fought the best we could.[22]

Col. ʿĀdil Sharārah reflected on the price paid by his unit for this success: "My squadron had twenty-four pilots, and after the war there were eight left. About four were able to eject and were captured, but the others were all dead. We lost sixteen aircraft to fighters and air defenses."[23]

The Egyptian offensive had shattered the status quo with Israel, restored its military self-respect, and created a climate that was to lead to peace. But the price had been high.

23

# Ramadan War Impact (1973)

When Israel continued to batter Egyptian forces following the cease-fire of 22 October, Soviet leaders threatened to send in Russian troops to end the fighting. The United States responded by demanding a halt to Israeli attacks and increasing the readiness worldwide of American military units, including nuclear forces. In spite of these events, Israeli air and ground forces kept pressing forward against the Egyptians. Israel now had the initiative when the second cease-fire took effect at sundown on 24 October. The final situation on the Sinai front saw the Egyptian Second and Third Armies still holding a large strip of the Sinai. But during two days of heavy fighting, Israeli forces had seized considerable Egyptian territory and surrounded the Third Army. One immediate impact of the 1973 War was the political and economic mobilization of the Arabs against the West. The Arab oil-producing nations had declared an oil embargo against the United States in retaliation for its support of Israel. This embargo and the dramatic rise in the price of oil created economic shock waves throughout the world.

## The Final Shots

On 22 December 1973, MiG-21s and Phantoms again clashed over the Gulf of Suez. Both sides claimed that they had inflicted a single loss in this brief air battle. It was to be the last confrontation between Egyptian and Israeli aircraft. Syria refused to end hostilities and continued to battle with the Israelis through the fall and winter of 1973. Fighting continued until May 1974 when a cease-fire was signed. The surprise Egyptian and Syrian attacks had, however, shattered the status quo. The success of the Arab offensive enabled Egypt and, to a lesser extent, Syria to negotiate from a position of strength.

The 1973 War had both a positive and negative impact on the Egyptian Air Force. Like the other military services, the air arm had suffered serious losses. However, these sacrifices had a reward in enhanced morale and self-confidence. With such a boost in confidence came a subtle change in attitudes. While relations with the Soviet Union improved slightly following the conflict, Egypt and the EAF increasingly turned to the West for political, economic, and military support. Links were reestablished with Great Britain, the United States, France, and other countries. These moves led to a revolutionary change in direction and orientation.

## 1973 Air War Lessons

Air Marshal Husni Mubarak stated shortly after the war that "Six years of training on our part paid off." Since the aircraft and equipment used by Egypt were inferior to Israel's U.S.-built systems, during the 1973 War the EAF used hit-and-run tactics, surprise, and numerical superiority to seize the initiative from the Israelis as often as possible. According to Air Marshal Mubarak, the EAF did well during the first part of the 1973 War. However, he admitted that the air arm suffered heavy casualties when it was forced to shoulder the responsibily for defense over the Deversoir Gap.[1]

The fighting in 1973 had again reinforced the lessons of the War of Attrition: when properly coordinated, SAMs, guns, and interceptors formed a very effective air defense shield. The Egyptian Air Defense Force and EAF fighters shot down many IAF aircraft and forced the Israelis on the defensive. Israeli pilots and navigators were initially shocked by their heavy losses and concentrated more on evading Arab defenses than delivering their weapons. This considerably reduced the accuracy and effectiveness of Israeli strikes.

American sources credited Egyptian ground-based air defenses with shooting down forty IAF aircraft.[2] The IAF admitted to flying 5,443 attack sorties on the Suez front during which fifty-two jets and helicopters were lost.[3] Reportedly thirty-three of these jets were shot down, and twenty-one aircrew killed while flying ground support missions.[4]

Egypt's air defense shield was also lethal to Arab aircraft. Col. Aḥmad Wafāʾī remarked: "One of our planes had to break off [from a dogfight with Israeli jets on 17 October] due to a hydraulic problem. When he tried to land at a small air base, our ground troops armed with an SA-7 Strella didn't realize he was an Egyptian. They shot him down, and the pilot was killed."[5] Due to the failure of IFF systems, inefficient command and control procedures, and the "fog" of war, many Egyptian, Iraqi, and Algerian jets were shot down by accident. Western estimates list forty-five to sixty Arab losses to friendly fire on both fronts. Many of these were lost over Syria where Iraqi and other Arab jets had different IFF systems.[6]

## Blunting Israeli Airfield Attacks

Starting on 7 October and continuing throughout most of the conflict, Israeli jets flew hundreds of sorties in an attempt to knock out Egyptian airfields and suppress the EAF. In a postwar interview, Air Marshal Mubarak remarked:

> In the course of the October War, the Egyptian Air Force dispelled the myth of invincibility of the Israeli Air Force. From the second day on, for seven days, Israeli attacks were heavy. We estimated on average, attacking waves were forty aircraft strong, directed against five bases. Against the al-Manṣūrah base, for example, the Israelis were sending up to 18 aircraft each day. Our defensive system did very well indeed. Against the Israelis we sent in first our MiG-21s as interceptors, then our SAM missiles and AA artillery. From the second day on the morale of our pilots rose, that of the Israeli pilots, who were frustrated in every one of their actions, began to sink.
>
> Our warning systems worked very well indeed. These are based not only on a comprehensive radar network, but also a very much simpler network of observers equipped with radio, who immediately report anything which flies to the data collection centers who sort out the most important reports. The joint anti-aircraft/air force headquarters take immediate steps to counter the threat in the most effective manner. As I have already said, the first line of defense would normally be the MiG-21s, which have given very good results. These are relatively simple aircraft, with a short range and pretty basic electronics; but they are immune from ECM, and are more maneuverable than the Phantom at altitude.
>
> Israeli pilots were faced with the choice of coming in at low level and running the gauntlet of our AA fire [which was even more dense than that put up in Vietnam], or coming in at altitude where they would meet, first our MiG-21s, then our SAM missiles. Their third choice was, of course, to retire; and after the first few days this was frequently done by Israeli pilots.[7]

Following the initial air assault at the start of the war, the EAF initially planned to keep most of its aircraft safe in the shelters and let the IAF suffer serious losses to the air defense shield. Most major EAF bases and other potential targets were ringed by SA-2 and SA-3 missile batteries and anti-aircraft guns. Some bases also had detachments of barrage balloons that prevented Israeli jets from making low-level attacks.

Repeated Israeli strikes against airfields and attacks against the missile screen forced the EAF to maintain fighter patrols over Egyptian territory. These defensive missions kept the EAF interceptor force very busy. While Israeli attacks did damage runways and taxiways at several airfields, these were quickly repaired and back in action within several hours. Aircraft shelters dramatically reduced the vulnerability of EAF aircraft on the ground. The IAF claimed to have destroyed twenty-two Arab aircraft by bombing. Most of these victories were, however, achieved at Syrian air bases. Nevertheless, several Egyptian MiGs were destroyed in their shelters by lucky bomb hits. In contrast, at least a dozen Israeli jets were lost during airfield strikes to SAMs, anti-aircraft fire, EAF interceptors, and fuel exhaustion.

Egyptian military leaders were fully aware of Israel's air power advantages. That is why they gave the air defense shield primary responsiblity to protect the ground forces and rear area positions from Israeli attacks. EAF MiG-21 interceptors filled in gaps and supported this air defense screen and protected reconnaissance aircraft and fighter bombers during raids into the Sinai.

Following the 1973 War an assessment of air operations was presented to the U.S. Senate. Lt. Col. John Corder, a U.S. Air Force intelligence officer who studied the lessons of the conflict, remarked:

> The (Egyptian) MiG-21 pilot was assigned to defend his homeland . . . essentially he was governed by Soviet doctrine, which says the controller on the ground will tell him what to do . . . when to take off, what altitude to fly at, what field to fly to, what weapons to attack [with]. He is not trained to any degree to initiate attacks on his own, although he may do so. The Egyptians had good radar coverage in most of the battle areas, and they did know when the Israelis were coming. However they did not know with sufficient accuracy to vector a big force to a long range intercept. This forced them into a point defense concept.
>
> So we found large formations of airplanes flying over the key target areas, trying to defend and intercept Israeli airplanes as they came in. The speeds at which the Israelis were penetrating made it very difficult. What this boiled down to was the Arab pilots were told to attack when they saw these Israeli

planes popping up . . . this is when we saw the large air battles of 24 MiGs versus 4 F-4s or 16 versus 8, things of that nature.

The Arab pilots, although in this kind of war they were exceptional, still had not received the kind of training in combat of this nature. He was aggressive, wanted to fight, but he was not trained to do the kind of things you have to do in order to succeed with the kind of weapons he had. It is this problem of being told what to do by the controller. As a result, you did not see the guy performing very well in the air to air combat arena.

So the situation was the Arab pilot was very eager; he stayed and fought. In fact, the Israelis, when they disengaged because of low fuel, were chased on the way out.[8]

## Air Combat

The EAF reportedly flew a total of 6,815 operational missions during the 1973 War. Perhaps a third of this total, or about 2,200 sorties, were estimated to be air defense missions. The remainder were flown by transports, helicopters, ground attack aircraft, and reconnaissance jets.[9] The IAF said it flew some 3,961 air defense, escort, and top cover missions. About 2,000 of these sorties were flown on the Suez front.[10]

EAF and Arab fighter pilots fought fifty-two air engagements with IAF jets during the Ramadan War. Egyptian interceptors hounded Israeli jets on deep-strike missions and fought many intense air battles near the Suez Canal. Most frequently air battles occurred near the front lines either just behind or beyond the Egyptian air defense screen. Many of these engagements began at low altitude when Israeli Mirages and F-4s tangled with Arab fighter-bombers and their MiG-21 escorts.

There were also many large air battles between EAF MiG-21s, Mirages, and F-4s near Port Said and Suez. For many days early in the war, the IAF tried to bomb holes in the Egyptian air defense screen at its northern and southern ends. The EAF reacted to these raids by maintaining regular interceptor patrols and attacking Israeli formations before they could roll in to drop their bombs. After suffering serious losses to Egyptian defenses, the Israeli fighters set up ambushes to bring EAF interceptors to battle over water, away from SAM defenses and at the limit of ground control coverage.[11]

Israeli superiority in pilot experience, aircraft performance, and weapons led to lopsided results. During air battles over Egypt and the

Sinai more than a hundred Egyptian, Iraqi, and Algerian jets and helicopters were shot down during the eighteen days of fighting. The IAF acknowledged the loss of only six aircraft in air combat on both fronts, but Arab sources claimed that there were more than fifty losses.[12]

Mirage pilots achieved a majority of the Israeli air combat victories, but F-4 crews also downed many Egyptian and Arab aircraft. Most Israeli air victories were achieved with infrared homing Shafrir or Sidewinder missiles, but some Arab planes were downed by cannon fire and a small number were hit by radar-guided Sparrow missiles.[13]

Egyptian pilots not only had to deal with a dangerous adversary but overcome technical problems with Russian-supplied weaponry:

> The main problem of Arab pilots was the ineffectiveness of their Soviet-built Atoll missile, primarily in locking onto the target. The Atoll is roughly compatible to the early model Sidewinders in performance and provides the pilot with an aural buzz when the infrared seeker head locks onto an aircraft tailpipe. Many Egyptian pilots particularly who flew far better than in any previous combat, simply could not get an Atoll target lock-on buzz when they were in firing position. . . . Israeli Mirage pilots conceded that Egyptian pilots showed better flying ability in aerial combat than previously but still lacked air discipline and tactics.[14]

Despite these problems, Egyptian fighter pilots achieved numerous air combat victories against Israeli jets. While a few gun camera films of Egyptian air victories have been published, no official listing of EAF "kills" from the 1973 War has been released. Israeli raids were frequently engaged by ground defenses and interceptors. Dozens of Israeli F-4s and A-4s were forced to dump their bombs and fuel tanks to defend themselves against missile and cannon attacks by Arab MiG-21 and MiG-17 pilots. This was a successful tactic because the Israeli jets were forced to withdraw without hitting their targets.

EAF MiG-21 pilots who spoke with the authors admitted that their units had suffered serious losses in dogfights with Israeli jets during the 1973 War. However, these pilots were sure that Israeli air combat losses were higher than the figures admitted by the IAF. One EAF MiG-21 regiment commander claimed that the pilots of his unit had achieved twenty-two victories during the war.[15]

Egyptian pilots suspected that, because of their pride, Israeli pilots often reported having been shot down by a SAM or AAA, rather than admit being beaten by an Arab pilot. It is also possible that many Israeli pilots never saw what hit them. Many American jets lost over North

Vietnam were ambushed by MiG-17s or MiG-21s and hit before the pilots knew they were under attack. The EAF used similar GCI-directed hit-and-run ambush tactics.

Within the EAF community there is acknowledgment of at least twenty confirmed Egyptian air victories during the 1973 War. Some Arab sources claim that several Egyptian fighter pilots achieved ace status by the end of the conflict, while one such Egyptian ace has been recognized in America.[16]

## Attack Operations

Air Marshal Husni Mubarak remarked following the war: "As everyone knows, our air force is mainly equipped with defensive aircraft, apart from a certain number of Tu-16s and Su-7s. For strike missions we were obliged to use, in addition to the Su-7s, MiG-21s with an Egyptian modification to the wing pylons. We also used the L-29 quite extensively for the less difficult missions, with excellent results; on 100 missions we lost only 3 of these aircraft. The Tu-16 was vulnerable, due to its low speed and size, and we therefore used it mainly for night bombing."[17]

The principal mission of EAF and Arab fighter-bombers was to fly hit-and-run air strikes against Israeli targets to inflict casualties and disupt Israeli operations. With the exception of the small number of Mirages, Su-20s, and Tu-16s available, Arab strike aircraft did not have range/payload performance or weapons delivery systems that could equal Israeli A-4s and F-4s. The limited experience of most Arab pilots also reduced the ability of the EAF and allied air arms to provide effective close air support for ground forces and accurately hit rear area Israeli positions.

Strike aircraft were effectively employed on the first day of the war when the EAF flew some three hundred strike sorties against Israeli airfields, command centers, and artillery batteries. EAF and allied Arab squadrons continued to fly hit-and-run strikes each day against Israeli targets in the Sinai until late in the war. Following the Israeli assault into Egypt, EAF and Arab fighter-bombers flew nearly 2,000 attack sorties against the Israeli bridgehead.

Arab MiG-17, Su-7/20, MiG-21, Hunter, and Mirage squadrons employed a variety of attack profiles against the Israelis. The preferred tactic was to approach at high speed and low altitude. Rockets, gunfire, and bombs were delivered in low-angle attacks to minimize exposure. These hit-and-run attacks frequently surprised the Israelis and caused serious damage and casualties.

However, many sorties were ineffective because no targets were detected, attacks were poorly flown, or munitions failed to work. Losses during these missions were somtimes heavy, at times reaching thirty-five per hundred sorties, since strike aircraft lacked radar warning devices and ECM self protection systems, and pilots had little training in how to defend against Israeli fighters.[18] EAF leaders were nevertheless pleased with the success achieved by the attack elements of the air force during the 1973 War. As Air Marshal Mubarak remarked: "The work done by the Su-7s and MiG-21s was excellent; we believe that some 400 enemy tanks were destroyed by these aircraft."[19]

## Helicopter Operations

More than a hundred EAF Mi-8 helicopters were sent into the Sinai on the first day of the war to deliver commando teams. Israeli jets intercepted several of these formations, and some twenty Mi-8s were shot down. Additional helicopters were lost to anti-aircraft fire and artillery. At least thirty-five EAF helicopters were shot down or lost in action during the 1973 War. There is no evidence that Egyptian helicopters were used in the attack role, even though some Mi-8s were modified to carry 57 mm rocket pods and machine guns.[20] Despite heavy losses, helicopter-delivered commandos caused considerable problems for the Israelis during the early part of the October War. EAF helicopters also played a major role in supporting the army in the Sinai. Senior EAF leaders felt that, in a future conflict, helicopters could play an even greater role.

## Air War Losses

American and Israel sources list total EAF losses in the Ramadan War as 220 aircraft and 35 helicopters.[21] No official Egyptian listing of 1973 War equipment or personnel losses has been released, although Egyptian President Sadat said in a speech given in January 1974 that Egypt had lost about 120 aircraft during the Ramadan War.[22]

Since the late 1950s the EAF has nearly always been short of experienced aircrew. Losses during the 1973 War continued this problem. According to American sources, the EAF lost about one hundred pilots during the conflict.[23] Many of these were squadron commanders and experienced pilots who had been flying since the 1956 Suez conflict. Aircrew training was given increased emphasis to make good this situation, but it would take many years before 1973 losses were replaced.

## Superpower Involvement

The Unites States and the Soviet Union were drawn into the 1973 conflict. Both countries were forced to send large quantities of weaponry and spare parts to their respective clients through massive air and sea lifts. Egypt, Syria, and Iraq received 12,500 metric tons of weaponry during 935 Soviet resupply flights. About 6,000 metric tons of material was flown to Egypt.[24] Late in the conflict three Soviet ships unloaded their cargos of crated aircraft, missiles, ammunition, and tanks at Alexandria. Following the cease-fire, Soviet ships brought more than 125 further aircraft, vital spare parts, and other weaponry to Egypt.[25]

The American airlift had a dramatic impact on IAF effectiveness late in the conflict. During the last week of the war, some thirty-five F-4s were flown to Israel along with hundreds of advanced weapons such as TV-guided Maverick missiles, laser-guided bombs, and ECM systems. These systems were immediately pressed into action. By the time the U.S. airlift ended, 22,300 metric tons of material was delivered by 421 C-141 sorties and 147 C-5 missions.[26]

Additional weaponry including fifty A-4s, hundreds of tanks, and other critical systems were delivered later by sea.

The dramatic confrontation between Soviet and American leaders toward the end of the war resulted from Israeli efforts to better its position before a cease-fire. The Soviet Union called for an immediate halt to protect the Egyptian gains and threatened to send in troops to stop the fighting. The strong American reaction could have led to further deterioration of the Middle East situation; however, both sides had the sense to compromise and allowed the United Nations to send in an emergency force to separate the Egyptians and Israelis.

## Isreali Lessons

The IAF was surprised by the heavy losses suffered to Arab air defenses and aggressive actions of the Arab air forces. The blow to Israeli morale was serious. Israel admitted that more than 110 of its aircraft had been shot down and more than fifty aircrew were killed, wounded, or captured.[27] More than 225 IAF aircraft were reportedly heavily damaged. Arab sources listed Israeli losses as much higher, in excess of 200 aircraft, but an itemized listing of these victories has not been released.

After the war a full review of Israeli military strategy led to many changes. Tactics were modified, and the IAF and Israeli Army strengthened. By the end of 1973 the massive American resupply program and Israeli reorganization resulted in an IAF that was 125 percent stronger than before the October War.[28] The effectivness of the IAF was also increased by the tactical lessons of war and advanced weapons and systems supplied by the United States. This included AH-1Q Cobra and MD500 attack helicopters and Lance surface-to-surface missiles.

On the ground, Israeli battle strategy was also changed. Additional artillery and rocket launchers were purchased, and the army assumed much of the close air support role previously assigned to the IAF. The success of Arab fighter-bombers and helicopter-borne commandos prompted the Israelis to increase its inventory of SAM and gun defenses. New weapons purchased for anti-aircraft defense included the Redeye shoulder-fired missile, Vulcan 20 mm gun system, and Chaparral mobile missile system.

An *Aviation Week and Space Technology* writer best summed up the EAF attitude following the 1973 War: "The Egyptian Air Force may not believe it won the air war of October, 1973, but it does believe that it was able to stay in the air and slug it out with the best the Israeli Air Force had to offer, that it denied the enemy freedom of operation over Egyptian forces to which it had become accustomed, that it carried the attack to enemy ground forces in crucial periods and that it has knocked considerable chrome off the Israeli Air Force's halo of invincibility."[29]

24

# New Directions Again (1974–1981)

During the decade after the 1973 War, President Sadat and the Egyptian government were forced to contend with serious domestic and international challenges. A peace treaty was negotiated with Israel, but President Sadat did not have the luxury of time to relax after this victory. He had to face pressing economic problems at home.

## The Legacy of War

Decades of high military spending and the drying up of foreign investment had starved Egypt's economy. The country's rapidly growing population and high unemployment also generated problems. Many educated Egyptians took jobs in Persian Gulf countries or in Europe because of a lack of opportunity at home. Limited opportunity, housing, resources, and religious differences led to tension between Muslims and Coptic Christians. In early 1977 major food riots broke out in many Egyptian cities following an announcement that the price of basic foods and fuel was being increased and that other economic reforms were being made. The Egyptian Army had to be called out to put down the disturbances because the situation went beyond the ability of the internal security forces.

President Sadat saw that he had to cut military expenditures and the size of his military in order to meet his modernization goals and improve Egypt's dire economic situation. By the late 1970s President Sadat had significantly reduced military spending. He had also begun to overhaul the socialist system installed by President Nasser and encouraged free enterprise development. By the late 1970s Egypt's economy was growing at an 8 percent rate due to increased tourism, oil exports, foreign investment, and overseas workers who returned money to the economy.

Closer relations with the United States also led to a new strategic outlook. In 1974 and 1975 Egypt and Israel signed a series of disengagement agreements that brought UN forces to the Sinai. However, the sacrifices of the 1973 War did not bear fruit until President Sadat made a dramatic move. With progress toward peace stalled, Sadat broke the deadlock. On 19 November 1977 he traveled to Jerusalem and spoke before the Israeli Knesset. Intense negotiations led to the 1978 Camp David Accords which set the terms for peace. On 26 March 1979 President Sadat and Israeli Prime Minister Begin signed a peace treaty that returned the Sinai to Egypt and ended forty years of confrontation. But it remained a "cold peace" since the basic problem between Israel and the Palestinians was not resolved.

The peace brought closer relations between Egypt and the United States. As a result billions of dollars of foreign aid, military assistance, and loans to purchase American equipment flowed between the two countries. Despite this new relationship, President Sadat and senior military leaders developed other sources of supply and built up Egypt's domestic arms industries to maintain independence.

The peace came at a considerable price: Egypt was isolated and deprived of economic and political support from Arab countries. Yet President Sadat was determined to play a leading role in the Middle East and the international arena. Despite many economic and political challenges, Egypt evolved from a country preoccupied with defense against Israel into a moderate Arab state. Egypt's military moved forward with a challenging program to trade Soviet systems for western equipment and internalize modern tactics and support concepts.

## Post-1973 War EAF Status

Following the 1973 War, Soviet ships brought more than 30,000 metric tons of military hardware to Egypt. These shipments included dozens of fighter aircraft, trainers, transports, helicopters, and an inventory of vital spare parts. The Egyptian Army and Air Defense Force also received large amounts of Soviet weaponry. This allowed Egypt to make good most of the losses suffered in the Ramadan War. Nevertheless, improved relations with the Soviet Union did not last long.

In 1974 the EAF had a personnel strength of some 23,000 and fielded more than 700 aircraft. It was divided into four air divisions and the

Southern Air District. Two air divisions were responsible for support aircraft (the 119th at Almāẓah took care of helicopters, the 129th at Cairo had transports), while the other two supported air defense (the 139th Division was at al-Manṣūrah and the 149th Division at Inshāṣ). Air Force Headquarters controlled the operations of all aircraft, but the air divisions provided administrative and logistics support. The Southern Air Division was the only unit to exercise direct control over the aircraft in its area. This Division was independent because of the need to protect the important Aswan Dam complex.[1]

The air brigade (formerly known as a regiment) was the primary EAF unit. An air brigade was composed of from two to four squadrons. EAF air brigades and squadrons were also regularly moved to meet changing tactical requirements and for security reasons.[2] In 1974 the EAF had six MiG-21 air brigades. Two fighter/bomber air brigades flew Su-7/Su-20s and two MiG-15/17s. Aswan was the home of the Tu-16 bomber air brigade. An air brigade of Il-14 transports and another of helicopters were stationed at Almāẓah. A second helicopter air brigade was headquartered at al-Khaṭāṭibah, and a brigade of An-12 transports was stationed at Cairo International. Squadrons that were attached to the above-mentioned units were stationed at more than forty primary and dispersal air bases located throughout Egypt.[3]

## Training Intensified

Following the 1973 War, the Air Force Command stated that there would be no major changes in overall doctrine. The EAF and EADF cooperated to defend Egyptian air space, and EAF interceptors and attack aircraft had scored numerous victories against the Israelis. Many tactical lessons had been learned, and the EAF moved quickly to incorporate these into its training programs.

In response to a question regarding the future objectives of the EAF, Air Marshal Husni Mubarak responded: "Our first objective must be the continual improvement in training. The October War is the best illustration of the difference in training between 1967 and 1973; we must continue along the same road if we wish to achieve even better results."[4]

The EAF training program combined the best elements of Soviet and western techniques with adaptations for the Middle East environment and created a uniquely Egyptian program. The three-year academic and flight training program at the Air Force Academy at Bilbays was shortened to eighteen months to generate additional pilots. This emphasis on pilot training was important because following the 1973 War the EAF pilot/aircraft ratio dipped to less than one to one.[5] Of some 200 cadets who started a typical training course, only 80 or fewer usually received their wings. In the mid-1970s about 100–130 new pilots graduated from the EAF academy each year. Following initial ground instruction, cadets flew 30–40 hours on the Jumhūrīyah, 120 hours on the L-29 Delfin, and 70 hours on the MiG-15UTI and MiG-17F. Navigators trained on Il-14 flying classrooms, while helicopter pilots flew 100 hours on the Mi-4 and Mi-8 for training. In the fall of 1974, Air Marshal Mubarak paid a visit to the Royal Air Force Central Flying School in England to observe RAF training methods and aircraft. The EAF was in the process of assessing positive aspects of foreign training programs as part of its long-term goal of converting to western systems.

Egypt still faced a challenge filling and maintaining its pilot, aircrew, and technical ranks as the growing economy drew away experienced technicians. The EAF pilot training program experienced a high washout rate, and serving pilots had to attend to staff and technical responsibilities in addition to their flying duties. To develop aircrew the EAF regularly sent pilots on exchange tours with foreign air forces and attended the Egyptian Air War Institute and overseas training programs. In the late 1970s EAF officers began attending military educational programs in the United States, France, and India.

EAF officers were sent to the French École de Guerre each year to study the strategy and tactics of air warfare. American military educational programs attended included the USAF Air Command and Staff course plus various technical and pilot training courses. The Indian Air Force assisted the EAF in training pilot instructors and test pilots. Sadly the old relationship with the British RAF was not revived to any great extent. The EAF also held joint exercises with the Egyptian Air Defense Force, Army, and Navy and hosted squadrons from Iraq, Algeria, and Persian Gulf countries.

As a result of budget reductions and a declining inventory of spare parts for Russian-built aircraft, during the 1970s EAF fighter pilots had to reduce their overall flying hours. On average interceptor and fighter-bomber pilots flew ten to fifteen hours a month, while bomber, transport, and helicopter crews received fifteen or more flying hours.[6]

Air Marshal Husni Mubarak, described by President Sadat as one of the, "five big heroes" of the Ramadan War, soon relinquished command of the EAF and became the nation's vice president. Command of the EAF was then assumed by Maj. Gen. Muḥammad Shākir ᶜAbd al-Munᶜim. Despite the improvement in relations, only one small detachment of Soviet Air Force MiG-25 reconnaissance aircraft was allowed to return to Egypt. Located at Cairo West, this unit only had some four aircraft and a small number of pilots and support personnel. It was not long before even this small Soviet detachment was sent home due to further deterioration in relations.[7]

On the first anniversary of the Ramadan War, 6 October 1974, tight formations of MiG-21s, Su-20s, and Antonov transports flew over Cairo. Meanwhile a similar anniversary fly-past over Damascus was highlighted by a formation of late-model Soviet-built MiG-23s. The absence of these new aircraft over the skies of Cairo betrayed the fact that Egypt and the Soviet Union were again at odds.

## MiG-23 Flogger Swing-Wing Fighter

In an attempt to improve their strained relationship, in late 1974 and early 1975 the Soviet Union provided Egypt with a modest supply of military spare parts and MiG-23 fighters ordered before the 1973 War. A ship carrying the first four MiG-23s docked at Alexandria in late 1974, and a team of Soviet technicians assembled these aircraft at Marsá Maṭrūḥ. By the summer of 1975 the EAF was flying a squadron of MiG-23s in the intercept role. Egypt was supplied with the MiG-23MS Flogger E, which was a downgraded export version of the standard Soviet fighter. Later a small number of MiG-23BN Flogger F aircraft optimized for ground attack and four MiG-23U Flogger-C two-seat trainers were delivered to Egypt.

On the plus side, the MiG-23 had greater range and speed than the older MiG-21. However, the MiG-23 was far less maneuverable than the MiG-21. The new MiG had a slightly better avionics suite and an internal 23 mm cannon but was armed only with the disappointing AA-2 Atoll air-to-air missile.[8] Some fourteen of these aircraft appeared in public for the first time during the October War fly-past over Cairo in 1976. Egypt's combat-experienced pilots had little trouble in converting to the MiG-23 since this new fighter was not really much more advanced than the late-model MiG-21MFs and Su-20s that had long been flown by the EAF. Because of economic and political problems, by the late 1970s the Soviet Union reduced deliveries of spare parts and support to a trickle.

## Diversification and Independence

After throwing out the Soviets in 1972, President Sadat had committed to a long-term program to diversify Egypt's military inventory with western equipment. Egyptian leaders never again wanted to be in a situation in which an embargo could totally disrupt military actions and limit political options. The move from Soviet to more advanced systems was to start as soon as possible. Egypt began to purchase aircraft and weaponry from the West even before the 1973 War.

Initially western nations would only agree to offer unarmed helicopters and defensive systems. However, after the Ramadan War this situation changed. "We are going from the Soviet era of vacuum tubes to the Western technology of integrated circuits," said one Egyptian Air Force General, "and we will never again be dependent on a single source of supply."[9]

## Mirage V: Return to the West

In October 1974 the EAF received the first of thirty-eight French-built Mirage 5SDE fighter-bombers and Mirage 5SDD trainers ordered before the Ramadan War. These aircraft were funded by Saudi Arabia and several other Arabian Gulf nations. These were the first western-built tactical aircraft delivered to the EAF since the Vampires and Meteors of the 1950s. This was also a symbolic purchase since the Mirage had for so long been closely associated with the Israeli Air Force.

Over the next decade the EAF went on to purchase more than eighty Mirages. The delta-wing jet was used in the intercept, fighter-bomber, reconnaissance, and trainer roles. Ṭanṭá air base has been the primary home for the Mirage fleet, but squadrons and detachments have also operated from Jiyānklīs and other bases. The multi-role Mirage V has similar air combat performance to the late-model MiG-21 but better avionics and weaponry. EAF Mirage Vs were armed with up to four

Matra 550 infrared-homing missiles and two 30 mm cannon. In the air-to-ground role the Mirage could better the Su-20 in payload, performance, and bombing accuracy.[10] EAF Mirage Vs were camouflaged and painted with large orange sections on the wing and tail to differentiate them from Israeli Mirages and similar Kfirs.

The EAF helicopter force also received some of the first western systems. Saudi Arabia funded the purchase of six British Westland Sea King helicopters to support the Egyptian Navy and twenty-four Commando helicopters to supplement the Mi-8 in 1974. Deliveries began in 1975 and were completed in 1978. In 1975 President Nixon presented President Sadat a Sikorsky S-61 helicopter as a gift for use as his personal transport.

The 6 October 1975 military parade and flyover demonstrated that Egypt was making rapid progress toward its goal of introducing western military hardware. The military parade featured American trucks and British Land Rovers carrying Egyptian soldiers armed with Russian weapons, while overhead French-built Mirages and British-produced Commando helicopters flew alongside various Soviet-built aircraft. Late in 1975 Egypt ordered four American C-130 Hercules transport aircraft. By the end of the decade the EAF had replaced its fleet of thirty An-12 transports with eighteen C-130s.

The Egyptian Air Defense Force, which accounted for a quarter of Egypt's military manpower, also profited from the move to western equipment. In the mid-1970s the service began to receive radios, command and control systems, and radars from the West. EADF crews tested several European air defense weapons and placed an order for the French Crotale SAM system.

## Industrial Investment

The ambitious military modernization plan was also used as a vehicle to develop Egypt's arms and aerospace industries. The Arab Organization for Industrialization (AOI) was organized in 1975 to pursue this goal. Consisting of Egypt, Saudi Arabia, Qatar, and the United Arab Emirates, this multinational company had a capital fund of more than $1 billion. Egypt, hoped that this new organization would develop into a regional industrial consortium that would supply the military needs of the entire Arab world.

The AOI negotiated with major British and French firms to establish factories in Egypt and produce up to 150 Alpha Jet aircraft, 200 Westland Lynx helicopters, 100 Mirage fighters plus associated weapons, support equipment, and vehicles. Considerable progress was made until overseas financial support was withdrawn following the signing of the Egyptian-Israeli Camp David Peace Accords.[11] Nevertheless, Egypt went ahead with several of the programs planned for the AOI using domestic funding and western investment.

## Progress toward Peace

In 1977 President Sadat shocked the world when he traveled to Israel. This brave act set the scene for negotiations between Egypt and Israel which resulted in a peace treaty and the return of the Sinai to Egyptian control. Before the peace treaty, Egypt and the United States were already developing closer ties. In 1978 President Jimmy Carter offered to sell Egypt one hundred F-5E/F fighters to modernize the EAF. At this same time Saudi Arabia was also buying this aircraft. President Sadat refused this offer and asked for more advanced F-4, F-15, and F-16 jets. Following the 1979 Camp David Accords, which set the terms for peace, the United States became Egypt's principal supporter and supplier of military systems.

To support the peace process, during the summer of 1979 the United States flew thirty-five F-4E Phantom fighter-bombers to Egypt. These were withdrawn from U.S. Air Force squadrons and delivered to Egypt under Operation Peace Pharaoh. President Carter ordered immediate delivery of these aircraft so that they could participate in the 6 October 1979 National Day Military ceremonies to highlight the new Egyptian-U.S. relationship.[12] The delivery of F-4s to Egypt was an important event. It showed that Egypt was on an equal level with Israel in American eyes. Since the War of Attrition the F-4 had been known and hated by Egyptians as a symbol of Israeli air power and aggression.

During this same period Air Vice Marshal Luṭfī Shabānah assumed command of the EAF. His mission was to introduce American and European equipment and western tactical concepts.

## Upgrading EAF Technical Skills

Out of the limelight, the EAF was working hard to improve its aircraft and the standards of training and maintenance. Problems caused by limited spare parts, battleworn aircraft, and old equipment forced the EAF to reduce its flying hours. This situation was made worse by the increased demands on the technical force when new western systems such as the Mirage, Phantom, C-130s, and Commando helicopters were phased into service.

Egypt also had difficulties in rapidly maintaining its cadre of trained support personnel. Large number of draftees were taken in each year, and an intensified training program helped improved the situation, but only slowly. The Air Force Technical Training Institute at Ḥulwān set out to increase the number of skilled support personnel. New facilities that opened in the mid-1970s allowed more airmen to attend these important courses. A new expanded training syllabus prepared airmen to support the new aircraft and equipment purchased from France, England, and other western countries. These systems demanded a broader theoretical background and a greater knowledge of advanced technology. Thousands of aircraft mechanics and NCO technicians, including many foreign students from Libya, the Sudan, and other Arab countries, passed through this training program at Ḥulwān.[13]

One of the most immediate problems facing the EAF and its technical support personnel was the limited supply of spare parts. Egypt turned inward as well as to many international sources for assistance. The Egyptian government instructed state-run factories and private firms to manufacture parts for Russian systems and develop aircraft and systems overhaul programs. Increasingly strained relations between Egypt and the Soviet Union also prompted the EAF to turn to China, India, and eastern Europe for parts and technical assistance. Egypt also approached private firms in the United States and Britain for assistance. Soviet pressure cut off sources of vital spares in India and eastern Europe.

An EAF commission was sent to Britain to study the possibility of fitting the Rolls Royce Spey turbojet engine into Egypt's MiG-21 interceptors. While this change would have extended the aircraft's range and performance, this modification was found to be too expensive given the jet's limited remaining life. In 1975 the EAF initiated a modest upgrade of its MiG-21 fleet. Firms in Egypt, the United States, France, and Britain supported these efforts. Several British firms including Ferranti, Smiths

Industries, and Marconi wound up with substantial contracts to upgrade the avionics of EAF aircraft.[14]

## Chinese Connection

However, in 1976 China agreed to provide the EAF with parts to support the MiG-17, MiG-21, Tu-16, and various trainers and transports. This relationship set the scene for EAF purchase of fighters from China. In 1979 the first of forty F-6 fighter and FT-6 two-seater conversion trainers arrived in Egypt. Egyptian and Chinese technicians reassembled the new F-6 fighters at several EAF air bases after their delivery by ship to Alexandria. The F-6 is a copy of the Soviet MiG-19 fighter built by Shenyang in China. These aircraft cost only about $1 million each and allowed the EAF to maintain the flying skills of its pilots at a relatively low cost.[15]

## Egypt versus Libya

Following the 1973 conflict, relations between Egypt and Libya deteriorated. Libyan leader Muammar Qaddafi was actively fostering terrorism and dissent in Egypt because of his opposition to President Sadat's new orientation toward the West and talk of peace with Israel. The long-simmering dispute between President Sadat and Muammar Qaddafi intensified in early 1977, and forces of both countries were placed on alert. On 19 July 1977 Libyan artillery units shelled Egyptian settlements along the coast near Sallūm. Two days later both sides again traded artillery fire, and then Libyan and Egyptian armor clashed. EAF Su-20 and MiG-21 jets struck at Libyan radar sites and artillery positions, and Libyan Mirages bombed border villages. Both sides suffered air and ground losses.

Egyptian pilots flew numerous fighter and strike sorties over Libya; the air base at al-ʿAdam near Ṭubruq was blasted with anti-runway bombs, and several radars and other targets were destroyed. During these strikes the EAF lost several aircraft, including at least two Su-20s, to fierce anti-aircraft fire. At least six Libyan jets were lost to EAF fighters and ground fire. Air strikes and ground combat continued until 25

July.[16] Arab League Secretary-General Maḥmūd Riyāḍ and other Middle East diplomats arranged a cease-fire, and the shooting came to an end by early August. President Sadat was determined to quiet down the problems on his western border so that he could concentrate on international and domestic challenges.

## EAF Modernization

The ambitious Egyptian military modernization program moved forward within the limits set by economic and political realities. Within a seven-year period Egypt had ended its twenty-year relationship with the Soviet Union and turned to the United States, France, Great Britain, and China for advanced air and ground weaponry.

Due to the large amount of Soviet weaponry and the high cost of new systems from the West, this was a slow transition. But the modernization plan was being fulfilled. Older Soviet systems would be maintained or modernized as long as possible. High-performance western systems including aircraft, command and control networks, plus training and maintenance systems would be emphasized.

By 1980 the Mirages and American-supplied Phantoms formed the first line of Egyptian air defense and attack force. Three squadrons of Mirages formed an air brigade based at Ṭanṭá. The two squadrons of F-4s were assigned to an air brigade at Cairo West. Both these aircraft could perform multiple missions. Phantoms were assigned the all-weather, beyond visual range interception mission because of their long-range radar and AIM-7 Sparrow radar-guided air-to-air missiles. EAF Phantoms received a two-tone grey paint job and sported orange panels like the Mirages to ease identification problems.

Initially Egyptian support personnel had difficulty maintaining the complex F-4s. However, U.S. Air Force training teams and contractors overhauled the planes and provided additional training to EAF support personnel. As EAF maintenance personnel and aircrew gained experience with the Phantom, the situation improved. Even so, the F-4 Phantom was never a popular aircraft in the EAF. Maj. Aḥmad ʿĀṭif, the first Egyptian pilot to shoot down an Israeli Phantom (9 December 1969) witnessed the arrival of the first EAF Phantoms at Cairo West in October 1979. When President Sadat asked the major what he thought of the F-4,

ʿĀṭif replied, "Excellent bird but it maneuvers like a loaded truck, Mr. President."[17]

In early 1980 the United States offered to sell Egypt advanced F-15 and F-16 fighters. The EAF ordered thirty-four F-16A Fighting Falcon fighter-bombers and six F-16B trainers later in the year. During 1982 and 1983 these aircraft were delivered, and subsequently they were formed into two squadrons to form a wing at Inshāṣ air base. EAF F-16s were armed with AIM-9P Sidewinder air-to-air missiles and free-fall bombs. A second order for forty F-16C and D aircraft was placed in 1981.

On 6 October 1981, during the National Day parade, President Sadat was assassinated. Vice President Mubarak assumed the position of president of the country. As a former commander of the EAF, President Mubarak has a great appreciation for the importance of air power. While serious economic and political challenges slowed the pace from previous plans, President Mubarak continued to modernize the Egyptian armed forces and to support the development of Egypt's aerospace and high technology industries.

## EAF Air Order of Battle 1974[18]

| Combat | Support | Helicopter | Trainer |
|---|---|---|---|
| MiG-21 × 200 | Il-14 × 50 | Mi-4 × 35 | Jumhūrīyah × 100 |
| Su-7/20 × 100 | An-12 × 20 | Mi-6 × 20 | L-29 × 100 |
| MiG-15/17 × 100 | | Mi-8 × 70 | Yak-18 × 50 |
| Tu-16 × 25 | | Commando × 10 | MiG-15UTI × 20 |
| Il-28 × 10 | | | Su-7U × 15 |
| | | | MiG-21UTI × 25 |

25

# Moving Forward (1981–1994)

The assassination of President Sadat by Islamic revolutionaries on 6 October 1981 demonstrated the internal resentment over Egypt's peace with Israel and frustration concerning the government's economic, religious, and political direction. During the month before the attack, President Sadat had cracked down on the Muslim Brotherhood, expelled the Soviet ambassador and his mission, and enacted economic reforms aimed at revitalizing the Egyptian economy. On 14 October 1981 Husni Mubarak was named president of Egypt. He acted quickly to root out potential threats to the state. At the same time President Mubarak was also willing to extend the olive branch and released many of the political and religious dissidents jailed during the Sadat period.

## The Challenge of Change

The Egyptian economy continued to grow into the mid-1980s, but soon thereafter it stagnated. The collapse of oil prices in 1986, decline of tourism due to terrorism, and expulsion of expatriate workers during the Gulf conflict created a difficult situation. Continuing population growth, budget deficits, and rising foreign debt continued the economic decline. Cancellation of debts by the United States and Gulf states and difficult economic reforms stabilized the situation but created internal pressures.[1]

President Mubarak pursued a moderate course and carefully balanced his actions. Egypt strongly supported Iraq during the Iran-Iraq War, even though Egypt was, at its start in 1980, politically an outcast in the Arab world. Military advisors were sent to Iraq during the conflict, and Egyptian military and civilian factories were kept busy producing weaponry to support the war effort.

While Egypt followed through on commitments made in the Camp David Accords, military investment under the new president increased. As a former commander of the EAF, President Mubarak had full appreciation of the importance of the military to domestic security and regional influence. The 1982 Israeli invasion of Lebanon and defeat of Syrian forces underscored that Egypt still faced a turbulent region. Egypt also faced a threat from Libya along the border and in Sudan to the south. The Egyptian-U.S. security relationship rapidly broadened, and American aid to Egypt soon nearly equaled that provided to Israel. The two countries cooperated closely on defense matters. Egyptian bases were used during the Iran hostage crisis to support rapid deployment force activities and other military activities in the Middle East.

Since the brief 1977 border skirmish with Libya, the Egyptian Air Force has enjoyed the longest period of peace in its tumultuous history. During these years the EAF, along with the other services and the entire country, has seen considerable changes. By 1989 the country had resumed diplomatic relations with most Arab states. President Mubarak demonstrated his leadership and support for the Persian Gulf states by sending more than 30,000 troops to fight in the war to reclaim Kuwait, though no air force units were committed.

## Egyptian Force Modernization (1980–1994)

Military aid from the United States, which has averaged about $1.1 billion annually since 1980, plus domestic investment have allowed Egypt to make the transition from a mostly Soviet to a mixed-force structure. Due to economic limitations, this transition has been gradual. Aging Soviet systems have been maintained far longer than expected. However, assistance from the United States, Europe, China, Poland, Rumania, and Yugoslavia and the creativity of the Egyptians have meant that Soviet tanks, missiles, and aircraft have been maintained in operational condition. But as the 1990s progress this equipment will inevitably reach the end of its service life.

Five-year defense plans, which began in 1983, carefully controlled investments to ensure that the needs of the military, domestic industries, and political constituents were fulfilled. Egypt gradually reduced its military force structure and followed a plan to best capitalize on American aid and national funding. Defensive weapons and systems that were force multipliers received first priority. This included Egyptian purchase of U.S.-made F-16 fighters and M-60A3 and M-1 tanks. A new air defense network comprised of both ground and airborne radars and air defense weapons such as the Improved Hawk and European Skyguard and Crotale were purchased. These weapons were all tied together with the new "Project 776" command and control system.

European, Chinese, and other foreign systems were produced under license to maintain the Egyptian defense industrial base. New western high-technology systems were mixed with Soviet and foreign equipment to form a hi/low mix. Western tactical, training, support, and maintenance concepts were adopted. This was a slow and often painful process that was expensive to implement. This unique approach did, however, allow Egypt to attain its goal of modernization while still retaining a high degree of independence.[2]

## International Cooperation and Training

Libyan involvement in Sudan prompted President Mubarak to ask the United States to send E-3A airborne warning and control aircraft (AWACS) to Egypt to monitor air movements in February 1983. Several of these aircraft operated from EAF bases and maintained a presence until after the 1983 Bright Star exercise.

The August 1983 Bright Star exercise saw close coordination between the armed forces of the United States and Egypt. American forces and Egyptian officers and enlisted men benefited from this training and communications. Exercises with American forces and foreign units continued. In 1984 U.S. and Egyptian forces cooperated to remove mines from the Suez Canal, Gulf of Suez, and Red Sea. French Air Force squadrons participated in war games with EAF and EADF units from Egyptian bases in late 1986. While these programs were valuable for training, Egypt's return to the Arab community has prompted the government to maintain a low profile concerning these activities.

The EAF learned many lessons during these operations through exposure to western weaponry and tactical concepts. Increased Egyptian attendance at U.S. technical schools and European military programs has enhanced the education of Egyptian officers and enlisted personnel.

Air Vice Marshal Muḥammad Luṭfī Shabānah, former commander of the EAF, remarked: "We are getting acquainted with modern technology which is used by the United States and we are also acquiring some of it. It is essential to know how it is used by different air forces. We have studied the tactics of the U.S. Air Force and their way of controlling their airplanes. Our fighter controllers have been trained on AWACS early warning planes. We planned and carried out joint maneuvers in which a great number of planes of different types took part. This was useful and a great success for us. All new experience and knowledge like this is indispensable."[3]

## EAF Hi/Low Mix

The "high" side of the EAF tactical force is comprised of some 165 F-16s and 20 Mirage 2000 fighters. Both are single-engine, highly maneuverable fighters that were optimized for the air superiority mission but are also capable attack aircraft. With their high-performance, advanced avionics and varied complement of missile and gun armament, these new fighters are excellent replacements for the aging MiG-21. The advanced systems in the F-16 and Mirage 2000 allow EAF pilots to switch easily to the strike role using bombs and guided missiles.

## F-16 Fighting Falcon

The first EAF air brigade of F-16A/Bs was formed at Inshāṣ in 1982, and by 1983 the unit was operational. The positive Egyptian-U.S. relationship has enabled the EAF to secure sophisticated systems that the Soviets never would have supplied. Additional orders for F-16C/Ds enabled the EAF to equip additional air brigades with the aircraft at Banī Suwayf and Abū Ṣuwayr. In the 1990s, F-16s assembled in Turkey further expanded the EAF's F-16 fleet.

Along with the delivery of the F-16C/Ds, the EAF received the all-aspect AIM-9L Sidewinder and AIM-7 Sparrow air-to-air missiles, ALQ-119 and 131 jamming pods, and AGM-65 Maverick air-to-ground missiles. EAF F-16s are assigned both air defense and ground attack missions. Like Egyptian Phantoms, the F-16s are painted in a grey two-tone camouflage.

## Mirage 2000

The EAF became the first air arm outside of France to buy the Mirage 2000 fighter. A squadron with sixteen single-seat Mirage 2000EN fighters and four Mirage 2000BM trainer aircraft became operational in 1987. These fighter bombers were armed with Matra 530 and 550 air-to-air missiles for the intercept mission. EAF Mirage 2000 aircraft can also perform ground attack missions using the precision AS-30 laser-guided air-to-surface missile and free-fall bombs. France supplied Egypt with an integrated ECM system for its Mirage 2000 jets. Originally Egypt planned to buy a second squadron of these jets, but financial limitations ended this hope. The purchase of advance French fighters again demonstrated the Egyptian desire to remain independent of a single source of supply for critical military hardware.

## Backbone of the EAF Tactical Force

Several air brigades of Mirage V and F-4 Phantom multi-role aircraft formed the cornerstone of the EAF tactical force during the 1980s and into the 1990s. These aircraft were upgraded and maintained in good condition through a mix of EAF, Egyptian industry, and contractor support. Egypt relied on a mix of high-performance western jets and relatively low-cost Chinese fighters to replace the large numbers of Soviet aircraft that had to be retired during the 1980s.

Some eighty F-6s (the Chinese-built version of the Russian MiG-19 supersonic fighter) were brought by ship to Egypt. Two air brigades of F-6s and TF-6s served in the air defense, ground attack, and advanced trainer roles. More than 100 Chinese-built F-7 fighters were also purchased by Egypt during the 1980s to replace wornout or unsupportable MiG-21s, MiG-17s, and Su-7/Su-20s. The Chinese F-7 is very similar to the Russian MiG-21F-13. Initial versions of this Mach-two jet were bought for a cost of only $3 million each. Egyptian and Chinese technicians reassembled the new F-6 and F-7 fighters at several EAF air bases after their delivery by ship to Alexandria. Working with suppliers in Europe and the United States, Egypt upgraded the avionics, weaponry, and ECM systems of its large fleet of MiG-21 aircraft which included new F-7s and decreasing numbers of Soviet-built Fishbeds. The F-6, F-7, and surviving MiG-21s were armed with French 550 Magic and U.S. AIM-9

Sidewinder air-to-air missiles and a variety of Egyptian and western bombs and rockets.[4]

## Egyptian Industrial Development

Egypt has the largest industrial base in the Arab world and a long history of manufacturing military products. President Mubarak emphasized development of Egypt's aerospace and high technology industries in order to expand domestic jobs and create new business opportunities. Even though its partners in the AOI withdrew their support in retaliation for the Camp David peace treaty, Egypt went ahead on its own with this development program. During the 1980s Egypt became a major exporter of army weapons such as rifles, mortars, and ammunition. Much of these munitions were exported to Iraq and other Arab nations.

In September 1982 the first Alpha Jet trainer was test flown in Egypt after being assembled at the Ḥulwān factory near Cairo. By the early 1990s some thirty Alpha Jet trainers had been delivered to the EAF Training School at Bilbays, and fifteen light attack versions of the aircraft were also in service. AOI factories also assembled nearly 100 Gazelle helicopters and 130 EMB-312 Tucano turboprop trainers and built parts for the Mirage 2000, the F-16, and the Falcon business jet. EAF and Chinese technicians, with assistance from civilian firms, assembled large numbers of F-6 and F-7 aircraft in Egypt.

Egyptian factories build most of the bombs, ammunition, rockets, drop tanks, and many of the spare parts used by the EAF. Many Chinese jets assembled in Egypt were provided to the EAF, but more than one hundred were shipped to Iraq. Likewise more than half of the Tucano trainers assembled at the Ḥulwān factory were exported to Iraq.[5]

Air force technical units and Egyptian industry worked aggressively to maintain the EAF fleet of Soviet-built aircraft. Technicians from the United States, Europe, and China assisted Egypt in keeping MiGs, Sukhois, Mil helicopters, plus Antonov and Ilyushin transports flying as long as possible. Hundreds of Egyptian firms were given contracts to produce spare parts, drop tanks, and munitions and to overhaul engines. Highly skilled EAF and contractor technicians proved that they could improvise and create critical spare parts to keep MiG-21s and F-7s flying.[6]

## EAF Fiftieth Anniversary, 1982

In May 1982 the EAF celebrated the fiftieth anniversary of the first training flight in Egypt which took place at Almāẓah. During this period Air Vice Marshal Muḥammad ʿAbd al-Ḥamīd Ḥilmī assumed command of the EAF. During a display and air show, the EAF demonstrated that considerable progress was being made toward its goals of diversifying its inventory and modernization.

During 1982 Israel completed its withdrawal from the Sinai. Egypt moved forward with its plan, established by the Camp David Accords, to deploy additional mechanized units to the Sinai. UN peacekeeping forces remained to ensure a smooth transition. There were no plans to station EAF units in the Sinai except for helicopters and liaison aircraft. The 1982 Israeli invasion of Lebanon has, however, remained in the mind of Egyptian defense planners for many years. Over the next several years Egyptian forces practiced on several occasions a combined air-ground defense of the Sinai against a potential Israeli attack.

## Force Modernization, End of the First Five-Year Plan

While Egypt has remained at peace, unrest in Sudan and Ethiopia has been a worry. EAF units have deployed to southern bases to send signals that Egypt was prepared to act if things got out of control. EAF E-2Cs, reconnaissance aircraft, and RPV drones have been used to monitor events in the Middle East region which could impact Egyptian interests.

While the EAF has traditionally served as Egypt's first line of defense and attack, the proliferation of ballistic missiles has changed the balance of power in the Middle East. Israel is known to have fielded the Jericho medium-range missile, perhaps with nuclear warheads, as well as the Lance tactical rocket. Saudi Arabia purchased medium-range missiles from China while Iran and Iraq used SCUDs in action during their eight-year war, and this weapon was fired regularly at Israel and Saudi Arabia during the Persian Gulf War. Syria also fielded a collection of FROG and SCUD missiles.

The army has control of Egypt's rocket and missile forces. During the 1980s Egypt pursued a program to manufacture its own ballistic missiles. The first to be fielded was the Ṣaqr 80, an upgraded version of the Soviet FROG rocket. While funds were invested in several domestic and international missile development programs, funding limitations and international political pressure brought these programs to an end. An effort to upgrade and expand the existing inventory of SCUD missiles was started in the 1980s, and this could result in a fielded missile in the future.[7]

Egyptian military modernization plans moved according to plan despite severe budget challenges. By the end of President Mubarak's first Five-Year Defense Plan (1987), the EAF was well into its planned upgrade program. Some 20 Mirage 2000s were operational along with 80 F-16s and 120 Chinese F-6s and F-7s. Command of the EAF had been passed along to Air Vice Marshal ʿAlāʾ Barakah.

By the early 1990s Egypt had retired all of the Su-7 and Su-20 fighter-bombers and MiG-23s because it was impossible to maintain a reliable source of spare parts for the aging aircraft. The MiG-17 had also been phased out of service after flying with the EAF for more than thirty years. The EAF's high level of experience helped smooth the transition to these new systems.

Maj. Gen. Nabīl Shuwakrī commented:

> The strong point of the EAF is its tremendous combat and training experience. For example we may send a MiG-21 to fight against a Mirage, since we are completely familiar with it. And we did modify our tactics with the new generation of aircraft. Command and control is a field where we have great experience, not just of the Air Force itself but also the control of manpower and the engineers. It's very important to be experienced in using a combination of aircraft. And of course it you've got a combination of Mirage 2000 and F-16, you can use both aircraft to make a very strong team. You can also use the F-16 in conjunction with the MiG-21 to improve the work of the latter aircraft, which had such a poor weapon delivery system. I can train pilots to shoot the F-16 with the MiG-21. It depends on how you use the aircraft. As we have many types of aircraft—Chinese, Russian, French, British, Brazilian, American and Canadian—we have the best "Red Flag" in the world.[8]

## Air Defense Force Upgraded

The Egyptians have also gone to great lengths to ensure that the Air Defense Command was equipped with the latest equipment. Many antiquated Russian radars and control systems have been replaced by U.S. and European systems, parts of which have been built in Egypt. Under "Project 776," a new nationwide air defense radar network and command and control system network have been established to ensure that Egyptian airspace is secure.

An air defense expert commented: "The Air Defense Command is structured so that air defense interceptor aircraft become operationally controlled by [headquarters] as do E-2C aircraft. They are succeeding in imprinting U.S. doctrine over Soviet organizational doctrine and making it all play together, and that's no mean task."[9]

The EAF operates six E-2C Hawkeye aircraft with the APS-125 radar to support the country's large network of ground-based radars and monitor activities well into the Mediterranean or over neighboring airspace. These important aircraft operate from Cairo West air base. The folding wings of the E-2Cs allow these aircraft to be housed in large concrete shelters for added protection. These excellent systems have proven a challenge to maintain. However, EAF and contractor personnel working together keep them in high readiness.

Hundreds of new air defense systems such as the U.S.-built Improved Hawk and Caparral, French-supplied Crotale, and Swiss Skyguard have entered service with the Egyptian Air Defense Force. These have replaced aging Soviet-built systems. Egypt armed the Swiss Skyguard missile system with U.S.-built AIM-7 Sparrow missiles and uses these to defend air bases and other vital positions.

Egyptian industry produced parts for U.S. and European air defense systems as part of offset programs and to ensure that follow-on support could be provided in-country. In the 1980s Egypt developed and fielded its own air defense equipment including the SAKR EYE (an improved version of the Soviet SA-7 shoulder-fired missile) and the Sinai 23 mobile air defense system which combined an American M113 armored vehicle with a French-built radar and Egyptian-built 23 mm cannon and SAKR missile system. One of the lessons of the 1973 War was the need for mobile air defense systems to move with armored forces and be able to displace if attacked.

## Reconnaissance and ECM: Force Multipliers

The EAF has acted on the many lessons of warfare since the 1973 conflict. The 1973 War, the 1982 Israeli invasion of Lebanon, and the Gulf War highlighted the need for force multipliers such as capable reconnaissance systems, electronic warfare, and command and control networks. Egypt has invested heavily in these areas.

Manned tactical reconnaissance is performed by a mix of Mirage V and MiG-21 aircraft fitted with sensor pods. Intensified training and doctrine changes have rectified the weakness in this area revealed during the 1973 War. Unmanned systems have assumed an increasing role during the past decade. Egypt has been a leader in the Arab world in the development, testing, and operational use of remotely piloted vehicles (RPV). In the late 1970s the EAF gained experience with RPV operations using the U.S.-made Firebee system and CT 20 drones from France. In the 1980s more sophisticated Skyeye and Scarab remotely piloted vehicles entered service to monitor activities along the border areas without the risk of losing a manned aircraft. RPV systems can be used in peace and in war to monitor activities and movements. Israel has been a leader in this area, and the tactical benefits of even first generation RPVs were demonstrated during the Israeli invasion of Lebanon. Egypt quickly learned these lessons and purchased several U.S.-built reconnaissance systems. The utility and effectiveness of the Skyeye and Scarab systems have been proven during many exercises and tactical operations.[10]

The EAF has also vastly upgraded its capabilities in the areas of electronic warfare and intelligence collection. During the 1980s Egypt purchased two specialized EC-130 aircraft from the United States and four 1900C aircraft with the Guardrail system for electronic intelligence gathering. Four Westland Commando 2E helicopters, fitted with Selenia IHS-6 ELINT and jamming systems, can also be used for intelligence.

Egypt has a large inventory of self-defense ECM systems for its fleet of tactical aircraft. All tactical aircraft and many transports and helicopters are equipped with radar warning receivers. Some of these systems are European, while U.S. aircraft typically are delivered with standard equipment used by American forces. EAF F-7 and Mirage fighters are equipped to carry the Selenia SL/ALQ-234 jamming pods which can disrupt both pulse and continuous wave radars. The F-4 can carry the AN/ALQ-119 jamming pod and the AN/ALE-40 chaff/flare dispenser, while the F-16 has the more advanced AN/ALQ-131 pod.[11]

## Training

Maj. Gen. Nabīl Shuwakrī, Chief of the Operations Department, EAF remarked:

> Primary training starts with the Egyptian-made Jumhurreyah; then we use the Brazilian Tucano for basic training and the Alpha Jet or the Czech L-29 for

> advanced training. When the pilots have accumulated between 200 and 300 flying hours, they convert to fighter. Because we do not manufacture our own aircraft, we treat the new generation of aircraft with great care. From my point of view, there are two things to teach a pilot: how to fly and how to fight. When the pilot fights, he has to forget the flying aspect and concentrate on the combat. After many flying hours on the MiG-19 [F-6] and MiG-21 [or F-7], pilots can fly an F-16. But everything depends on the pilot: we can move a graduate pilot straight to the Mirage V or F-16.
>
> After 1973, we decided to buy some Chinese F-6 and F-7s, which are equivalent to the MiG-19 and MiG-21 respectively, just to keep a generation of fighter pilots so that we would be able to convert them to any other modern aircraft. Most of our qualified F-16 pilots are former F-6 pilots. The F-6 is a tough aircraft. An experienced F-6 pilot will find it very easy to fly the F-16, though of course he will have to study the technology and all the systems which are different. The most important thing is to have skilled pilots, and it costs me nothing on the F-6 [to keep them trained].[12]

The EAF has streamlined its large and diverse inventory of training aircraft. The Ḥulwān Jumhūrīyah fleet has been modernized with bubble canopies and uprated 220 hp Continental engines. More than fifty EMB-312 Tucano turboprop trainers have been produced to replace the aging L-29 Delfin. In 1990, as a demonstration of improved relations, Libya gave the EAF ten Aero L-39 advanced trainers. An additional batch of forty-eight L-39s was purchased from Czechoslovakia in 1992 for use as advanced trainers and light attack aircraft.

A squadron of thirty Alpha Jets are also used for advanced and armament training. After flying the Alpha Jet or L-39, new EAF pilots usually fly a tour on the F-6 or F-7 before converting to high-performance F-4s, F-16s, or Mirages. Helicopter pilots first train on the Heller UH-12E and then fly the Gazelle before converting to larger rotorcraft. Navigators are instructed on four specially modified DHC-5 Buffalo aircraft.

Egypt has a long history of training foreign students. In the recent past the EAF has provided pilot training to students from Chad, Zimbabwe, Somalia, Sudan, and several Arab countries. Students from more than a dozen countries have attended EAF technical training courses.

A fleet of twenty-five C-130H Hercules long-range transports form the backbone of the Egyptian transport force. These turboprop aircraft are supported by five DHC-5D Buffalo, three Mystère Falcon, two

Gulfstream 3, four Boeing 707s, and one Boeing 737 jet. The EAF has a number of aerial refueling tankers: former airline 707s with tanks and refueling systems. These enhance the EAF's ability to deploy and maintain an airborne alert of fighters.

The EAF continues to maintain a sizable inventory of helicopters. Four squadrons with more than eighty SA-342 Gazelle light helicopters are in service. About half of these agile craft are armed with HOT wire-guided anti-tank missiles or 20 mm cannon. Fifteen CH-47Cs serve as the EAF's large lift helicopters, supported by twenty Westland Commandos and some fifty surviving Mi-8s. A small number of U.S. UH-60 Black Hawks have been purchased for VIP and special mission duties. The EAF will significantly upgrade its attack helicopter force through the addition of twenty-four AH-64 Apache helicopters during the 1990s.

As the force moved into the 1990s and Air Vice Marshal Aḥmad ʿAbd al-Raḥmān assumed command, the EAF faced new challenges. The air force is a reflection of the society in general, and a large percentage of its manpower is conscripts. The trend away from the study of technical subjects such as engineering and science and emphasis on Islamic studies has, however, reduced the pool of trained talent. Many of the experienced technicians that served as the backbone of the EAF's support groups are now retiring. Also, high-technology F-16s, E-2Cs, and other systems require new levels of training. To keep a MiG-21 flying you needed skills in hydraulics and metal work; for an F-16 it is writing software and maintaining electronic systems.

Egypt still maintains one of the largest and most diverse air forces in the Middle East. Since the 1973 War the EAF has transitioned from a force supplied and trained by the Soviet Union to one that operates a mix of Russian, American, French, British, and Chinese equipment. While the EAF has declined in size from 700 to about 400 aircraft, most of these are new generation systems. This is the same trend seen in nearly all western air forces and neighboring Israel. With fewer aircraft, pilot and support skills have become increasingly important. The experience level in the EAF is high due to regular investment in flight time, intensified training, and regular operational exercises. The air force has selected the best tactical concepts, aircraft, weaponry, and support systems to meet its unique needs.[13]

## Gulf Conflict (1990–1992)

President Mubarak committed more than 30,000 men and two Egyptian divisions to the defense of Saudi Arabia and other Persian Gulf countries. Egyptian commandos flown in by EAF C-130s were the first Arab troops to arrive in Saudi Arabia. EAF F-4s and F-16s were offered to support the war effort, but none were sent because of limited air base facilities. However, EAF fighters patrolled over allied shipping in the Red Sea. Egyptian forces were among the first to enter Kuwait during the invasion that recovered the country and pushed out the Iraqis. The EAF played only a supporting role during the conflict. A small number of Gazelle and Mi-8 helicopters operated with Egyptian forces, and EAF C-130s flew regular missions to carry troops and supplies.

Since then Egypt has continued to be very active in support of the United Nations. The Egyptian armed forces, including the EAF, already wear the "blue berets" in Bosnia and may do so in support of the UN elsewhere in the coming years.

As the EAF moves into its sixth decade, the force continues to develop and expand its capabilities. It is clearly important for Egypt to maintain its air capability in order to contribute to the defense of the country and keep its position as a leader in the Middle East.

## EAF Air Order of Battle 1994[14]

| Tactical | Recce | Helicopter | Transport | Training |
|---|---|---|---|---|
| F-16A/C × 110 | Mirage VR × 6 | AH-64 × 24 | C-130H × 22 | F-16B/D × 20 |
| Mirage 2000 × 20 | MiG-21R × 15 | UH-60 × 3 | DHC-5 × 5 | Mirage VSDD × 6 |
| F-4 × 32 | EC-130 × 2 | Gazelle × 50 | Mystère-Falcon × 4 | F-6/TF-6 × 20 |
| Mirage V × 54 | Beech 1900 × 4 | Mi-8 × 50 | Gulfstream III × 2 | Alpha Jet × 10 |
| MiG-21/F-7 × 100 | | Commando × 23 | Boeing 707 × 1 | L-29 × 20 |
| Alpha Jet × 36 | | Sea King × 5 | Boeing 737 × 1 | L-39 × 10 |
| F-6 × 25 | | CH-46C × 15 | | EMB-312 Tucano × 20 |
| | | UK-12E × 12 | | Jumhūrīyah × 80 |

# Appendix 1

## Aircraft Flown by the Egyptian Air Force

First figures are for length in meters, second figures are span; figure in parentheses is number of crew, followed by armament load, engine type, and thrust or horsepower (hp) and maximum takeoff weight.

*Aero L-29 Delfin*
12.4 m, 11.8 m; (2); 2 × 100 kg bombs or 8 rockets; 1 × M701, 892 kg thrust; 3,636 kg

*Airspeed Oxford 1*
12 m, 18.6 m; (3); 1 × 7.7 mm machine gun, 12 small bombs; 2 × AS Cheetah X, 355 hp; 3,454 kg

*Antonov An-12*
42.47 m, 43.63 m; (6); 2 × 23 mm cannon; 4 × Ivchenko AI-20K, 4,015 hp; 61,136 kg

*Avro 618 "Ten" (license-built Fokker Trimotor)*
16.625 m, 24.9 m; (3); —; 3 × AS Lynx, 215 hp; 4,648 kg

*Avro 626*
9.275 m, 11.9 m; (2–3); —; AS Cheetah V, 270 hp; 1,212 kg

*Avro 652*
14.79 m, 19.775 m; (3); —; 2 × Cheetah IX, 270 hp; 3,409 kg

*Avro 674 (license-built Hawker "Egyptian" Audax)*
9.9 m, 13 m; (2); AS Panther VI, 750 hp; 1,993 kg

*Avro Anson I*
14.79 m, 19.78 m; (4); 2 × 7.7 mm machine guns, 2 × 45.5 kg bombs, 8 × 9 kg bombs; 2 × AS Cheetah XV, 350 hp; 3,636 kg

*Avro Commodore*
9.54 m, 13 m (2); —; AS Lynx, 215 hp; 1,590 kg

*Avro Lancaster B1*
24.325 m, 35.7 m; (5); 8 × 12.7 mm Browning machine guns, 1,818 kg bombs; 4 × RR Merlin 24, 1,460 hp; 31,818 kg

*Beech D-18S*
15.37 m, 16.65 m; (2); —; 2 × P&W R 985-BS, 450 hp; 3,966 kg

*Beech 1900*
15.31 m, 19.075 m; (2); —; 2 × P&W PT6A, 850 hp; 3,955 kg

*Boeing 707/320*
53.5 m, 51 m; (4); —; 4 × P&W JT3D-7, 8,182 kg; 116,818 kg

*Boeing 737*
35 m, 32.55 m; (2); —; 2 × P&W JT8D, 6,591 kg; 45,454 kg

*Boeing Vertol CH-47C (built by Agusta)*
15.3 m, 21 m (rotor diameter); (3); —; 2 Lycoming T55, 3,750 hp; 20,909 kg

*Convair PBT-5 Catalina*
36.4 m, 22.34 m; (7); 2 × 12.7 mm and 2 × 7.6 mm machine guns; P&W R-1830, 1,200 hp; 15,455 kg

*Curtis C-46D Commando*
26.7 m, 39.8 m; (3); —; 2 × P&W R2800–75, 2,000 hp; 20,455 kg

*Curtis P-40 Tomahawk IIA*
11 m, 13 m; (1); 4 × 7.7 mm, 2 × 12.7 mm machine guns; Allison V-1710–33, 1,160 hp; 3,636 kg

*Dassault-Dornier Alpha Jet*
14,09 m, 10.47 m; (2); 1 × 30 mm cannon, 1,818 kg bombs; 2 × SNECMA Larzac 04, 1,352 kg; 6,818 kg

*Dassault Falcon*
19.69 m, 18.725 m; (2); —; 2 × GE CF 700–2D-2, 1,864 kg; 11,818 kg

*Dassault Mirage IIISDE*
17.24 m, 9.45 m; (1); 2 × 30 mm cannon, 909 kg bombs; SNECMA Atar 9C, 6,227 kg; 12,272 kg

*Dassault Mirage V SDD*
17.85 m, 9.45 m; (1); 2 × 30 mm cannon, 1,818 kg bombs; SNECMA Atar 9C, 13,636 kg

*Dassault Mirage 2000*
17.18 m, 10.3 m; (1); 2 × 30 mm cannon, 1,818 kg bombs, SNECMA M53, 9,090 kg; 13,636 kg

*De Havilland 60T*
8.37 m, 10.5 m; (2); —; DH Gipsy II, 120 hp; 795.5 kg

*De Havilland Canada DHC 1 Chipmunk*
8.9 m, 12 m; (2); —; DH Gipsy Major 8, 145 hp; 909 kg

*De Havilland Canada DHC-5 Buffalo*
27.65 m, 33.6 m; (3); —; 2 GE CT-64, 3,095 hp; 22,364 kg

*De Havilland DH 104 Dove*
13.77 m, 19.95 m; (2); —; 2 × DH Gipsy Queen Mk 3, 335 hp; 3,864 kg

*De Havilland Dragon-Rapide*
12.075 m, 16.8 m; (2); —; 2 × DH Gipsy Six, 200 hp; 2,500 kg

*De Havilland Vampire FB52*
10.76 m, 14 m; (1); 4 × 20 mm cannon, 8 × rockets; DH Goblin D, 1,523 kg; 5,636 kg

*De Havilland Vampire T55*
12,075 m, 14 m; (2); 2 × 20 mm cannon; DH Goblin 35, 1,590 kg; 5,455 kg

*Douglas C-47 Dakota*
22.575 m, 33.25 m; (3); up to 1,818 kg bombs internally; 2 × P&W R-1830, 1,200 hp; 11,818 kg

*EMB-312 Tucano*
11.32 m, 12.95 m; (2);——; PT-6A, 750 hp; 2,555 kg

*Fairey Gordon*
22.3 m, 16 m; (2); —; AS Panther II A, 520 hp; 2,682 kg

*Fiat G55 A1*
10.7 m, 13.59 m; (1); 4 × 12.7 mm machine guns, 2 × 160 kg bombs; Daimler Benz 601, 1,475 hp; 3,682 kg

*General Dynamics F-16A*
17.3 m, 10.85 m; (1); 1 × 20 mm cannon, 6 Sidewinder missiles, 4,545 kg bombs; P&W F100, 10,681 kg; 15,909 kg

*General Dynamics F-16C*
17.3 m, 10.85 m; (1); 1 × 20 mm cannon, 6 Sidewinder missiles, 4,545 kg bombs; GEF110, 12,273 kg, 17,273 kg

*Gloster Gladiator I/II*
9.6 m, 11.3 m; (1); 4 × 7.7 mm machine guns; Bristol Mercury IX, 840 hp; 2,159 kg

*Gloster Meteor F4/F8*
14.35 m, 13 m; (1); 4 × 20 mm cannon, 10 rockets; 2 × Rolls Royce Derwent 5, 1,590 kg; 8,682 kg

*Gloster Meteor T7*
15.225 m, 13 m; (2); —; 2 × Rolls Royce Derwent 8, 1,590 kg; 8,682 kg

*Gloster Meteor NF13*
17.47 m, 15 m; (2); 4 × 20 mm cannon; 2 × Rolls Royce Derwent 8, 1,659 kg; 9,313 kg

*Grumman E-2C Hawkeye*
19.72 m, 28.2 m; (6); 2 × Allison T-56A, 23,545 kg

*Grumman Mallard*
16,92 m, 23,33 m; (2); —; 2 × P&W R1340, 450 hp; 3,636 kg

*Ha Jumhūrīyah*
9.04 m, 12.16 m; (2); —; Continental C-145, 200 hp, 1,000 kg

*Ha-200 Al-Qāhirah*
10.3 m, 11.96 m; (2); 1 × 20 mm cannon; 2 × Turbomeca Marbore II, 400 kg; 3,355 kg

*Ha-300*
14.15 m, 6.77 m; (1); —; E-300, 4,500 kg; 5,455 kg

*Handley Page Halifax*
25.05 m, 36.46 m; (6); 4 × 7.7 mm machine guns; 4 × Bristol Hercules XVI, 1,650 hp; 29,545 kg

*Hawker "Egyptian" Audax*
9.89 m, 13 m; (2); 1 × 7.7 mm machine gun, 8 × 9 kg bombs; AS Panther X, 750 hp; 1,977 kg

*Hawker Fury prototype*
11.93 m, 13,42 m; (1); 4 × 20 mm cannon; Bristol Centaurus XII, 2,480 hp; 5,307 kg

*Hawker Hart*
10.27 m, 13 m; (2); 2 × 7.7 mm machine guns, 227 kg bombs; RR Kestrel 1B, 525 hp; 2,070 kg

*Hawker Hurricane IIC*
11.26 m, 14 m; (1); 4 × 20 mm cannon; RR Merlin 20, 1,650 hp; 3,682 kg

*Hawker Sea Fury*
11.99 m, 13.42 m; (1); 4 × 20 mm cannon, 910 kg rockets, bombs; Bristol Centaurus XVIII, 2,480 hp; 5,682 kg

*Heller UH-12E*
10.4 m, 12.4 m (rotor diameter); (2); —; Lycoming VO-504, 305 hp; 1,250 kg

*Ilyushin Il-14P*
24.4 m, 36.4 m; (5); —; 2 × Ash-82T, 1,900 hp; 18,000 kg

*Ilyushin Il-28*
20.27 m, 24.6 m; (4); 2 × 23 mm cannon, 3,000 kg bombs; 2 × Klimov VK-1, 2,727 kg; 20,909 kg

*Lockheed C-130H*
34.2 m, 46.4 m; (4) —; 4 × Allison T56, 4,500 hp; 77,273 kg

*Lockheed Loadstar*
22.925 m, 17.44; (2); —; 2 × Wright Cyclone 3, 1,200 hp; 7,955 kg

*Macchi MB 308*
7.47 m, 11.49 m; (2); —; Continental 65, 60 hp; 551 kg

*Macchi MC205V*
10.09 m, 11.49 m; (1); 2 × 20 mm cannon and 2 × 12.7 mm machine guns; FIAT RC-58 (license DB601), 1,465 hp; 3,628 kg

*McDonnell Douglas F-4E Phantom II*
21.93 m, 19.125 m; (2); 1 × 20 mm cannon, 4,545 kg bombs; 2 × GEJ79, 8,136 kg; 25,455 kg

*McDonnell Douglas Helicopter AH-64 Apache*
19.95 m, 16.8 m (rotor diameter); (2); 1 × 30 mm cannon, 16 Hellfire missiles, rockets; 2 × GET700, 1,550 hp, 7,955 kg

*MiG-15*
12.72 m, 11.58 m; (1); 2 × 23 mm cannon, 1 × 37 mm cannon, 2 × 250 kg bombs; Klimov VK-1, 2,705 kg; 6,455 kg

*MiG-15UTI*
11.5 m, 11.58 m; (2); 1 × 12.7 mm machine gun; Klimov VK-1, 2,705 kg; 6,200 kg

*MiG-17F*
12.72 m, 11.1 m; (1); 3 × 23 mm cannon, 2 × 250 kg bombs or 16 × 57 mm rockets; Klimov VK-1F, 3,177 kg; 7,045 kg

*MiG-19SF*
14.99 m, 10.325 m; (1); 3 × 30 mm cannon, 16 × 57 mm rockets; 2 × Tumansky RD-9B, 3,273 kg; 10,227 kg

*MiG-21F-13*
18.08 m, 8.2 m; (1); 1 × 30 mm cannon, 2 × K-13 Atoll missiles, 2 × 250 kg bombs; Tumansky R-11, 5,682 kg; 8,545 kg

*MiG-21MF*
18,08 m, 8.2 m; (1); 4 × K-13 Atoll missiles, 4 × 455 kg bombs, 1 × 23 mm cannon; Tumansky R-13, 6,590 kg; 9,420 kg

*MiG-21PF*
18.08 m, 8.2 m; (1); 2 × K-13 Atoll missiles, 16 × 57 mm rockets; Tumansky R-11-F2S-300, 6,000 kg; 8,864 kg

*MiG-21UM*
18,08 m, 8.2 m; (2); 2 × K-13 Atoll missiles; Tumansky R-13, 6,590 kg; 9,318 kg

*MiG-23B*
17.88 m, 16.36 m; (1); 1 × 23 mm cannon, 1,818 kg bombs; Tumansky R-27, 12,539 kg; 15,909 kg

*Mil Mi-2*
13.59 m, 16.625 m (rotor diameter); (2); —; 2 × Isotov GTD-350, 437 hp; 3,500 kg

*Mil Mi-4*
13.59 m, 24.12 m (rotor diameter); (2); —; Ash-82v, 1,700 hp; 7,773 kg

*Mil Mi-6*
38.09 m, 40.19 m (rotor diameter); (5); 1 × 23 mm cannon; 2 × Soloviev D-25V, 5,500 hp; 42,591 kg

*Mil Mi-8*
21.03 m, 24.44 m (rotor diameter); (4); 1 × 12.7 mm machine gun, 16 × 57 mm rockets; 2 × Isotov TU-2–117A, 1,900 hp; 13,000 kg

*Miles M14A Hawk*
8.4 m, 11.9 m; (2); —; DH Gipsy Major, 130 hp; 864 kg

*Miles M19 Master II*
10.325 m, 12.52 m; (2); —; Bristol Mercury XX, 870 hp; 2,579 kg

*Morane 502 (license-built Fiesler Fi156 Storch)*
11.35 m, 16.36 m; (2); —; Argus AS10C, 240 hp; 1,364 kg

*Mraz Sokol M-1D*
8.43 m, 11.46 m; (2); —; Walters Minor 4–111, 105 hp; 782 kg

*Muegyetemi Sportrepulo Egyesulet M-24*
9.1 m, 11.64 m; (2) —; Hirth HM 504, 100 hp; 382 kg

*Norduyn Norseman*
11.32 m, 18.05 m; (2); —; P&W Wasp, 600 hp; 3,364 kg

*North American T-6 Harvard*
10.12 m, 14.7 m; (2); 227 kg bombs; P&W R-1340, 600 hp; 2,409 kg

*Percival Q6*
11.29 m, 16,3 m; (2); —; 2 × DH Gipsy Six II, 205 hp; 2,500 kg

*Piper Super Cub*
7.82 m, 11.26 m; (2); —; Lycoming 0–290, 150 hp, 682 kg

*Republic P-47*
12.63 m, 14.175 m; (1); 8 × 12.7 mm machine guns, 909 kg bombs; P&W R-2899, 2,300 hp; 8,818 kg

*SA-342 Gazelle*
10.9 m, 12.04 m (rotor diameter); (2); 4 × HOT missiles, 1 × 20 mm cannon; Turbomecca Astazou, 592 hp; 1,705 kg

*Short Stirling IV*
31.675 m, 34.68 m; (4); —; 4 × Bristol Hercules, 1,650 hp; 31,818 kg

*Sikorsky S-61A*
25.43 m, 21.7 m (rotor diameter); (3); —; 2 × GE T-58, 1,500 hp; 10,000 kg

*Sikorsky UH-60*
17.79 m, 18.55 m (rotor diameter); (2); —; 2 × GET700, 1,536 hp; 10,000 kg

*Sukhoi Su-7BMK*
19.95 m, 10.24 m; (1); 2 × 30 mm cannon, 4 × 500 kg bombs, 4 × 16 57 mm rockets; Lyulka AL-7F, 7,402 kg; 13,468 kg

*Sukhoi Su-20*
21.525 m, 15.84 m; (1); 2 × 30 mm cannon, 7 × 500 kg bombs, 4 × 16–57 mm rockets; Lyulka AL-21F, 11,227 kg; 17,727 kg

*Supermarine Sea Otter*
13.91 m, 16.1 m; (3); —; Bristol Mercury 30, 855 hp; 4,545 kg

*Supermarine Spitfire V*
10.47 m, 12.89 m; (1); 2 × 20 mm cannon, 4 × 7.7 mm machine guns; RR Merlin, 1,440 hp; 2,917 kg

*Supermarine Spitfire Mk 9*
10.97 m, 12.89 m; (1); 2 × 20 mm, 4 × 7.7 mm machine guns; RR Merlin, 1,710 hp; 3,318 kg

*Supermarine Spitfire Mk 22*
11.43 m, 12.92 m; (1); 4 × 20 mm cannon; RR Griffon 61, 2050 hp; 4,091 kg

*Supermarine Spitfire T. Mk. 9*
10.97 m, 12.89 m; (2); —; RR Merlin, 1,580 hp; 3,545 kg

*Tupolev Tu-16*
39.96 m, 37.8 m; (6); 7 × 23 mm cannon, 3,000 kg bombs, AS-5 missile; 2 × Mikulin AN-3, 9,295 kg; 34,545 kg

*Westland Commando*
19.54 m, 21.7 m (rotor diameter); (2); 1 × machine gun; 2 × RR Gnome, 1,590 hp; 9,091 kg

*Westland Lysander*
10.675 m, 17.5 m; (2); 3 × 7.7 mm machine guns, 16 × 9 kg bombs; Bristol Mercury, 840 hp; 2,682 kg

*Westland Sea King*
19.54 m, 21.7 m (rotor diameter); (4); homing torpedoes; 2 × RR Gnome, 1,590 hp; 9,091 kg

*Westland Wessex IV*
13.3 m, 20.125 m; (3); —; 3 × Genet Major 140 hp; 2,864 kg

*Westland-Sikorsky S1*
19.95 m, 16.8 m (rotor diameter); (2); —; P&W Wasp Junior

*Yak 11*
9.74 m, 10.74 m; (2); 1 × 12.7 mm machine gun; Shvetsov Ash-21, 730 hp; 2,636 kg

*Yak 18*
9.57 m, 12.162 m; (2); 6 × 57 mm rockets; Ivchenko AI-14RF, 260 hp; 1,273 kg

*Zlin 226T*
8.95 m, 11.81 m; (2); —; Walter Minor 6-III, 160 hp; 818 kg

# Appendix 2

## Extracts from Annual Report No. 1 by the British Air Attaché in Cairo on the REAF during the 1948–49 Palestine War (Doc. 168248, FO371/80474, PRO, London)

(On prewar expansion plans and overall view of the REAF's future role)

The expansion plan for the Royal Egyptian Air Force which had been put before the Egyptian authorities in 1947 by the Military Mission seems to have been forgotten early in 1948. The Royal Egyptian Air Force at that time were making many enquiries, both in America and in the United Kingdom for newer types of aircraft, but seemed totally unable to reach a firm decision. No idea of a well thought out plan for expansion and training could be found; in fact it appeared that there was no cohesion and coordination of effort on the part of various departments of the Air Force Headquarters. "Ad hoc" measures were adopted for obtaining small quantities of equipment which suddenly became necessary. The basic natural trait of living only for the present without regard to the future was thus effecting the organization and efficiency of the Air Force with the result that disorganization and inefficiency (as judged by British standards) was the order of the day in the early months of 1948.

(On effects of total arms embargo imposed on Egypt by Britain from 4 June 1948)

This seriously handicapped the Egyptian Air Force who then turned to other countries for purchase of aircraft and arms. A considerable number of Fiats and Macchis were bought from Italy, and Stirlings were bought from Belgium. At the same time a great effort was made to repair their aircraft, particularly Spitfire aircraft, even to the extend of undertaking extensive repair work which even during the "Battle of Britain" days would not have been carried out in the United Kingdom. For example, two American "Thunderbolt" fighter aircraft were built from wrecks recovered from a scrap heap. Many Commando transport aircraft were overhauled, repaired and fitted for bombing, and many Spitfire air-frames and Merlin engines were modified locally to give increased speed and performance for these aircraft. To assist the service personnel in this work a force of approximately 3,000 civilian mechanics was enlisted by the Royal Egyptian Air Force. These mechanics had considerable experience having worked with the Royal Air Force and American Air Force during the 1939/45 war. This indicates a very sincere desire to engage in battle and the Royal Egyptian Air Force showed commendable energy and initiative in the way they tackled the problem.

(On situation on 31 December 1948)

At the end of the year the Egyptian land forces were in a very serious position, and early in December there were indications of a mounting threat towards Gaza and the danger that El Arish, the forward air base in Sinai might become unusable. On December 31st. I was called to the Ministry of War and Marine by the Minister, Haidar Pasha, who gave me the military situation in detail, and implored me for help despite the arms embargo, so that the Egyptian forces, both land and air should not be routed. The Israeli forces had, on 29th. December, 1948, crossed the border and entered Egyptian territory and directly threatened El Arish. As a result the Royal Egyptian Air Force were forced to evacuate El Arish and having no base between that airfield and Cairo were both likely to have several forced landings and probable crashes due to aircraft running out of fuel, unless use could be made of Royal Air Force airfields in the Canal Zone. It was decided, and the Minister of War was told by me, that the Royal Air Force airfields in the Canal Zone could be used by the Royal Egyptian Air Force to save their aircraft from risk of serious damage or loss but that refuelling facilities only would be provided. This offer of assistance was accepted with warm appreciation by the Minister and the Royal Egyptian Air Force.

(On morale and efficiency of the REAF during the Palestine War)

The morale of the Air Force remained high throughout the war up to 31st. December, 1948, despite the handicap of the arms embargo, and the Egyptian Air Force showed commendable initiative and a keen desire to engage in battle. Judged by Egyptian standards, the Air Force was efficient and gave valuable support to their ground troops as well as carrying out independent action. An example of their support of the Army was the use of Spitfire aircraft carrying long range petrol tanks filled with arms and equipment, which were dropped on the Egyptian garrison beleaguered at Faluja. Another example was the operation by night on bombing missions of Stirling four engined aircraft, despite the limited experience of the type, and in night operations in general.

(On final phase of the Palestine War)

The Egyptian land forces had been defeated by more experienced Israeli forces and were not keen to continue the battle; the Egyptian Air Forces, however, were still keen to fight despite heavy losses during the final phase when they had to evacuate El Arish. At the same time the Royal Egyptian Air Force recognized the fact that their combat efficiency was of a low order.

# Appendix 3

## The Syrian Air Force (1946–1958)

Syrian independence was declared by the country's first freely elected parliament in August 1943, but Syria remained under foreign occupation until spring 1946. The Syrian Air Force (SAF) was founded in that year when France was still controlling much of the country. Clashes between French troops and a hastily assembled Syrian militia of police and civilians culminated in a three-day French artillery and air bombardment of Damascus. But the British then intervened and forced the French back to their barracks. A totally new and unexpected sight then appeared in the skies over the Syrian capital, a silver-painted Piper Cub, the first aircraft of the newly created Syrian Air Force.

When the French military eventually departed in April 1946, they left behind only a few runways and some empty buildings. Within little more than a year the infant SAF was involved in the Palestine War against Israel, during which a small number Harvard trainers armed with machine guns and light bombs were used to support Syrian ground forces. Some were flown by Syrian pilots, others by volunteers from Arab countries, and some by European mercenaries. One Harvard was shot down by an Israeli fighter, while an Israeli Avia that disappeared over the Golan Heights in 1948 was destroyed by a Syrian Harvard.

After the Palestine War a new Syrian military regime came to power in a coup later claimed to have been helped by the American CIA.[1] This military government worked hard to build up an effective air force, a task made more difficult by the continuing conflict with Israel as well as confusion and political intrigue within Syria during the early 1950s. At first most of the SAF's new equipment came from Italy, including Fiat G59–2A, ex-Egyptian Macchi MC 205 fighters,[2] and some Spitfire F22s from Britain. The Fiat G59 was a Rolls Royce Merlin-powered version of the G55 Centauro then flown by the Egyptian Air Force. Other aircraft acquired during this period included De Havilland Canada Chipmunk trainers, French-built versions of the Junkers Ju-52 transport, C-47, Beech C-45s, and Fairchild F-24s. Syria also purchased some thirty Italian-built De Havilland Vampire jets in 1952, but these were transferred to Egypt.

The first jets to serve with the SAF were Gloster Meteor F8s, which, with some Meteor T7 trainers, were ordered from Britain in January 1950. This order was held up by a British arms embargo in October 1951. Finally, in September 1952, the Meteors were delivered. Additional Meteor F8 fighters, plus some Meteor NF13 night-fighters, were supplied to Syria in the mid-1950s.

Syria, like Egypt, then turned to the Soviet Union for modern combat aircraft and other weaponry. In 1955 twenty-five MiG 15 fighters and a small number of MiG 15UTI conversion trainers were ordered. A training mission composed of Russian and Warsaw Pact personnel was sent to Syria to instruct the SAF to fly and maintain these new jets. The Syrian MiGs were actually shipped to Alexandria, Egypt, and trucked to Almāẓah air base where they were assembled. Meanwhile Syrian aircrew and maintenance troops trained alongside the Egyptians at courses in the USSR and eastern Europe.

The tripartite assault on Egypt in 1956 caught Syria in the middle of this process of reequipment. SAF personnel inevitably became involved in the fighting as Syrian pilots were training alongside their Egyptian comrades at Abū Ṣuwayr near the Suez Canal. As the Anglo-French air assault began on the night of 31 October, a number of MiG-15s and Il 28 jet bombers destined for the SAF were flown to safety by Egyptian, Syrian, Soviet, and Czech pilots. However, a number of SAF training mission MiG 15s and MiG 15UTIs were destroyed on the ground at Abū Ṣuwayr by British and French aircraft. During the Suez conflict RAF Canberra and French RF-84 reconnaissance aircraft overflew Syria on several occasions to monitor activity at SAF bases. On 6 November 1956 a Canberra of No. 13 Squadron RAF was shot down by a Syrian Meteor while flying over Syria. The aircraft crashed in Lebanon, two of the crew bailing out safely, though a third was killed. Among the Syrian Meteor F8 pilots of this time was a certain Hafez al-Assad, the future president of Syria. He was involved in an attempted interception of a British Canberra which resulted in his crashing after landing in the dark at Nayrab airfield near Aleppo, a base not equipped for night flying. Assad came down near a Palestinian refugee camp and, having leapt from the now burning Meteor, he heard people running about among the trees looking for him. "No doubt they expected to find a corpse hanging from a branch," the Syrian president later recalled.[3]

Following the Suez War, the Soviet bloc sent a significant amount of aid to Syria to strengthen its defenses. Toward the end of 1956 Syria ordered sixty MiG 17s, and in December of the same year a group of SAF pilots went to the Soviet Union and Poland for training. The first batch of twelve MiG 17s arrived in Syria in January 1957. Egyptian pilots flew some of these over Damascus on Syrian National Day in April 1957 because of a shortage of Syrians qualified to fly the new jets.[4] All sixty MiG 17s had been delivered by late 1957. When the SAF pilots returned from training in Russia and eastern Europe, they manned the new fighter squadrons and began flying operational air defense missions over Syria. Jet training was resumed in Syria itself in 1957 when a fighter conversion course was established at Hama under the direction of Soviet instructors.

By the end of 1957 the SAF had two operational MiG 17 squadrons defending the capital from their base at Damascus-Mazzah. Two additional MiG 17 squadrons were being established, and advanced training at Hama on night and blind interception techniques were intensified to prepare for the delivery of MiG 17PF night-fighters which replaced the Meteor NF 13s in 1958. Some thirty pilots a year passed through initial instruction at an air college outside Aleppo where support personnel were also trained at the SAFs Aeronautical Technical Institute. Other pilot candidates continued to be sent to the USSR and East bloc countries for training.

# Notes

## Introduction

1. Wing Comdr. A. B. Dicken, letter to D. Nicolle, February 1993.

2. Jacques Berque, *Egypt: Imperialism and Revolution,* trans. J. Stewart (London, 1972), 553.

3. David C. Isby, September 1992.

## 1. Background

1. Brig.-Dr. ʿAbd al-Raḥmān Zakī, first curator of the Egyptian Army Museum, in conversation with D. Nicolle, Cairo, 1971.

2. A. B. Theobald, *Ali Dinar, Last Sultan of Darfur, 1898*–1916 *(London, 1965),* 188–89, 193, 195.

3. Letter in Great Britain Public Records Office (PRO), London.

> Ras el Tin Palace, June 3rd 1916.
> My Dear General,
> I have today received a letter from the Sirdar giving me details of the great services rendered by the Royal Flying Corps in the recent operations which led to the taking of El Fasher.
> I wish to convey to you without delay both my congratulations and my gratitude for the valuable help you have accorded us, and which has very largely contributed to the success of my Army's operations. . . .
> Believe me,
> My dear General,
> (signed) Hussein Kamil

4. Y. Kansu, S. Şensöz, and Y. Öztuna, *Havacılık Tarihinde Türkler,* vol. 1 (Ankara, 1971), 257–59.

5. David Nicolle, *Lawrence and the Arab Revolts* (London, 1989).

6. Kansu et al., 257–59, 323–31, 390–408.

7. Ibid., 361.

8. Letter from RAF Headquarters (HQ), Middle East Command, concerning the proposed establishment of an Egyptian air force (PRO, Doc. 1922, AIR 2, no. 1066, London).

9. Report from RAF HQ, Middle East Command (PRO, Doc. 1925, AIR 2, no. 1066, London).

## 2. A Start

1. Muslims are basically divided between the majority Sunni and minority Shīʿa groups. These two "sects" have much more in common than do Catholic and Protestant Christians. On the other hand, only Sunni recognized the authority of the caliphs. Comparisons between the Muslim caliph and the Christian pope are, however, misleading. Islam has no priests, no established ecclesiastical hierarchy, and religious questions are settled by the will of the *ummah* (Muslim people as a whole).

2. Air Vice Marshal Mīqātī (ret.), EAF, interview with D. Nicolle, Cairo, May 1973.

3. Muḥammad Ṣidqī later joined Egypt's civil airline, Misrairwork, as a captain, while his son subsequently joined the Egyptian Air Force. Nevertheless, Muḥammad Ṣidqī remains something of a mystery. According to Air Vice Marshal Tait, he was a civilian who learned to fly in Germany, but according to J. Berque (*Egypt*), he had served in the German Air Force. Air Commodore Jazzārīn offered a third version, suggesting that Ṣidqī was trained by German instructors but had served in the Ottoman Turkish Air Force. There certainly was a Lieutenant Ṣidqī in the Ottoman 4th Bölük (Squadron) who flew as an observer-gunner in southern Turkey and on the Syrian front in 1917–18 (Kansu et al.), 290, 320–22, 396–98.

4. Air Vice Marshal V. H. Tait, RAF, interview with D. Nicolle, London, 1974.

## 3. Fledgling Years

1. Tait.

2. ʿAli Muḥammad Labīb, *Al-Qūwah al-Thālithah: Tārīkh Al-Qūwah al-Jawwīyah al-Miṣrīyah (The Third Power: A History of the Egyptian Air Force)* (Cairo, 1977).

3. Tait.

4. Ibid.

5. T. W. Russell, *Egyptian Service 1902–1946 (London, 1949).*

6. *The Aeroplane,* 17 May 1933.

7. Tait.

8. Tait; Mīqātī.

9. Tait; *The Aeroplane,* 27 December 1933.

10. Air Commod. Ibrāhīm Ḥasan Jazzārīn EAF (ret.), EAF, interview with D. Nicolle, Cairo, May 1973.

## 4. A New Treaty and a World Crisis

1. Egypt did not yet have a navy, only a coastguard and fishery protection services with a handful of lightly armed vessels.

2. The first prototype Panther-powered Audax, designed specifically for Egypt and known as the "Egyptian Audax," was built by the Hawker Company itself. The remainder were constructed under license by the Avro Company with the designation Avro 674.

3. Report to the Foreign Office by the British Ambassador in Cairo, supplementary note by Major General Marshall-Cornwall, 1937 (PRO, Doc. 1937, FO 371, no. 20912, London).

4. Report by the British Advisory Mission to the Foreign Office in London, 24 May 1938 (PRO, Doc. 1938, FO 371, nos. 21939, 21944, London).

5. According to the Report by the British Advisory Mission to the Foreign Office in London, 9 November 1938 (PRO, London), the strength of the REAF consisted of twenty-one Avro 626s, four DH Moths, six Hawker "Egyptian" Audax Mark VIs and eighteen Mark Xs, nineteen Miles Magister primary trainers, one Avro 652, one Westland Wessex, and one Avro 642 Commodore.

6. Ibid.

7. Jazzārīn; correspondence with Sq. Ldr. I. Blair (ret.), RAF, 1990.

8. Report by the British Advisory Mission to the Foreign Office in London, 14 May 1939 (PRO, Doc. 1939, FO 371, no. 23331, London).

9. Ibid.

10. Jazzārīn.

11. Report No. 10 on the Egyptian Army by the British Advisory Mission to the Foreign Office in London, August 1939 (PRO, Doc. 1939, FO 371, no. 23334, London).

## 5. Neutrals at War

1. The REAF now had "on strength" eighteen Lysanders and thirty-two Gladiators, most of which were still in storage, backed up by a reduced number of Avro 626s and DH Moths, the Audaxes, the Avro 652, Wessex, Commodore, and an increased trainer force of Magisters plus six Gordons.

2. Report of the British Advisory Mission, 30 November 1939 (PRO, Doc. 1939, FO 371, no. 23337, London).

3. "Combined Plan for the Defence of Egypt," report to the Foreign Office, London, August 1939 (PRO, Doc. 1939, FO 371, no. 23326, London).

4. Report on the REAF by the British Advisory Mission, 21 April 1940 (PRO, Doc. 1940, FO 371, no. 24612, London); Labīb.

5. Jazzārīn.

6. Report on the REAF by the British Ambassador in Cairo, 12 September 1940 (PRO, Doc. 1940, FO 371, no. 24612, London).

7. British Foreign Office comments added to Report No. 11 on the Egyptian Army by the British Advisory Mission for the Foreign Office, London, December 1939 (PRO, Doc. 1940, FO 371, no. 24610, London).

8. Report No. 14 on the Egyptian Army by the British Advisory Mission for the Foreign Office, London, July 1940 (PRO, Doc. 1940, FO 371, no. 24610, London).

9. Labīb.

10. Report No. 13 on the Egyptian Army by the British Advisory Mission for the Foreign Office, London, April 1940 (PRO, Doc. 1940, FO 371, no. 24610, London).

11. Report No. 14 on the Egyptian Army.

12. Labīb.

13. Report on the REAF by the British Advisory Mission, 27 December 1940 (PRO, Doc. 1940, FO 371, no. 24612, London); Labīb.

14. This Mobile Brigade was the best integrated force in the Egyptian Army and was viewed as something of an elite. It included a light armored car regiment, a squadron of light tanks, a light artillery battery, an anti-tank battalion, a machine-gun company, a field engineering section, a detachment of ordinance maintenance personnel, six motor transport companies from the Egyptian Army Service Corps, and the flight of six REAF Lysander aircraft. Report on the Egyptian Army by the British Advisory Mission, to the Foreign Office, September 1940 (PRO, Doc. 1940, FO 371, no. 24626, London).

15. Jazzārīn.

16. Of the eight Italian submarines in the region, one was eventually captured, two sunk, and a third wrecked; the remainder were withdrawn to Europe in February 1941.

17. Confidential Memo from the British Advisory Mission, to the Foreign Office, London, 6 December 1940 (PRO, Doc. 1940, FO 371, nos. 24612, 24613, London).

18. A. Emiliani, G. F. Ghergo, and A. Vigna, *Immagini e storia dell'aeronautica italiana 1935–1945*: Regia Aeronautica, Il settore mediterraneo *(Milano, 1976);* I fronti africani *(Parma, 1979).*

19. Report No. 15 on the Egyptian Army by the British Advisory Mission for the Foreign Office, London, October 1940 (PRO, Doc. 1949, FO 371, no. 24610, London).

20. Reports to the Foreign Office, London, 31 October 1941 to 27 July 1942 (PRO, Doc. 1941, FO 371, no. 27383; Doc.1942, FO 371, nos. 31561, 31562, London).

21. A. W. Sansom, *I Spied Spies* (London, 1965); A. Sadat, *In Search of Identity* (London, 1978); Jazzārīn; Mīqātī; ʿAbd al-Raḥmān Zakī.

22. Sadat, *Identity.*

23. Reports by the British Advisory Mission for the Foreign Office, London, May and June 1942 (PRO, Doc. 1942, FO 371, nos. 31561, 31562, London).

24. The British Intelligence reports cited in note 23 listed the following REAF officers as being involved in anti-British activities. The most active was said to be Ḥaqqī Hārūn, one of the REAF's first Avro 626 pilots, who was believed to lead the REAF group; Aḥmad Saʿīd al-Shalabī; Ḥasan ʿĀkif, who became King Fārūqs ADC and personal pilot before being exiled following Colonel Nasser's Revolution in 1952; and Muḥammad ʿAbd al-Munʿim Aḥmad, who had also flown Avro 626s in the 1930s. Those listed as less active members were Muḥammad Ibrāhīm Abū Rabīʿah, Ibrāhīm Ḥasan Jazzārīn, Saʿīd ʿAlī Zaytūn, Kamāl al-Dīn Ḥamādah, Aḥmad Shawqī, ʿAbd al-Ḥamīd Abū Zayd, Aḥmad Nājī and Muḥammad ʿAbd al-Ḥalīm Khalīfah, who were both early Avro 626 pilots, Midḥat Muḥammad Qaṣdī, and Muḥammad Ṣabrī Sharābah. The royalist "secret organization" in the Egyptian Army was said to have been directed by Col. ʿAbd al-Raḥmān Zakī, director general of the Ministry of National Defense and a personal friend as well as admirer of General al-Mīṣrī.

25. ʿAbd al-Raḥmān Zakī.

26. Jazzārīn.

27. Sadat, *Identity;* C. Shores, *Fighters over the Desert* (London, 1969).

28. Confidential Memo from the British Advisory Mission to the Foreign Office, London, December 1942 (PRO, Doc. 1942, FO 371, no. 31573, London); Report on the REAF by the British Advisory Mission to the Foreign Office, London, 18 January 1946 (PRO, Doc. 1946, FO 371, no. 53268, London).

29. Confidential Memo from the British Advisory Mission, December 1942.

## 6. Forgotten Allies

1. Report by the British Advisory Mission to the Foreign Office, London, 21 December 1942 (PRO, Doc. 1942, FO 371, no. 31561, London).

2. Ibid.

3. Report by the British Advisory Mission to the Foreign Office, London, 25 July 1943 (PRO, Doc. 1943, FO 371, nos. 35546, 35549, London).

4. Early in 1944 the chief of the Advisory Mission reported on the REAF's uncertainty as to its future role, and suggested that: "It appears that, whereas Egypt can look forward to a time when there will be no very strong enemy within reach by land, there will always be an air threat to be taken into consideration." Report by the British Advisory Mission to the Foreign Office, London, 28 February 1944 (PRO, Doc. 1944, FO 371, 41313, London). Some months later a British Foreign Office representative stated that: "It is clear that under the guidance of the British Advisory mission, the Royal Egyptian Air Force have [*sic*] now begun to make a necessarily small but nevertheless direct contribution to the Allied war effort." The writer was none other than Anthony Eden who, as the British prime minister in 1956, was to have a less sanguine attitude toward Egypt's armed forces. Memo attached to a Report by the British Advisory Mission to the Foreign Office, London, 18 August 1944 (PRO, Doc. 1944, FO 371, no. 41314, London).

5. ʿAbd al-Raḥmān Zakī.

6. Reports of the British Advisory Mission to the Foreign Office, London, 23 July 1943, 28 February 1944, 18 August 1944, 15 January 1945 (PRO, Doc. 1943, FO 371, nos. 35546, 35549; Doc. 1944, FO 371, nos. 41313, 41314; Doc. 1945, FO 371, nos. 45477, 45945, London).

7. Ibid.

8. The Egyptian troops who undertook these balloon-defense duties were trained by No. 206 Wing RAF at Ismāʿīlīyah and eventually took over Nos. 971, 974, and 980 Squadrons balloon defenses in Port Said, Suez, and Alexandria, becoming in the process Nos. 1, 2, and 3 Balloon Squadrons REAF. While the basic work was largely carried out by Egyptian Army troops, technical tasks were carried out by REAF NCOs and warrant officers under Squadron Leader Ṣāliḥ and Flying Officers Makkāwī and Ṣalāḥ al-Dīn.

9. Among the most experienced radio operators in the REAF was Yūsuf Wāṣif, the first Copt (Christian Egyptian) to join the air force after having served in it as a civilian clerk for several years. Wāṣif was trained by Marconi and, after his retirement, was the senior representative of Egyptian Airlines, then UAA, in London. Tait.

10. Report by Wing Commander Burgess of the British Air Ministry, 18 August 1943 (PRO, Doc. 1943, FO 371, no. 35550, London).

11. Official information about the earliest stages of the "Egyptian Hurricane Saga" is not available, but this probably does not mean that sensitive documents are being withheld by the British Public Records Office. It is more likely that the relevant papers were among those known to have been destroyed at the British embassy in Cairo when, during the darkest months of 1942, Rommel's Afrika Korps was expected in the Egyptian capital at any moment.

12. Reports of the British Advisory Mission to the Foreign Office, London, 9 August 1944, 21 June 1945 (PRO, Doc. 1944, FO 371, nos. 41313 and 41314, and Doc. 1945, FO 371, nos. 45477 and 45945, London); letter from Wing Comdr. P. F. Rogers, 208 Sq., RAF, 14 July 1974.

13. Report of the British Advisory Mission, 9 August 1944.

14. Labīb.

15. Ibid.

16. Sq. Ldr. Muṣṭafá Māhir, and Pilot Officers Rifāʿī, Bakr, Izzat, Ḥashshād, and Saʿd al-Dīn Sharīf, the latter having joined the REAF early in 1940.

17. Reports of the British Advisory Missions, 25 July 1943, February 1944, August 1944, June 1945; Labīb.

18. Air Marshal Saᶜd al-Dīn Sharīf, acting chief ADC to President Sadat, interview with D. Nicolle, London, 1973, and subsequent correspondence in 1973. Saᶜd al-Dīn Sharīf remained in the Egyptian Air Force longer than any of the other ferry pilots. During the Suez Crisis of 1956 he flew as one of President Nasser's personal pilots and was wounded by British bombing. Another ferry pilot to reach prominence was ᶜAlī Ṣabrī, a member of the Free Officers movement who was chief of air force intelligence in 1952. He played a key role in the Revolution of that year and subsequently became one of President Nasser's closest confidants.

19. Saᶜd al-Dīn Sharīf.

20. Labīb.

21. Report of the British Advisory Mission, 9 August 1944.

22. Ibid.

23. The Egyptian FTS kept these Miles Masters flying for several months, a feat described as "a great credit to the maintenance personnel" by the British Advisory Mission.

24. British Advisory Mission Report, 21 June 1945.

25. Memos of the British Advisory Mission to the Foreign Office, London, 12 December 1944, 24 January 1945 (PRO, Doc. 1944, FO 371, nos. 40072, 41313, 41314, and Doc. 1945, FO 371, nos. 45930, 45988, London).

26. Memos from the British Embassy in Cairo to the Foreign Office, London, 27 March, 21 June, August 1945 (PRO, Doc. 1945, FO 371, nos. 45930, 45948, 45949, 45988, London).

27. Ibid.

28. Letter from the British Ambassador in Egypt to the Foreign Office, London, 21 March 1946 (PRO, Doc. 1946, FO 371, nos. 53268, 53269, London); Memo from the British Advisory Mission to the Foreign Office, London, 4 April 1945 (PRO, Doc. 1945, FO 371, nos. 45945, 45946, London); Report of the British Advisory Mission to the Foreign Office, London, 18 January 1946 (PRO, Dod. 1946, FO 371, no. 53268, London).

29. Mīqātī; Labīb.

30. Report by the British Advisory Mission, London, 21 December 1942.

## 7. Threatening Horizons

1. The following aircraft were supplied to the REAF between 1943 and 1946:

*In 1943:* 26 × Curtiss Tomahawk (loaned and returned); 20 × Hawker Hurricane Mk. II; 4 × North American Harvard (on loan)

*In 1944:* 9 × Avro Anson Mk. I; 1 × Avro Anson Mk. 22 (gift to King Fārūq); 12 × Hawker Hurricane Mk. II (on loan)

*In 1945:* 20 × Supermarine Spitfire Mk. V; 1 × Avro Anson Mk. 22; 20 × Hawker Hurricane Mk. II (on loan)

*In 1946:* 40 × Supermarine Spitfire Mk. 9; 3 × Airspeed Oxford (on loan).

Report by the British Advisory Mission to the Foreign Office, London, 19 February 1947 (PRO, Doc. 1947, FO 371, nos. 63074, 63076, London).

2. Report by the British Advisory Mission to the Foreign Office, London, 27 September 1946 (PRO, Doc. 1946, FO 371, nos. 53268, 53269, London).

3. The only exception was No. 2 Squadron based at Dākhilah outside Alexandria.

4. Report of the British Advisory Mission to the Foreign Office, London, 27 June 1946 (PRO, Doc. 1946, FO 371, no. 53268, London).

5. Tait.

6. Confidential assessment by the British War Office, 26 November 1946 (PRO, Doc. 1946, FO 371, no. 53268, London).

7. The aircraft on offer were Supermarine Spitfire Mk. V, Mk, VIII, and Mk. 9 fighters, Bristol Beaufighter Mk. X long-range fighter-bombers, and De Havilland Mosquito fighter-bombers and night-fighters; Foreign Office Memo, London, 7 June 1946 (PRO, Doc. 1946, FO 371, nos. 53268, 532629, London); Report of the British Advisory Mission to the Foreign Office, London, 2 September 1946 (PRO, Doc. 1946, FO 371, no. 53260, London). The Egyptian government wanted Supermarine Spitfire Mk. 24 fighters, Vickers Wellington Mk. XIV, or Vickers Warwick Mk. V bombers.

8. Reports by the British Advisory Mission to the Foreign Office, London, 19 February, 17 June 1947 (PRO, Doc. 1946, FO 371, nos. 63074, 63076, London).

9. Report by the British Advisory Mission, 19 February 1947.

10. Reports of the British Advisory Mission to the Foreign Office, London, 19 February, 14 April 1947.

11. Labīb.

12. Foreign Office Memorandum, 18 July 1947 (PRO, Doc. 1947, FO 371, nos. 63074, 63076, London).

13. Foreign Office notes added to the Report by the British Advisory Mission to the Foreign Office, London, 17 June 1947 (PRO, Doc. 1947, FO 371, nos. 63074, 63076, London).

14. Report of the British Advisory Mission to the Foreign Office, 14 April 1947.

15. Foreign Office Memorandum, London, January 1948 (PRO, Doc. 1848, FO 371, no. 69188, London).

16. Foreign Office Memoranda, London, January 1948 (PRO, Doc. 1948, FO 371, no. 69223, London).

17. Among these were three aircraft, owned by the Belgian Contracting & Trading Company, Hornton Airways, and Cairo Aviation, impounded by Sq. Ldr. Timurtāsh of the REAF: a Miles Aerovan (G-AIDJ), a Dakota (G-AKLL), and a Beechcraft Traveller (G-AIHZ). The Aerovan and Dakota were subsequently released. The Beechcraft was commandeered, but paid for. Foreign Office Memorandum, London, January 1948 (PRO, Doc. 1948, FO 371, no. 69188, London).

18. The Hawker Fury was the highlight of an air display that also included formation fly-pasts in groups of three by REAF Magister and Harvard trainers, C-47 transports, and Spitfire fighters. Three De Havilland Dove light transports, which had been flown from England by REAF crews, also arrived during the display.

19. W. Humble, in correspondence and conversations with D. Nicolle, London, 1971; report in the Cairo daily newspaper *Al-Ahrām*, 25 April 1948.

20. Memoranda from the British Embassy in Cairo to the Foreign Office, London, April and May 1948 (PRO, Doc. 1948, FO 371, no. 69188, London); Biography of Wing Comdr. ʿAbd al-Ḥamīd Abū Zayd, supplied by the EAF Historical Department, 1972.

21. Memo from the British Embassy in Cairo to the Foreign Office, 14 June 1948 (PRO, Doc. 1948, FO 371. no. 69223, London).

22. Reports of the British Advisory Mission to the Foreign Office, London, 19 February, 14 April 1947.

## 8. The First Offensive

1. Report of the British ambassador's meeting with the Egyptian prime minister, in a letter to the Foreign Office, 14 June 1948 (PRO, Doc. 1948, FO 371, no. 69223, London).

2. According to a report on Zionist terrorism drawn up by the British Foreign and Colonial Office in June 1946, Zionist forces in Palestine already numbered 67,000 men:

Haganah and Palmaḥ militia: 40,000
Mobile and fully trained field army: 16,000
Palmaḥ mobile force: 6,000
Irgun, Stern, and Leḥi terrorist groups: 5,000

PRO, Doc. 1946, FO 371, no. 52542, London.

3. By 14 May 1948 the total number of Arab troops available for war in Palestine was 14,426 regulars supported by 6,000–7,000 irregular volunteers, most of whom were Palestinian Arab militiamen. The Arab regular troops initially committed were:

Egypt, plus small supporting units from Sudan, Saudi Arabia, and North Africa: 5,000 (basically three infantry battalions)
Transjordan: 4,550 (basically four infantry battalions)
Iraq: 2,000 (basically infantry battalions number unknown)
Syria: 1,876 (basically two infantry battalions)
Lebanon: 1,000 (basically one infantry battalion).

From a typescript account of *The Faluja Battle,* supplied by the Egyptian Defence Attaché's Office, London, 1975.

4. Air Brig. Jabr ʿAlī Jabr's book, *Al-Qūwat al-Jawwīyah bayn al-Siyāsat al-Miṣrīyah wā Īsrāʾīlīyah, vol. 1: 1922*–1952 (Cairo, 1993), is a general strategic and politicomilitary analysis rather than a blow-by-blow operational history.

5. Munīrah Kafāfī, *Fī Dhikrá 15 Māyū 1948: ʿIndamā Ustushhida Abī! (In Memory 15 May 1948: When My Father Was Killed in Action)* (Cairo, 1975), 144–46.

6. Annual Report No. 1 by the British Air Attaché, Cairo, 23 March 1950 (PRO, Doc. 1950, FO 371, no, 80474, London).

7. Munīrah Kafāfī; Labīb.

8. Biography of Wing Comdr. ʿAbd al-Ḥamīd Abū Zayd.

9. Until Egypt releases a full official account of the REAF's operations over Palestine in 1948–49, this event remains a mystery. The Israelis released a press account of the event which was repeated without denial in Egypt's English-language newspaper, the *Egyptian Gazette.* The Israelis also released several photographs of a severely damaged Spitfire LF9 on a Mediterranean beach, but this aircraft carried neither correct Egyptian national markings nor an REAF serial number. Its pilot was also said to have been captured by Israeli troops who included at least one woman soldier. A man claiming to be Maḥmūd Barakah, "second pilot" to Squadron Leader Abū Zayd and the son of a Cairo police chief, was subsequently interviewed by a journalist whose commitment to the Zionist cause was without question. But recent research among REAF personnel who saw service in 1948 produced the following emphatic claim that the whole episode was propaganda designed to raise Israeli morale at a difficult time:

> Regarding the story of Barakah of which you mentioned in your letter, I've done some asking around (contacts among retired and serving members of the Egyptian Air Force) and the answer I got was the following: If they claim the Spitfire crash landed then, why have not the Israelis showed a picture of the papers all Egyptian pilots carried? Why no picture of the pilot? . . . Then they said: We had no prisoner returning by the name of Barakah simply because there was no officer by the name Barakah in our Air Force then. We had few pilots then and it was easy to know them all by their names and faces too! Most of the people I asked for proof said that you can't give proof that Barakah *never existed.* I was completely silenced and fell into laughter. It is Israeli propaganda as usual and I think it is about time they know that!

Mr. Sharīf Sharmī, correspondence with D. Nicolle, 1992.

10. Foreign Office memoranda of 23 May (PRO, Doc. 1948, FO 371, nos. 69188, 69223, London).

11. Munīrah Kafāfī, 147–48.

12. *Egyptian Gazette,* 30 May 1948.

13. Mīqātī; Jazzārīn; Biography of Wing Comdr. ʿAbd al-Ḥamīd Abū Zayd.

14. For a detailed account of the fighting in this part of Palestine, see D. Nicolle, "The Faluja Pocket," parts 1 and 2, *The Army Quarterly and Defence Journal,* October 1975 and July 1976.

15. Foreign Office Memoranda on the Ramat Dawid Incident (PRO, Doc. 1948, FO 371, nos. 69188, 69223, London); Annual Report No. 1 of the British Air Attaché, Cairo, 23 March 1950; Air Marshal Sir David Lee, *Wings in the Sun: A History of the Royal Air Force in the Mediterranean 1945*–1986 (London, 1989); Munīrah Kafāfī.

16. Munīrah Kafāfī, 187–88.

17. Mīqātī; Jazzārīn.

18. Labīb.

19. Biography of Wing Comdr. ʿAbd al-Ḥamīd Abū Zayd.

20. Annual Report No.1 of the British Air Attaché, Cairo, 23 March 1950.

## 9. Fighting the Hydra

1. Letter from Colonel Ryan to the British Embassy in Cairo, 27 May 1948 (PRO, Doc. 1948, FO 371, no. 69188, London).

2. N. Arena, *I Caccia della serie 5: Parte terza, Macchi 205* (Modena, 1976); information from Dr. Ing. G. Cattaneo, November 1994.

3. The REAF was undertaking repair work which even during the Battle of Britain would not have been carried out in the United Kingdom. For example two American [P-47] Thunderbolt fighter aircraft were built from wrecks recovered from a scrap heap. Many Commando transport aircraft were overhauled, repaired and fitted for bombing, and many Spitfire airframes and Merlin engines were modified locally to give increased speed and performance for these aircraft. To assist the service personnel in this work a force of approximately 3,000 civilian mechanics was enlisted by the Royal Egyptian Air Force. These mechanics had considerable experience having worked with the Royal Air Force and American Air Forces during the 1939/45 war. This indicated a very sincere desire to engage in battle and the Royal Egyptian Air Force showed commendable energy and initiative in the way they tackled the problem.

Annual Report of the British Air Attaché, Cairo, 23 March 1950.

4. *Egyptian Gazette,* 19 July 1948.

5. Tait; Jazzārīn.

6. Mīqātī.

7. On 9 and 10 July the REAF struck Tel Aviv airfield, factories along the Yarqon river north of Tel Aviv, Dayr Haim, Dorot, and Kefar Am. Troops and armor concentrations were also attacked in the Ruhāmah area. On 10 July REAF Spitfires again bombed and strafed Israeli positions at Ruḥāmah, Nirʿam, Negba, and Beʾer Tuviyya. On 12 July the Tel Aviv area was attacked again; *Egyptian Gazette,* 11 and 12 July 1948.

8. Nicolle, "The Faluja Pocket."

9. Munīrah Kafāfī, 150–51.

10. *Egyptian Gazette,* 19 July 1948.

11. Annual Report of the British Air Attaché, Cairo, 23 March 1950.

12. Israel had so far seized 201 of the 219 Arab villages that then existed within the UN-designated Jewish state. Israel had also taken a further 112 Arab villages within the UN-designated Arab state. The Arab armies had captured 13 Jewish settlements within the UN-designated Arab state and in the UN-designated International Zone around Jerusalem. They had also taken one Jewish settlement within the UN-designated Jewish state.This left Israel in control of 1,300 sq. km of territory allotted to the Arabs by the UN, while 330 sq. km of land allotted to the Jews had fallen to Arab control.

13. After being refurbished, the Stirlings were resold to a Belgian charter company called Trans-Air of Melsbroeck, six being fitted out for cargo, six for passengers. For some time they flew a sporadic service from Blackbushe in England to Shanghai in China, one crashing on takeoff at Kunming in December 1947. In 1948 five of these transports were sold to the Tangiers Charter Co., four then being sold to Egypt.

14. Munīrah Kafāfī, 154–55.

15. N. Arena, *I Caccia della serie 5: Parte seconda, Fiat G.55* (Modena, 1976), 149–50.

16. Foreign Office Memoranda of 5 January to 17 February and 4 May 1949 (PRO, Doc. 1949, FO371, no. 73573, London).

17. *Tall:* small artificial hill resulting from thousands of years of human habitation on one site, typical of many parts of the Middle East.

18. Labīb.

19. *Al-Ahrām,* 17 October 1948; *Egyptian Gazette,* 17 October 1948.

20. Information passed from British air attaché in Cairo to the Foreign Office, 19 October 1948 (PRO, Doc. 1948, FO 371, no. 69188, London).

21. The westernmost part of the Egyptian line was at Isdūd north of al-Majdal (held by the 2d Infantry Brigade). The line then curved south and east through Jūlis (held by the 7th Infantry Battalion), ʿIrāq Suwaydan (held by the 2d Infantry Battalion), Faluja (held by the 1st Infantry Battalion), ʿIrāq al-Manshīyah, and Bayt Jibrīn (an extended position held by the 6th Infantry Battalion). East of this the Arab front line was held by a mixture of Transjordanian Arab Legion regulars, Palestinian militias, and some mixed irregular forces still apparently under Egyptian command. "The Faluja Battle," Egyptian Army Historical Division; D. Nicolle, "The Faluja Pocket."

22. "A Biographical Note on the Late Squadron Commander Mohamed Abdul Hamid Abu Zeid," EAF Historical Department, via the Egyptian Air Attaché, London (1974).

23. According to the EAF's biographical note on Abū Zayd:

> His confidential reports always cited him as "exceptional," "highly experienced and notably active." He was an excellent officer, highly skilled in fighters' operations, noticeably daring and valorous in active field service. He made a record of 72 air sorties during the operations from 15 May to 19 October 1948, when he was announced missing in action. He was always the first to inaugurate air operations following the expiry of every truce, and always made more than one air sortie at rush times when the enemy's air and land activities were increased.
>
> On May 15, 1948, flying a Spitfire, he led the first raid on Tel Aviv airport, destroying a landing Dakota and setting the hangers aflame.
>
> He also took part in supporting the Army during land battles and in protecting the ground forces against enemy aircraft. Besides he took part in bombing the Israeli vital targets in Tel Aviv. In the first week of June 1948 he dropped an Israeli light passenger aircraft which had been used as a bomber.

Shortly after the end of the Palestine War the Egyptian authorities proposed naming one of the REAF's main air bases after Sq. Ldr. Abū Zayd, though nothing seems to have come of this idea.

"Biographical Note on Abu Zeid," EAF Historical Dept.

24. Munīrah Kafāfī, 187.

25. Ibid.

26. Information passed from British Air Attaché in Cairo to the Foreign Office, 19 October 1948 (PRO, Doc. 1948, FO 371, no. 69188, London).

## 10. A Losing Battle

1. Sayyid Ṭāhā, from an undated extract from an Egyptian magazine account of the Faluja battles, including extracts from Sayyid Ṭāhā's diary (supplied by Brig.-Dr. ʿAbd al-Raḥmān Zakī, first director of the Egyptian Army Museum, Cairo, 1974).

2. Ibid.

3. Jazzārīn.

4. Munīrah Kafāfī, 148–49.

5. G. Cattaneo, letter to *Aeroplane Monthly,* June 1988.

6. Jazzārīn; Mīqātī.

7. Munīrah Kafāfī, 160–61.

8. Muṣṭafá Shalabī al-Ḥinnāwī rose to become commander of the Egyptian Air Force after the catastrophic Arab-Israeli war of June 1967. Maj. Gen. Muṣṭafá al-Ḥinnāwī (ret.), interview with L. Nordeen, 26 March 1989, Cairo.

9. *Egyptian Gazette,* 29 December 1948.

10. Munīrah Kafāfī, 188.

11. *Egyptian Gazette,* 30, 31 December 1948; 2, 4, 5 6, 7 January 1949.

12. At the beginning of January, the remainder of No. 208 Squadron's Spitfires were also flown into Fāʾid air base from Cyprus.

13. Annual Report of the British Air Attaché, Cairo, 23 March 1950.

14. Ibid.

15. Foreign Office memorandum concerning IAF flights over the Canal Zone (PRO, Doc. 1948, AIR 20. no.6906, London); Foreign Office memorandum on Egyptian requests for long-range fuel tanks (PRO, Doc. 1949, FO 371. no. 69287, London).

16. Annual Report of the British Air Attaché, Cairo, 23 March 1950. This arrangement was originally meant to last only while Israeli forces remained inside Egyptian territory, but in fact the REAF continued to use al-Ballāḥ throughout most of 1949.

17. L. Boyce, May 1990, E. Thomason, June 1991, and F. Adkin, July 1990 (RAF ret.) in private correspondence with D. Nicolle; F. Adkin, "Mossies over the Med," *Aeroplane Monthly* (August, September, October 1989), 466–69, 546–84, 634–36.

18. Air Marshal Sir David Lee, *Wings in the Sun: A History of the Royal Air Force in the Mediterranean 1945–1986* (London, 1989). The Spitfires of No. 208 Squadron and the Mosquitos of No. 13 Squadron, based at RAF Fāʾid in the Suez Canal Zone, undertook these missions with cover being provided by Tempests of Nos. 6 and 213 Squadrons.

19. Lee.

20. Annual Report of the British Air Attaché, 23 March 1950.

21. Ibid.

22. Ibid.

23. "Report by Squadron Leader J. R. Baldwin on the Training Given to the Royal Egyptian Air Force by the Royal Air Force—April/May, 1949," Appendix B to Air Attaché Report No. 1 on Royal Egyptian Air Force (PRO, Doc. 1950, FO 371, no. 80474, London).

24. Ibid.

25. Ibid.

26. Information from Dr. Ing. G. Cattaneo, November 1994.

27. Annual Report of the British Air Attaché, 23 March 1950.

28. The air attaché's report was probably wrong in referring to this "unnumbered" bomber squadron. The single Halifax probably formed part of No. 8 (Bomber) Squadron. The C-47s were shortly reported in No. 3 (Transport) Squadron, the C-46s in No. 7 (Trans.) Squadron, and the Bonanzas joined other assorted aircraft in a new No. 10 (Comm.) Squadron.

## 11. Reactionaries and Revolutionaries

1. The last Macchi was flown to Almāẓah by the Macchi representative and World War II Italian fighter ace, Ettore Foschini, in June 1951.

2. Foreign Office memorandum, February 1949 (PRO, Doc. 1949, FO 371, no. 74984, London).

3. Annual Report of the British Air Attaché, Cairo, 23 March 1950.

4. Foreign Office correspondence of September 1949 (PRO, Doc. 1949, FO 371, 73574, London).

5. Annual Report of the British Air Attaché, Cairo, 23 March 1950.

6. A. Sabit, *A King Betrayed* (London, 1990). Contrary to Mr. Sabit's belief, the British were aware of the presence of a German officer in Cairo and of the latter's proposal to recruit other German officers; Foreign Office Memoranda (PRO, Doc. 1950, FO 371, no. 80466; Doc. 1951, FO 371, nos. 90183, 90184, London).

7. The collection, housed at Almāẓah, reportedly included a DH Moth Trainer, Avro 626, "Egyptian" Audax, Hawker Hart, Hurricane, Gloster Gladiator, Westland Lysander, and a Spitfire; Mīqātī.

8. Foreign Office Memorandum, July 1949 (PRO, Doc. 1949, FO 371, no. 73574, London); Annual Report of the British Air Attaché, Cairo, 23 March 1950.

9. Ibid. Nine reconditioned Lancaster B1 and Halifax A9 bombers were ordered to replace the unsatisfactory Short Stirlings. Three Gloster Meteor F4 jet fighters and a Meteor T7 trainer were requested to supplement the two F4s ordered earlier and now due to arrive. Seven additional Meteor F4s were ordered later in 1949.

10. Mīqātī; Foreign Office correspondence, January 1950 (PRO, Doc. 1950, FO 371, no. 80473, London).

11. One Halifax was soon lost due to maintenance problems. ʿAbd al-Laṭīf Baghdādī was taxiing to the runway with a complement of REAF mechanics on their way to England for a Vampire maintenance course when the aircraft burst into flames. No one was hurt, but the Halifax was a write-off; Mīqātī. The last of nine ex-RAF Halifax A9s to arrive in Egypt in 1949 was the final production machine (RAF serial number RT938), which had been accepted by the RAF in November 1946.

12. Unconfirmed intelligence reports stated that an unidentified jet fighter exploded in Israeli airspace in April 1950. The Syrians had no jets, and the Egyptians did not lose anything around this date. The British speculated that this unidentified aircraft was a Vampire, whereas REAF Intelligence reported that eight crated S 92 jet fighters (Czech-built Messerschmitt Me 262 A-1As) had been secretly delivered to Israel. Memorandum from the British Embassy in Cairo to the Foreign Office, London, 19 April 1950 (PRO, Doc. 1950, FO 371, no. 80474, London).

13. The REAF eventually acquired twelve Meteor F4 fighters and three T7 trainers. An even bigger order for Meteors had been sent to Glosters in October and December 1949, but it was never completed, the aircraft in question going to the RAF and to Denmark.

14. The first Vampire to reach Egypt was a standard FB5, although all subsequent Vampires were FB52s, plus twelve T55 trainers.

15. In the summer of 1949 work started on a Parachute Training School, and a dozen Airspeed Horsa transport gliders were purchased with a view to creating an airborne force in the Egyptian Army, but this was abandoned early the following year. In 1950 an Air Force College was also established at Bilbays, modeled on the famous RAF College at Cranwell. During 1949, sixteen additional Miles Magister trainers had arrived in Egypt by ship, having been purchased from the civil register.

16. Britain had provisionally agreed to supply Marconi AMES 13, 14, and 21 radars back in October 1948; Foreign Office Memorandum (PRO, Doc. 1948, FO 371, no. 69188, London). Two sets of Marconi AMES 21 radars had been ordered in May 1949, but Egypt was still trying to obtain these three years later; Foreign Office Memoranda of March 1952 and June 1952 (PRO, Doc. 1952, FO 371, no. 96993, London).

17. Annual Report No. 2 on the REAF by the British Air Attaché in Cairo, 24 January 1951 (PRO, Doc. 1951, FO 371, no. 96993, London).

18. Foreign Office Memoranda of 15 August 1952 (PRO, Doc. 1952, FO 371, no. 90175, London).

19. The REAF's inventory was large but remarkably varied at this time and included the following types of aircraft: Avro Anson Mk. 1 and Mk. 19, Beechcraft AT-7, Boeing B-17 Flying Fortress, Consolidated Catalina, Curtis C-46, De Havilland Dove, De Havilland Vampire FB5 and FB52, DHC Chipmunk, Douglas C-47, Fiat G55, Gloster Meteor F4 and T7, Grumman Avenger, Grumman Mallard, Handley Page Halifax C8 and A9, Lockheed Loadstar, Macchi MC205, Miles Magister, Mraz Sokol, Norduyn Norseman, North American Harvard, Republic P-47 Thunderbolt, Short Stirling Mk. IV, Stinson (probably AT-19 Reliant), Supermarine Sea Otter, Supermarine Spitfire LF9, Vultee Valiant, Westland Lysander, Westland-Sikorski S-51 Dragonfly; Annual Report No. 2 on the REAF by the British Air Attaché in Cairo, 24 January 1951.

20. Ibid.

21. In October 1949 the joint RAF-REAF "Exercise Gestic" tested the air defenses of the Canal Zone, but an accompanying joint British and Egyptian public air display was cancelled. In contrast, a big parade and fly-past over Cairo in November, which included the first appearance by REAF jets, commemorated the centenary of the death of Muḥammad ʿAlī, founder of modern Egypt, and was accompanied by an outburst of Egyptian nationalism.

22. Two Macchi MB 308 light aircraft had arrived before the end of the Palestine War as "gifts" following the conclusion of the Italian MC 205V order, though these did not officially form part of the REAF's Royal Flight. Two Westland-Sikorski S-51 Mark 1B Dragonflies were sold to the REAF in September 1949 and continued to be the subject of a tussle between the REAF and its Royal Flight. One of the REAF's Curtiss C-46s was also sent to Italy where it was fitted out with a bathroom and bedroom so that the king could fly nonstop, and in some style, to and from London. The king's favorites, however, remained his two Grumman Mallard amphibians. By the time the 1952 Revolution erupted, the Egyptian Royal Flight consisted of no less than thirteen aircraft; Mīqātī; Annual Report No. 2 on the REAF by the British Air Attaché in Cairo, 24 January 1951.

23. British Embassy in Cairo Memorandum to the Foreign Office, London, January 1950 (PRO, Doc. 1950, FO 371, nos. 73574, 80474, London).

24. Ibid; Mīqātī.

25. The list of aircraft ordered by Egypt but currently held up by the British authorities was: 20 × Balliol trainers, 34 × Chipmunk trainers (plus an unknown number of additional aircraft), 8 × Furies, 6 × Lancasters, 3 × Meteor T7s, 24 × Meteor F8s, 6 × Spitfire trainers, 20 × Spitfire 18s, 22 × Vampire FB52s, 18 × Vampire FB52s to be assembled in Egypt, 16 × Vampire night-fighters, 2 × Westland-Sikorski S51 helicopters; Foreign Office correspondence, 13 May 1952 (PRO, Doc. 1952, FO 371, no. 96968, London).

26. Annual Report No. 2 on the REAF by the British Air Attaché in Cairo, 24 January 1951.

27. Ibid.

28. Sabit; Sadat, *Identity.*

29. Ibid.

30. Sadat, *Identity.*

31. For example, General Najīb and Anwar Sadat flew from Cairo to Alexandria in a DH Dove of No. 10 Squadron on 25 July to arrange King Fārūq's departure from Egypt.

32. Sabit.

33. Mīqātī. The air vice marshal's son, Ramzī Mīqātī, was similarly unable to pursue his own flying career in Egypt. Instead he went to England, slightly altered his name to Ramsey McCarthy, and became a pilot with British Airways.

34. This C-47 continued to serve as President Nasser's personal transport until it was replaced by a new Ilyushin Il-14 presented by the USSR in 1955.

35. A. Sadat, *Revolt on the Nile* (London, 1957).

36. Tait.

37. Jazzārīn.

38. Annual Report No. 2 on the REAF.

## 12. "Czech" Arms

1. Memo from the British Embassy in Cairo to the Foreign Office, 7 August 1952 (PRO, Doc. 1952, FO 371, no. 96993, London).

2. Secret cipher from the Prime Minister to the Secretary of State at the Foreign Office, 27 January 1953 (PRO, Doc. 1953, FO 371, no. 102878, London). It should be pointed out that Winston Churchill had a personal antagonism toward the Egyptian armed forces dating back to his experiences in the Sudan in the late nineteenth century.

3. After considerable pressure from the British government, Italy admitted that 13 Vampires had been sold to Syria, 45 to another unidentified Middle Eastern country which was probably Egypt, and that 25 would soon go to Saudi Arabia. Two EAF squadrons are believed to have been equipped with these aircraft, probably Nos. 30 and 31. Foreign Office Memo, 18 November 1955 (PRO, Doc. 1955, FO 371, nos. 113707, 113708, 113709, London).

4. This EAF jet conversion unit is believed to have been No. 5/6 Combined Squadron. From this period on, the exact identities of Egyptian squadrons have rarely been released officially.

5. Letter from Wing Commander Bradley, RAF (ret.) to D. Nicolle, 15 February 1990.

6. Letter from Flight Lieutenant Bushe, RAF (ret.) to D. Nicolle, 21 March 1990.

7. Annual Report No. 6 for the year 1955, by the British Air Attaché in Cairo, 24 January 1956.

8. Article in the *Jewish Chronicle* (London), 30 July 1965.

9. Israeli radio broadcast taped by the BBC Monitoring Service, Cavesham, 1955.

10. Memo from the British Air Attaché in Cairo to the Foreign Office, 27 January 1955 (PRO, Doc. 1955, FO 371, no. 113707, London).

11. Various Foreign Office Memos of 4 April, 30 and 31 August, 3 and 7 September, and 1 November 1955 (PRO, Doc. 1955, FO 371, nos. 113707, 113708, 113709, London).

12. *Keesings Contemporary Archives,* February–March 1956.

13. Ibid., December 1956.

14. Annual Report No. 6 for the year 1955, by the British Air Attaché in Cairo, 24 January 1956 (PRO, Doc. 1956, FO 371, no. 119009, London).

15. Ibid.

16. Ibid.

17. Memo from the British Embassy in Cairo to the Foreign Office, 18 January 1956 (PRO, Doc. 1956, FO 371, nos. 119007, 119008, 119009, 119011, London).

18. Annual Report No. 6 for the year 1955, by the British Air Attaché in Cairo, 24 January 1956.

19. Ibid.

20. EAF Historical Department information via correspondence with Sharīf Sharmī.

## 13. The Other Side of Suez

1. Patrick Facon, "Trente ans plus tard: L'Armée de l'Air à Suez ou l'autopsie d'une victoire," *Air Fan,* November 1986, pp. 8–17.

2. Alfred Goldberg, "Air Operations in the Sinai Campaign 1956" (formerly secret study), USAF Historical Division; EAF Historical Department information via correspondence with Sharīf Sharmī.

3. "Air Superiority and Airfield Attack: Lessons from History," BDM Corporation Report prepared for the Defense Nuclear Agency, 1982; "Operation Kadesh, IDF/AF 1950–1956, Buildup and Operations," IDF/AF History Branch, 1986.

4. Air Vice Marshal R. A. Mason and John W. R. Taylor, *Aircraft, Strategy and Operations of the Soviet Air Force* (London, 1986), 65.

5. Paul Gaujac, *Suez 1956* (Paris, 1986); Facon.

6. EAF Historical Department information via Sharīf Sharmī: In 1956 Egyptian anti-aircraft weapons were the responsibility of the Egyptian Army's artillery branch. Egypt's anti-aircraft defenses were divided into three elements; Unit 1 (Cairo sector): 31 heavy and 11 light guns; Unit 2 (northern sector): 19 heavy and 14 light guns; Unit 3 (eastern sector): 13 heavy and 9 light guns.

7. Lee, *Wings in the Sun;* Victor Flintham, "The Suez Campaign, 1956," *Scale Aircraft Modeling,* November 1986, pp. 54–85.

8. Interview (D. Nicolle) and correspondence with Air Marshal Sa'd al-Dīn Sharīf, acting ADC to President Sadat, 14 December 1973, Cairo; Ehud Yonay, *No Margin for Error: The Making of the Israeli Air Force* (New York, 1993), 160–62.

9. Sir Anthony Eden, *The Memoires of R. H. Sir Anthony Eden,* vol. 2: *Full Circle* (London, 1960), 524.

10. Guy Ramon, "Pride of Place," *Israel Air Force Magazine,* 1989; Memorial plaque to the dead of the Suez War in the Military Museum, Cairo, Labīb.

11. Interview (L. Nordeen) with Brig. Gen. Y. Shavit (ret.), 17 November 1987, Tel Aviv; Memorial plaque, military Museum Cairo.

12. Brig. Gen. Fārūq al-Ghazzāwī (ret.), interview with author (L. Nordeen), 29 March 1989, Cairo.

13. Robert Jackson, *The Israeli Air Force Story* (London, 1970), 76.

14. "The Wild Horses, The Mustang Squadrons," *Born in Battle no. 44,* 1947. Well after the war an Israeli publication admitted that a Mustang pilot, "[named] K, forced down twice, by AA fire and an Egyptian Vampire, and thought dead, managed to hitch a ride back to Ekron and surprise his squadron commander."

15. Charles Christienne and Pierre Lissarague, *The History of French Military Aviation,* trans. Frances Kianka (Washington, D.C., 1986), 473.

16. Gaujac.

17. BBC-TV program, "Suez-Ten Years After" (London, 1966).

18. Lee, 75–76.

19. Flintman, "Suez Campaign."

20. Labīb.

21. Many books have been written about the Suez War, particularly by those intimately concerned. One excellent source is Anthony Nutting's *No End of a Lesson: The Story of Suez* (London, 1967). This book and many others make it clear that the safety of the Suez Canal was more of a pretext than a real cause for concern by the Anglo-French-Israeli Alliance.

22. Muḥammad H. Haykal, (Heikal) *Cutting the Lion's Tail: Suez through Egyptian Eyes* (London, 1986), 186.

23. Maj. Gen. Muṣṭafá al-Ḥinnāwī (ret.), interview with author (L. Nordeen), 29 March 1989, Cairo.

24. D. Lee, 76–79.

25. Al-Ḥinnāwī.

26. Maj. Gen. Muḥammad Nabīl al-Masīrī (ret.), interview with author (L. Nordeen), 28 March 1989, Cairo.

27. Al-Ghazzāwī, 29 March 1989.

28. Al-Ḥinnāwī.

29. Al-Masīrī, 28 March 1989.

30. Al-Ghazzāwī, 29 March 1989.

31. Gen. Lucien Robineau, "Les Portes-à-Faux de l'Affaire de Suez," *Revue historique des armes,* no. 4, 1986, 41–50.

32. Lee, 82.

33. Lee; Memorial plaque, Military Museum, Cairo, Egypt, 82–85.

34. Lee, 90.

35. Ibid.
36. Flintman, "Suez Campaign."
37. Ibid.

## 14. A Doubtful Anniversary

1. U.S. Navy, *Office of Naval Intelligence Review (ONI Review),* September 1959, vol. 14, no. 9, p. 406.
2. Nadav Safran, *From War to War: The Arab-Israeli Confrontation, 1948–1967* (New York, 1969), 210–12.
3. *ONI Review,* 401–3.
4. Ibid., 402–3, 408.
5. Sadat, *Identity,* 205.
6. Very little information has ever been published about the Yemeni and Syrian air forces. One of the few histories to be published on Syria appeared in an Italian military journal, "Syrian Air Force," *JP4,* Florence, October and November 1975. *ONI Review,* September 1959, also provides a good overview of the status of both the Syrian and Yemeni air arms in the late 1950s.
7. U.S. Navy, *ONI Review,* December 1958, vol. 14, no. 7, 253.
8. Safran, 209–12.
9. Information on the 14 February 1960 air victory over Syria came from the EAF History Branch; the authors have a copy of a gun-camera photo reportedly from the engagement.
10. *Far East Intelligence Roundup,* June 1954, issue 151, vol. 5, no. 6, secured through Freedom of Information act request.
11. Edward Luttwak and David Horowitz, *The Israeli Army* (New York, 1975), 223.
12. Nile Valley Fledgelings: Egypt's Aircraft Industry Described," *Air Pictorial* (London), January 1976.
13. Annual Report No. 1 for the period 1/1/1958–31/1/1950, of the British Air Attaché in Cairo, 23 March 1950 (PRO, London).
14. Annual Report for 1951 of the British Air Attaché in Cairo, 24 January 1952 (PRO, London).
15. Ibid.
16. Memo from the British Embassy in Cairo to the Foreign Office, 19 January 1955 (PRO, London).
17. Information supplied in confidence by the staff of the BBC Arabic Service to the author (D. Nicolle), London, 1971, and discussions with EAF personnel during a visit to Cairo by author (Nordeen) in 1989.
18. Letter from the Office of the Indian Air Attaché, London, 1975.
19. *Flight International* news archives.
20. BBC Arabic News Service, London.
21. Joseph S. Bermudez, "Ballistic Missile Development in Egypt," unpublished technical paper prepared for the Missile Proliferation Project, Monterey Institute of International Studies, 28 August 1992, 10–19.
22. Maj. Gen. ʿĀdil Naṣr, interview with author (L. Nordeen), 9 November 1987, Cairo.
23. Maj. Gen. Nabīl Shuwakrī (ret.), interview with author (L. Nordeen), 9 November 1987, Cairo.
24. U.S. Navy, *ONI Review,* December 1958, vol. 14, no. 7.

## 15. Wider Horizons

1. "The First Transplant . . . The Super Mystère in Action," *Born in Battle, Defense Update No. 88, Special 40th Annual IAF Issue,* June 1988, pp. 40–42.

2. Brig. Gen. Tamīm Fahmī ʿAbd Allāh (ret.), interview with author (L. Nordeen), 27 March 1989, Cairo.

3. Mason and Taylor, 28–32, 65.

4. Al-Ghazzāwī, 31 March 1989.

5. ʿAbd Allāh.

6. Maj. Gen. ʿAwaḍ Ḥamdī (ret.), interview with author (L. Nordeen), 9 November 1987, Egyptian Air Force Headquarters, Cairo.

7. ʿAbd Allāh.

8. Lon Nordeen, *Air Warfare in the Missile Age* (Washington, D.C., 1985), 46.

9. Col. Taḥsīn Zakī (ret.), letter to D. Nicolle via Sharīf Sharmī, May 1991.

10. J. N. Westwood, *The History of Middle East Wars* (Greenwich, Conn., 1984), 70–72; Patrick Brogan, *The Fighting Never Stopped: A Comprehensive Guide to World Conflict since 1945* (New York, 1990), 334.

11. ʿAbd Allāh.

12. David Holden and Richard Johns, *The House of Saud: The Rise and Rule of the Most Powerful Dynasty in the Arab World* (New York, 1981), 230–35.

13. "Free World Air Intelligence Brief, United Arab Republic, AP-240–3-1–67-INT, 1 January 1967," Defense Intelligence Agency (DIA), pp. 2–3, acquired through a Freedom of Information Act request.

14. Marquis W. Childs, "Egypt Using Gas in Yemen," *St. Louis Post Dispatch,* 19 June 1967.

15. Seymour Hersh, *Chemical and Biological Warfare: America's Hidden Arsenal* (Garden City, N.Y., 1969), 243–46.

16. DIA Air Intelligence Brief.

17. ʿAbd Allāh.

18. Al-Masīrī, 26 March 1989.

19. Zakī.

20. Safran, 210–12.

21. DIA Air Intelligence Brief. According to American sources, in early 1967 the EAF had an order of battle as follows: 6 all-weather fighter squadrons, 15-day fighter squadrons, 3 light bomber squadrons, 2 medium bomber squadrons, 1 medium bomber flight, 4 transport squadrons, 2 helicopter squadrons, and 1 headquarters support unit.

22. Ibid.

23. Gunther E. Rothenberg, *The Anatomy of the Israeli Army* (New York, 1979), 140.

24. DIA Air Intelligence Brief.

25. Ibid.

26. Ibid.

27. Nordeen, *Air Warfare,* 16, 225.

28. Zakī.

29. Brig. Gen. Muṣṭafá Ḥāfiẓ, interview with author (L. Nordeen), 29 March 1989, Cairo.

30. Edgar O'Ballance, *The Third Arab-Israeli War* (Hamden, Conn., 1972), 9–14.

31. Ḥamīd.

32. Shavit.

33. Ḥamīd.

34. An official air order of battle of the EAF on the eve of the 1967 War has not been released. This is the best assessment available. Sources included: 1967 DIA Air Intelligence Brief; conversations with the Israeli Air Attaché, London, 1968 (D. Nicolle); *Flight International* News Archives; BBC News Service; DBM Report, "Air Superiority and Airfield Attack: Lessons from History" and "The Six Day War," *Born in Battle,* no. 6, 1979; material from Israeli Defence Force archives supplied by Vajda Ferenc-Antal, November 1995 (D. Nicolle).

## 16. The Surprise Assault

1. *Military Balance* 1966–67, International Institute for Strategic Studies (IISS) (London, 1966).
2. "Free World Air Intelligence Brief, 1 January 1967, United Arab Republic, AP-240–3-1–67-INT," DIA.
3. O'Ballance, *The Third Arab-Israeli War,* 79.
4. D. K. Palit, *Return to the Sinai* (New Delhi, 1974), 23.
5. Al-Masīrī, 26 March 1989.
6. Ḥamīd, 29 March 1989.
7. Brig. Gen. Samīr ʿAzīz Mīkhāʾīl (ret.), interview with author (L. Nordeen), 29 March 1989, Cairo.
8. Brig. Gen. ʿAwaḍ Ḥamdī (ret.), interview with author (L. Nordeen), 27 March 1989, Cairo.
9. Ḥāfiẓ, 27 March 1989, Cairo.
10. "Airfield Attack—Lessons of Middle East Wars," *Born in Battle,* no. 37, p. 13.
11. Ḥamīd, 29 March 1989.
12. Mīkhāʿīl, 29 March 1989.
13. Shuwakrī (ret.), 28 March 1989.
14. Al-Masīrī, 26 March 1989.
15. Information from *Military Balance, 1968–1969, IISS; Trevor Dupuy,* Elusive Victory: The Arab-Israeli Wars, 1947–1974 (New York, 1978), 337. On the first day of the 1967 War, Egypt suffered the loss of an estimated 322 aircraft.
16. Naṣr, 9 November 1987. This victory was confirmed by the Israeli Air Force officer who commanded the one Super Mystère squadron during the 1967 War; Shavit.
17. Taḥsīn Zakī (ret.), letter to D. Nicolle, May 1991; David Nicolle, "1967—An Egyptian Testament," *Air Pictorial International* (London), August 1992, pp. 420–25.
18. Maj. Thomas J. Marshal, "Israeli Helicopter Forces: Organization and Tactics," *Military Review,* May 1972, pp. 94–98.
19. Nāṣr.
20. Al-Ghazzāwī, 26 March 1989.
21. Ibid.
22. Zakī.
23. Ibid.
24. Dupuy, 339.
25. Shuwakrī, 9 November 1987.
26. Al-Masīrī, 26 March 1989.

27. Dupuy, 333; "Massive Resupply Narrows Israeli Margin," *Aviation Week and Space Technology,* 19 June 1967, p. 16: EAF losses listed as 30 Tu-16, 29 Il-28, 24 Il-14, 8 An-12, 10 Mi-6, 1 Mi-4, 100 MiG-21, 29 MiG-19, 87 MiG-15/17, 14 Su-7, and 4 unidentified aircraft.

28. Zeʿev Schiff, *A History of the Israeli Army* (London, 1985), 155.

## 17. Egyptian Phoenix

1. Roger F. Pajak, "The End of Soviet Arms Aid to Egypt?" a paper presented at the 8th Annual International Affairs Symposium 1973–74.

2. A. J. Barker, *The Arab-Israeli Wars* (London, 1980), 104–8.

3. Dupuy, 343–60.

4. Ḥamīd, 29 March 1989.

5. Col. Taḥsīn Zakī (ret.), letter to D. Nicolle, May 1991.

6. Maj. Gen. Farīd F. Ḥarfūsh, interview with author (L. Nordeen), 27 March 1989, Cairo.

7. Zakī.

8. David A. Korn, *Stalemate: The War of Attrition and Great Power Diplomacy in the Middle East, 1967–1970, (Boulder, Colo., 1992), 79–82.*

9. Lawrence L. Whetten, "June 1967 to June 1971: Four Years of Canal War Reconsidered," *New Middle East,* June 1971, pp. 15–17.

10. Edgar O'Ballance, *The Electronic War in the Middle East, 1968–*1970 (Hamden, Conn., 1972), 73

11. Whetten.

12. Pajak.

13. "Air Forces Intelligence Study (AFIS), 1 January 1971, United Arab Republic," DIA-240–3-1–71-INT, secured through a Freedom of Information Act request from the Defense Intelligence Agency, 1991.

14. Al-Masīrī, 29 March 1989.

15. "Free World Air Intelligence Brief," 1 December 1967, AP-240–3-1–69-INT, United Arab Republic, secured through a Freedom of Information Act request from the Defense Intelligence Agency, 1991, p. 3.

16. Briefing and interview with technical representatives of the Ḥulwān Government Aircraft Factory, Egypt, 2 April 1989.

## 18. Fighting Back

1. Jean Zumbach, *On Wings of War: My Life as a Pilot Adventurer* (London, 1977), 284.

2. Shuwakrī, 11 November 1987.

3. Briefing to author (L. Nordeen) by Col. A. Badr, Research Department, Egyptian Air Defense Command, 10 November 1987, Cairo.

4. Ibid.

5. "Free World Air Intelligence Brief, 1 February 1969, United Arab Republic," DIA, secured through a Freedom of Information Act request, 1991.

6. Maj. Gen. Muṣṭafá al-Ḥinnāwī, interview with author (L. Nordeen), 26 March 1989, Cairo.

7. Ḥamīd, 29 March 1989.

8. Al-Masīrī, 26 March 1989.

9. Ḥamīd, 29 March 1989.

10. "UAR and Israel Wage Artillery Duel across Suez," *New York Times,* 9 March 1969, p. 1.

11. *Military Balance* (London, England, International Institute for Strategic Studies) 1968–1969, p. 45; "Free World Air Intelligence Brief, 1969." p. 6.

12. "Free World Air Intelligence Brief, 1969," 4.

13. Al-Masīrī, 26 March 1989.

14. Mīkhā'īl, 29 March 1989.

15. Chaim Herzog, *The Arab-Israeli Wars: War and Peace in the Middle East* (New York, 1982) 225.

16. O'Ballance, 75.

17. ʿAbd Allāh.

18. Gerald Astor, "The World's Toughest Air Force—It Keeps Israel Alive," *Look Magazine,* 30 May 1970, pp. 17–23.

19. Ibid.

20. "Free World Air Intelligence Brief, 1969," 5.

21. Raymond A. Anderson, "Spirit More Aggressive," *New York Times,* 24 July 1969; James Ferron, "Israelis Say Jets Again Hit Egyptians," *New York Times,* 26 July 1969.

22. "Israel Hits Egyptian Targets near Canal," *New York Times,* 20 August 1969.

23. Mīkhā'īl.

24. James Ferron, "Israel Reports Downing 11 Jets in Suez Clashes"; and Thomas F. Brady, "Egyptians Report Strikes," *New York Times,* 12 September 1969.

25. "A Talk with General Bar Lev," *Newsweek,* 24 November 1969, p. 52.

26. Mīkhā'īl.

27. *Military Balance* 1969–1970, IISS.

## 19. Attrition War

1. John Bentley, "Inside Israel's Air Force," *Flight International,* 19 March 1970, p. 427.

2. Col. Aḥmad ʿĀṭif, interview with author (L. Nordeen), 28 March 1989, Cairo.

3. "Flash on the Combat Scene of April 2," *Flash of Damascus* (Damascus), May 1970, pp. 2, 8.

4. Badry, Maj. Gen. Hassan el (ret.), Maj. Gen. Ṭāhā el Magdoub (ret.), and Maj. Gen. Muḥammad Ḍiā' el Din Zohdy, (ret.), *The Ramadan War 1973* (Dunn Loring, Va., 1978), 138–39.

5. "Israelis Say They Shot Down 3 Syrian MiGs," *New York Times,* 9 January 1970.

6. Al-Masīrī, 26 March 1989.

7. "Mrs. Meir Declares Israeli Raids Show That Nasser Is a Failure," *New York Times,* 6 February 1970, p. 6.

8. "Israelis Report Losing Jet in Dogfight at Canal," *New York Times,* 10 February 1970.

9. Col. Rifʿat Fatḥī, interview with author (L. Nordeen) Ṭanṭá air base, Egypt, 30 March 1989.

10. Peter Grose, "Nasser Concedes That Israelis Have Air Superiority in Mideast," *New York Times,* 9 February 1970.

11. Aleksey Basenko, "In the Air over Egypt: The Time Has Come to Discuss," *Krasnaya Zvezda* (in Russian), 20 September 1990.

12. Ibid.

13. "Israelis Seek to Thwart SAM-3s", *New York Times*, 25 March 1970, p. 8.

14. "Soviet Participation in the Arab-Israeli War of Attrition Recalled," *Krasnaya Zvezda* (in Russian), 25 March 1989, p. 2; Alan Georges, "Soviet Troops Fought in War of Attrition," *Defense*, October 1989, p. 750.

15. John Bentley, "Inside Israel's Air Force Part 2: The Enemy We Face," *Flight International*, 23 April 1970, p. 669.

16. Basenko.

17. Whetten, 19.

18. Al-Masīrī, 26 March 1989.

19. Edward H. Kolcum, "Soviets Shifting Middle East Balance," *Aviation Week and Space Technology*, 16 May 1970, pp. 18–21.

20. "Soviets Deploy New Suez Defenses", *Aviation Week and Space Technology*, 13 July 1970, pp. 14–16.

21. Merav Halperin and Aharon Lapidot, *G-Suit: Combat Reports from Israel's Air War* (London, 1990), 73–75.

22. Whetlen, 19.

23. Halperin and Lapidot.

24. Al-Masīrī, 26 March 1989.

25. "Air Forces Intelligence Study, 1 January 1971, United Arab Republic," Defense Intelligence Agency, secured through a Freedom of Information Act request 1991, pp. 4–7.

## 20. New Directions

1. "Air Forces Intelligence Study, 1 January 1971, United Arab Republic," DIA.

2. Whetten, 17–20.

3. Ibid.

4. 1971 DIA study.

5. Ibid.

6. Peter Young, "Phantom Shadow over the Suez," *Life Magazine*, 25 September 1970.

7. Edward H. Kolcum, "SAM Changes Force New Strategy," *Aviation Week and Space Technology*, 16 November 1970, 16–21.

8. "Soviet Naval Air Threat," *Defense Intelligence Digest*, August 1970, pp. 4–5.

9. Patrick Brogan, *The Fighting Never Stopped: A Comprehensive Guide to World Conflict since 1945* (New York, 1990), 103–4.

10. "MiG-25 Foxbat," *Warplane, no. 18*, 1985, 346–55.

11. Al-Masīrī, 26 March 1989.

12. Fatḥī.

13. Ibid.

14. *U.S. State Department Area Handbook, Egypt 1990*.

15. Maj. Gen. Ḥasan al-Badrī, Maj. Gen. Ṭāhā el Magdoub Maj. Gen. Mohammed Dia el Din Zohdy, *The Ramadan War, 1973*, Dunn Loring, Va. 1978. pp. 21–30.

16. *Military Balance 1972–73, IISS;* "Air Forces Intelligence Study," 1 October 1974; "The Yom Kippur Arab-Israeli War," *Warplane*, no. 93, 1983, pp. 1841–45.

17. "Air Forces Intelligence Study, (AFIS) DI-240-EG-74, Egypt (Arab Republic of), 1 October 1974, Defense Intelligence Agency, secured through Freedom of Information Act request 1990.

18. *Military Balance 1971–72*, IISS.

## 21. Ramadan War

1. General Muḥammad al-Jamasī, deputy prime minister, minister of war, and commander in chief of the Egyptian Armed Forces, "The Military Strategy of the October 1973 War," Cairo University International Symposium on the October 1973 War, Cairo, 27–30 October, 1975, Technical Paper distributed by the Egyptian government at the meeting.

2. Steven Zaloga, *Soviet Air Defense Missiles* (London, 1989), 65–68, 88–89.

3. Col. Muḥammad Ibrāhīm, interview with author (L. Nordeen), al-Manṣūrah air base, Egypt, 29 March 1989.

4. Al-Masīrī, 26 March 1989.

5. Badry, Maj. Gen. Hassan el (ret.), Badry et al., 156.

6. Col. ᶜĀdil Sharārah, interview with author (L. Nordeen), Cairo West air base, Egypt, 28 March 1989.

7. Ḥāfiẓ.

8. Col. Aḥmad Wafāʾī, interview with author (L. Nordeen), EAF Headquarters, Heliopolis, Egypt, 9 November 1987.

9. Jerry Asher and Eric Hammel, *Duel on the Golan: The 100 Hour Battle That Saved Israel* (New York, 1987), 54.

10. Robert Hotz, "Offense, Defense Tested in 1973 War," in *Both Sides of the Suez: Airpower in the Middle East,* a special reprint of *Aviation Week and Space Technology* articles, 1975, pp. 38–39.

11. Halperin and Lapidot, 92–94.

12. Sadat, *Identity,* 147–49.

13. Julian S. Lake and Richard V. Hartman, "Air Electronic Warfare," *U.S. Naval Institute Proceedings,* October 1976, p. 48.

14. Radio Baghdad, 9 October 1973, BBC Monitoring Service Report from Arabic, October 1973.

15. Halperin and Lapidot, 101–5.

16. *Insight on the Middle East War, Sunday Times* (London, 1974).

17. Dupuy, 18.

18. *Insight on the Middle East War.*

19. Ibid.

20. *Born in Battle, Defense Update International, Issue no. 42, 10 Years: The Yom Kippur War, The War in the Air,* p. 18.

21. Al-Ḥamīd, 29 March 1989.

22. Al-Masīrī,, 26 March 1989.

23. John F. Kreis, *Air Warfare and Air Base Defense* Special Studies, Office of Air Force History (Washington, D.C., 1988), 338.

24. Mīkhāʾīl.

25. Maj. Gen. Avraham Adan (ret.), *On the Banks of the Suez: An Israeli General's Personal Account of the Yom Kippur War* (Denver, Colo., 1980), 225.

26. Radio Cairo, 9 October 1973, BBC Monitoring Service Report from Arabic, October 1973.

27. Wafāʾī.

28. Lt. Gen. Muḥammad ᶜAlī Fahmī, chief of staff, Egyptian Armed Forces, "The Role of the Egyptian Air Defense Force in the October 1973 War," Cairo University International Symposium on the October 1973 War, 27–30 October 1975, papers distributed by the Egyptian government at the meeting.

29. "Soviet Aid Sparks Arab Gains," *Aviation Week and Space Technology,* 15 October 1973, p. 13.

## 22. Battles for the Bridgeheads

1. Lt. Col. Rifʿat Yusrī and Brig. Gen. Farīd Ḥarfūsh, interview with author (L. Nordeen), Ṭanṭá air base, Egypt, 30 March 1989.

2. "Soviet Aid Sparks Arab Gains," 12–13.

3. Col. ʿĀdil Sharārah, interview with author (L. Nordeen), Cairo West air base, Egypt, 28 March 1989.

4. Mark Lambert, "Middle East Market Report," *Flight International*, 13 March 1975; D. Nicolle, interview with members of the *Flight International* editorial staff, August 1975, London.

5. Fatḥī.

6. Henry Tanner, "Egyptian Forces Launch Major Offensive in the Sinai" and Craig Whitney, "Israel Puts Enemy Loss of Tanks at 220," *New York Times*, 15 October 1973.

7. Al-Ḥamīd, 29 March 1989.

8. William B. Quandt, "Soviet Policy in the October 1973 War," *Rand Paper R1864-ISA*, May 1976; Dupuy, 568–70.

9. Robert Hotz, "Egypt Plans Modernized Air Arm," *Aviation Week and Space Technology, Special Issue*, 1975.

10. Dupuy, 505.

11. Wafāʾī.

12. Badry et al., 82, 98, 107; Terrence Smith, "Hundreds of Tanks Clash in Struggle for Suez Area," *New York Times*, 18 October 1973.

13. David Nicolle, "The Holy Day Air War," *Air Enthusiast International*, May 1974, pp. 244–49.

14. Leslie H. Gelb, "Jets Flown by North Koreans Are Reported in Clash with Israel," *New York Times*, 19 October 1973; "Israeli Forces Say North Koreans Pilot Some Egyptian MiGs," *New York Times*, 16 August 1973.

15. Ḥāfiẓ.

16. Fatḥī.

17. Col. Muḥammad Ibrāhīm, interview with author (L. Nordeen), al-Manṣūrah air base, 29 March 1989.

18. Ḥāfiẓ.

19. Adan, 325; Hotz, "Offense, Defense Tested in 1973 War."

20. ʿIzzat.

21. ʿĀṭif.

22. Wafāʾī.

23. Sharārah.

## 23. Ramadan War Impact

1. Lambert, 419–24.

2. DIA, 1974.

3. Charles W. Corddry, "The Yom Kippur War: Lessons New and Old," *National Defense*, May–June 1974, p. 509.

4. "The War in the Air," *Born in Battle, Defense Update no. 42, Yom Kippur Special*, 1983, p. 18.

5. Wafāʾī.

6. Kreis, 336; "Yom Kippur War," *Warplane,* no. 93, 1983, pp. 1841–45.

7. "Air Power in the Middle East," *Aviation and Marine,* July–August, 1975, p. 55.

8. *FY75 U.S. Senate Armed Services Committee Report, Tactical Air Power Subcommittee,* 11–20 March 1974, pp. 4306–14.

9. Badry et al., 82, 112; Dupuy, 550.

10. Corddry.

11. Robert Hotz, "Israeli Air Force Faces New Arab Arms," *Aviation Week and Space Technology, Special Issue, Both Sides of the Suez: Airpower in the Mideast,* 1975, pp. 6–10.

12. "The War in the Air," *Born in Battle;* "Selected Readings in Tactics," *1973 Middle East War,* U.S. Army Command and Staff and General Staff College Publication, RB-100–2, Ft. Leavenworth, Kans., August 1976, pp. 5–11/5–13; data secured from a retired senior Israeli Air Force general officer listed 186 Egyptian air combat losses during the 1973 War; Herbert J. Coleman, "Israeli Air Force Decisive in War," *Aviation Week and Space Technology,* 3 December 1973, pp. 18–21, listed 200 EAF air combat losses vs. 3 for the IAF; Dupuy, p. 555.

13. Hotz, "Israeli Air Force Faces New Arab Arms."

14. Ibid.

15. Robert Hotz, "Egypt Plans Modernized Air Arm," *Aviation Week and Space Technology,* 30 June 1975, p. 12.

16. Letter from Sharīf S. Sharmī to D. Nicolle, 10 September 1991. According to Sharmī: "Five EAF pilots are officially aces—four with five kills and the fifth with more than six—true figures! As I said before, some pilots scored during 1967 and went on to score in 1973 while flying MiG-17s and MiG-21s. Also, two of these scored during the period from 1967 to 1970 when the War of Attrition ended. I'm sorry I cannot say more because of censorship." Interviews (by L. Nordeen) with Maj. Gen. Nabīl Shuwakrī (ret.) and Maj. Gen. ʿAwaḍ Ḥamdī (ret.), 9 November 1987. A letter from B. Tillman, of the *American Fighter Aces Association,* indicated that an EAF pilot named ʿAlī Wajaʿī, who flew MiG-21s during the War of Attrition and Ramadan War, reportedly achieved ace status.

17. *Aviation and Marine,* 55.

18. Kreis, 334; *1973 Middle East War,* U.S. Army Command and Staff Paper, 5–11/5–13.

19. *Aviation and Marine,* 55.

20. *1973 Middle East War,* U.S. Army Command and Staff Paper, "Airmobile Operations," pp. 5–6/5–8.

21. Nicolle, "The Holy Day Air War," 248; Corddry.

22. Radio Cairo, BBC Monitoring Service Report from Arabic, February 1974.

23. DIA, "Air Forces Intelligence Study, 1974."

24. Dupuy, 569; Quandt.

25. Anthony H. Cordesman, and Abraham R. Wagner, *The Lessons of Modern War, Volume 1, Arab-Israeli Conflicts 1973–1989* (Boulder, Colo., 1991), 102: "the USSR demonstrated that it could rapidly assemble and move large amounts of aircraft and armor to Egypt and Syria. Egypt and Syria got about 175 new fighters with about 50 going to Syria . . . In fact, the USSR provided roughly three times the number of tanks during the fighting that the U.S. eventually supplied to Israel during the fighting. It also supplied roughly twice the number of fighter aircraft."

26. Dupuy, 569.

27. Coleman, 18.

28. Lon Nordeen, *Fighters over Israel: The Story of the Israeli Air Force from the War of Independence to the Bakaa Valley* (New York, 1990), 149–52.

29. Hotz, "Offense, Defense Tested in 1973 War," 16.

## 24. New Directions Again

1. DIA, 1974.

2. Ibid.

3. *Military Balance 1974–75*, IISS (London, 1974).

4. "Air Power in the Middle East," *Aviation and Marine*, July–August 1975, p. 55.

5. DIA, 1974.

6. Hotz, "Egypt Plans Modernized Air Arm," 12.

7. DIA, 1974.

8. Piotr Butowski and Jay Miller, *MiG: A History of the Design Bureau and Its Aircraft* (Dallas, Tex., 1991), 102–5.

9. Hotz, "Egypt Plans Modernized Air Arm," 12.

10. Dale Tahtinen, *The Arab-Israeli Military Balance since October 1973*, Foreign Affairs Study, American Enterprise Institute for Public Policy Research (Washington, D.C., 1974), 11–13.

11. "The Egyptian Air Force after Fifty Years: Muscling Up to Meet the Missions of the Future," *African Defense*, May 1982, pp. 52–77.

12. Robert F. Dorr, *F-4 Phantom II* (London, 1984), 161.

13. Robert Ropelewski, "Technical Training Effort Growing," *Aviation Week and Space Technology, Special Issue, Both Sides of the Suez*, 1975, pp. 58–61.

14. "Massive Egyptian Arms Modernization Launched with Saudi Funds, U.K. Help," *Armed Forces Journal International*, October 1977, p. 32.

15. "Egypt—An Airpower in Transition," *Air International*, April 1982, pp. 163–201.

16. "Revenge in the Desert," *Time*, 1 August 1977, pp. 20–21; Dennis Chaplin, "Libya: Military Spearhead Against Sadat?" *Military Review*, November 1979, pp. 42–50.

17. ʿĀṭif.

18. *Military Balance 1975–76*, IISS.

## 25. Moving Forward

1. Clyde R. Mark, "Egypt-United States Relations," Congressional Research Service Issue Brief, 14 July 1993.

2. Cordesman and Wagner, 320–33; "Egypt—An Airpower in Transition."

3. Ezet and Louchet, 52–58.

4. "China Shipping F-7 Fighter Kits to Jiyanklis for Assembly," *Aviation Week and Space Technology*, 15 August 1983, pp. 160–62.

5. Ibid.

6. Interview by author (L. Nordeen) at AOI Factory at Ḥulwān, 12 November 1989; F. Cliffton Berry, "Egyptian Air Force: Ready When Needed," *Air Force Magazine*, January 1982, pp. 49–55.

7. Joseph Bermudez, "Ballistic Missile Development In Egypt," 28 August 1992, prepared for the Missile Proliferation Project at the Monterey Institute of International Studies.

8. Jacques Clostermann and Robert Salvy, "Strength in Diversity," interview with EAF Operations Chief, Maj. Gen. Nabīl Shuwakrī, *International Defense Review*, no. 1, 1990, pp. 63–65.

9. "Hawk Air Defense Brigade Operational," *Aviation Week and Space Technology*, 15 August 1983, pp. 162–64.

10. Michael A. Dornheim, "Egypt Using Unmanned Aircraft For Reconnaissance," *Aviation Week and Space Technology*, 23 January 1989, pp. 56–57.

11. Martin Streely, "Middle East Airborne EW," *Jane's Defense Weekly*, 3 February 1990, p. 42.

12. "Strength in Diversity," IDR.

13. Cordesman and Wagner.

14. *Military Balance 1993–94*, IISS.

## Appendix 3. The Syrian Air Force

1. Ms. Qūwatlī, daughter of Syrian ex-President Qūwatlī, in conversation with D. Nicolle, London, 1975.

2. Annual Report No. 2 of the British Air Attaché in Cairo, 21 January 1952 (PRO, Doc. 1952, FO 371, no. 96993, London).

3. President Assad, quoted in Patrick Seale, *Asad of Syria: The Struggle for the Middle East* (London, 1988).

4. Al-Ghazzāwī, 28 March 1989.

# Selected Bibliography

## Published Sources

Abdel-Malek, Anouar. *Egypt: Military Society,* trans. Charles L. Markmann. Random House, New York, 1968.

Abu Lughod, I. (ed.). *The Arab-Israeli Confrontation of June 1967: An Arab Perspective.* Arab Information Center, New York, 1968.

Adan, A. ("Bren"). *On the Banks of the Suez.* Arms & Armour Press, London, 1980; Presidio Press, San Francisco, 1980; original title: *On Both Banks of the Suez,* Edanim, Jerusalem, 1979.

Adkin, Fred. "Mossies over the Med." *Aeroplane Monthly,* London, August, September, October 1989, pp. 466–69, 546–49, 634–36.

*The Aeroplane.* News reports, London, 17 May and 27 December 1933.

*Al-Ahrām,* daily newspaper, Cairo, passim.

*Air International.* "Egypt—An Airpower in Transition." April 1982, pp. 163–201.

Alono, Scholomo. "The Jet Age," *Air Enthusiast,* No. 50, 1993, pp. 38–48.

Anderson, Raymond A. "Spirit More Aggressive." *New York Times,* 24 July 1969.

Arcanelis, Mario de. *Electronic Warfare.* Blandford Press, Dorset, 1985.

Arena, Nino. *I Caccia della serie 5: Parte Seconda, Fiat G.55.* Stem Mucchi, Modena, 1976.

———. *I Caccia della serie 5: Parte Terza, Macchi 205.* Stem Mucchi, Modena, 1976.

*Armed Forces Journal International.* "Massive Egyptians Arms Modernization Launched with Saudi Funds, U.K. Help." October 1977, p. 32.

Aruri, Naseer H. (ed.). *Middle East Crucible: Studies on the Arab-Israeli War of October 1973.* AAUG Monograph series No. 6, Medina University Press International, Wilmette, Ill., 1975.

Asher, Jerry, and Eric Hammel. *Duel on the Golan: The 100 Hour Battle That Saved Israel.* William Morrow and Co., New York, 1987.

*Aviation and Marine.* "Air Power in the Middle East." July–August 1975, 55–62.

———. "The Egyptians in the 1973 War." July–August 1975, p. 55.

*Aviation Week and Space Technology.* "China Shipping F-7 Fighter Kits to Jiyanklis for Assembly." 15 August 1983, pp. 60–62.

———. "First F-16 Wing at Inshas Nears Operational Status." 15 August 1983, pp. 88–89.

———. "Hawk Air Defense Brigade Operational." 15 August 1985, pp. 162–164.

———. "Soviets Deploy New Suez Defenses." 13 July 1970, pp. 14–16.

Badry, Hassan el, Maj. Gen., Maj. Gen. Taha el Magdoub, and Maj. Gen. Mohammed Dia el Din. *The Ramadan War 1973.* T.N. Dupuy Associates, Dunn Loring, Va., 1978.

Barker, A. J. *The Arab-Israeli Wars.* Ian Allen, London, 1980.

———. *Suez: The Seven Day War.* Faber & Faber, London, 1964; Praeger, New York, 1965.

Basenko, Aleksey. "In the Air over Egypt: The Time Has Come to Discuss." *Krasnaya Zvezda,* 20 September 1990 (in Russian).

Beaufre, A. *The Suez Expedition 1956.* trans. Richard Barry. Faber, London, 1969; Praeger, New York, 1969; original title, *L'Expédition de Suez,* Grasset, Paris, 1967.

Be'eri, Eliezer. "Arab Officers and Politics." *The Jerusalem Quarterly,* No. 6., Jerusalem, Winter 1978, pp. 125–38.

———. *Army Officers in Arab Politics and Society.* Praeger, New York, 1970.

Bentley, John. "Inside Israel's Air Force." *Flight International,* 19 March 1970, 427.

Berger, Morroe. *Military Elite and Social Change: Egypt since Napoleon.* Research monograph 6, Center for International Studies, Woodrow Wilson School of Public and International Affairs, Princeton University, February 1960.

Berque, Jacques. *Egypt: Imperialism and Revolution,* trans. Jean Stewart. Faber and Faber, London, 1972.

Berry, F. Clifton, Jr. "Egyptian Air Force: Ready When Needed." *Air Force Magazine,* 65, no. 1, January 1982, pp. 40–55.

*Born in Battle.* "The Six Day War." No. 6, Eshel Dramit, Tel Aviv, 1979.

———. "The Wild Horses, The Mustang Squadrons." No. 44, Eshel Dramit, Tel Aviv, 1983.

*Born in Battle, Defense Update International.* "Yom Kippur War—10 Years." No. 42, Eshel Dramit, Tel Aviv, 1983.

*Born in Battle, Defence Update, Special 40th Annual IAF Issue.* "The First Transplant: The Super Mystère in Action." No. 88, Eshel Dramit, Tel Aviv, June 1988, pp. 40–42.

Boutros-Ghali, Boutros. "The Foreign Policy of Egypt in the Post-Sadat Era." *Foreign Affairs,* 60, no. 4, Spring 1982, pp. 769–88.

Brady, Thomas F. "Egyptians Report Strikes." *New York Times,* 12 September 1969.

Brogan, Patrick. *The Fighting Never Stopped: A Comprehensive Guide to World Conflict since 1945.* Random House, New York, 1990.

Byford-Jones, W. *The Lightning War.* Hale, London, 1967; Bobbs-Merrill, Indianapolis, 1968.

Carterman, Jacques, and Robert Salvy. "Strength in Diversity." Interview with EAF operations Chief-Maj. Gen. Norbil Shuᶜakri. *International Defense Review.* January 1990, p. 1143.

Cattaneo, G. Letter published in *Aeroplane Monthly,* London, June 1988.

*Christian Science Monitor.* "Ex-Pilot Says US Jets Spied for Israel in 1967." 15 March 1984.

Christienne, Charles, and Pierre Lissarague. *The History of French Military Aviation,* trans. Frances Kianka. Smithsonian Institution Press, Washington, D.C., 1986.

Churchill, Randolf, and Winston Churchill. *The Six Day War.* Heinemann, London, 1967; Houghton Mifflin, Boston, 1967.

Copley, Gregory (ed.). "Egypt Revitalizes Its Defense Industry." *Defence and Foreign Affairs,* 7, no. 9, September 1979, pp. 32–40.

Corddry, Charles W. "The Yom Kippur War: Lessons New and Old." *National Defence,* May–June 1974, p. 509.

Cordesman, Anthony H., and Abrham R. Wagner. *The Lessons of Modern War, Volume 1: The Arab Israeli Conflicts.* Westview Press, Boulder, Colo., 1991.

Cremeans, Charles D. "Nasser's Approach to International Politics." In Benjamin Rivlin and Joseph Szliowicz (eds.), *The Contemporary Middle East: Tradition and Innovation,* Random House, New York, 1965, pp. 507–19.

Cull, Brian, Shlomo Aloni, and David Nicolle. *Spitfires over Israel,* Grub Street, London, 1993.

Dawisha, A. I. *Egypt in the Arab World: The Elements of Foreign Policy.* John Wiley and Sons, New York, 1976.

Dawisha, Karen. *Soviet Foreign Policy Towards Egypt.* St. Martin's Press, New York, 1979.

Dawson, J. "The Air War in the Middle East." *Air Force & Space Digest,* August 1967, pp. 26–29.

Dayan, M. *Diary of the Sinai Campaign.* Weidenfeld and Nicolson, London, 1966; Harper and Row, New York, 1966.

*Defense Intelligence Digest*. "Soviet Naval Threat Growing in Mediterranean," vol. 8, August 1970, pp. 4–5.
Dornheim, Michael A. "Egypt using Unmanned Aircraft for Reconnaissance." *Aviation Week & Space Technology*. 23 January 1989, p. 27.
Dorr, Robert F. *The F-4 Phantom II*. Osprey Publishing, London, 1984.
Duncan, Andrew. "The Military Threat to Israel." *Survival*, 24, no. 3, London, May–June 1982, pp. 98–107.
Dupuy, Col. T. N. *Elusive Victory: The Arab-Israeli Wars, 1947–1974*. Macdonald and Janes, London, 1978; Harper and Row, New York, 1978.
Dyer, Gwynne. "Egypt." In John Keegan (ed.), *World Armies*, Facts on File, New York, 1979, pp. 190–204.
Eden, Sir Anthony. *Eden Memoirs*, 3 vols. Cassell, London, 1960–65.
*Egyptian Gazette*, daily newspaper, Cairo, passim.
Emiliani, Angelo, Giuseppe F. Ghergo, and Achille Vigna. *Immagini e storia dell'aeronautica italiana 1935–1945: Regia Aeronautica, I fronti africani*. albertelli, Parma, 1979.
———. *Immagini e storia dell'aeronautica italiana 1935–1945: Regia Aeronautica, Il settore mediterraneo*. Intergest, Milan, 1976.
Entelis, John P. "Nasser's Egypt: The Failure of Charismatic Leadership." *Orbis*, 18, no. 2, Summer 1974, pp. 451–64.
Facon, Patrick. "Trente ans plus tard: L'Armée de l'Air à Suez ou l'autopsie d'une victoire." *Air Fan*, Paris, November 1986, pp. 8–17.
Ferron, James. "Israelis Say Jets Again Hit Egyptians." *New York Times*, 26 July 1969.
———. "Israel Reports Downing 11 Jets in Suez Clashes." *New York Times*, 12 September 1969.
Fisher, Eugene M., and M. Cherif Bassiouni. *Storm over the Arab World: A People in Revolution*. Follett, Chicago, 1972.
*Flash of Damascus*. "Flash on the Combat Scene of April 2," Damascus, May 1070, pp. 2, 8.
Flintman, Victor. *Air Wars and Aircraft: A Detailed Record of Air Combat 1945 to the Present*. Arms and Armour Press, London, 1989.
———. "The Suez Campaign, 1956." *Scale Aircraft Modeling*, November 1986, pp. 54–85.
———. "Suez 1956: A Lesson in Airpower." *Air Pictorial*, August–September 1965, p. 270.
Freeman, Robert O. *Soviet Policy Towards the Middle East since 1970*, rev. ed. Praeger, New York, 1978.
Gaujac, Paul. *Suez 1956*. Charles-Lavauzelle, Paris, 1986.
Georges, Alan. "Soviet Troops Fought in War of Attrition." *Defence*, October 1989, p. 750.
Green, Stephan. *Taking Sides*. William Morrow and Co., New York, 1984.
Grose, Peter. "Nasser Concedes That Israelis Have Air Superiority in Mideast." *New York Times*, 9 February 1970.
Hadar, Moshe, and Yehuda Ofer. *Heyl Haʿavir: The Israeli Air Force*. Ministry of Defence Publishing House, Tel Aviv, n.d.
Halperin, Marav, and Aharon Lapidot. *G-Suit: Combat Reports from Israel's Air War*. Sphere Books, London, 1990.
Halpern, Manfred. "Egypt and the New Middle Class: Reaffirmations and New Explorations." *Comparative Studies in Society and History*, 11, no. 1, Cambridge Mass, January 1969, pp. 97–108.
Heikal, Mohammed H., *Cutting the Lion's Tail: Suez through Egyptian Eyes*. Andre Deutsch, London, 1986.
———. *The Road to Ramadan*. Collins, London, 1975; Quadrangle/New York Times Book Co., New York, 1975.
———. *The Sphinx and the Commissar: The Rise and Fall of Soviet Influence in the Middle East*. Harper and Row, New York, 1978.

Henriques, R. *A Hundred Hours to Suez: An Account of Israel's Campaign in the Sinai Peninsula.* Collins, London, 1957; Viking Press, New York, 1957.

Hersh, Seymour M. *Chemical and Biological Warfare: America's Hidden Arsenal.* Doubleday Anchor Books, Garden City, N.Y., 1969.

Herzog, Chaim. *The Arab-Israeli Wars: War and Peace in the Middle East.* Random House, New York, 1982.

———. "Middle East War 1973." *RUSI, Journal of the Royal United Services Institute for Defence Studies,* London, 1975.

———. *War of Atonement.* Weidenfeld and Nicolson, London, 1975; Little, Brown, Boston, 1974; Edanim, Jerusalem, 1975.

Hewish, Mark, et al. "Egypt." In Ray Bonds (ed.), *Air Forces of the World,* Simon and Schuster, New York, 1979, pp. 148–51.

Hinnebusch, Raymond A. "Egypt under Sadat: Elites, Power Structures, and Political Change in a Post-Populist State." *Social Problems,* 28, no. 4, April 1981, pp. 442–64.

Hirst, D., and I. Beeson. *Sadat,* Faber & Faber, London, 1981.

Holden, David, and Richard Johns. *The House of Saud: The Rise and Rule of the Most Powerful Dynasty in the Arab World.* Holt, Reinhart and Winston, New York, 1981.

Hotz, Robert. "Egypt Plans Modernized Air Arm." In *Both Sides of the Suez: Airpower in the Middle East,* special reprint of *Aviation Week and Space Technology* articles, 1975, pp. 30–32.

———. "Offense, Defense Tested in 1973 War." In *Both Sides of the Suez: Airpower in the Middle East,* special reprint of *Aviation Week and Space Technology* articles, 1975, pp. 38–43.

Hurewitz, J. C. *Middle East Politics: The Military Dimension.* Praeger, New York, 1969.

International Institute for Strategic Studies. *The Middle East and the International System, I. The Impact of the 1973 War, Adelphi Papers, No. 114,* London, Spring 1975.

———. "The Middle East War." *Strategic Summary,* London, 1974, pp. 16–27.

———. *The Military Balance 1966–67.* London, 1966, pp. 45–46.

———. *The Military Balance 1972–73.* London, 1971, p. 44.

———. *The Military Balance 1975–76.* London, 1975, p. 33.

———. *The Military Balance 1981–82.* London, 1981, p. 57.

———. *Strategic Survey 1973,* London, 1974, pp. 26–27.

Jabr, Jabr ʿAlī Air Brig. *Al-Qūwah al-Jawwiyah bayna al-Siyāsat al-Miṣrīyah wa-al-Isrāʾīlīyah (Air Power in Egyptian and Israeli Policy), Volume I: 1922–1952.* Al-Maktabah al-Akādīmīyah, Cairo, 1993.

Jackson, Robert. *The Israeli Air Force Story.* Tandem Books, London, 1970.

*Jewish Chronicle,* news item, London, 30 July 1965.

Jindī, Muḥammad Fakhrī al-, Air Brig. *Nusūr Miṣṣrīyūn fawq al-Jūlān (Egyptian Eagles over the Golan).* Al-Ḥayʾah al-Miṣrīyah al-ʾĀmmah lil-Kitāb, Cairo, 1992.

Kafāfī, Munīrah. *Fī Dhikrá 15 Māyū 1948: ʿIndamā Ustushnida Abī! (In Memory of 15 May 1948: When My Father Was Killed in Action!),* Dār al-Maʿārif bi-Miṣr, Cairo, 1975.

Kansu, Yavuz, Sermet Şensöz, and Yılmaz Öztuna. *Havacılık Tarihinde Türkler 1 (History of Turkish Aviation 1).* Hava Kuvvetleri Basım, Ankara, 1971.

Kedourie, Elie, and Sylvia G. Haim (eds.). *Modern Egypt: Studies in Politics and Society.* Frank Cass, London, 1980.

*Keesings Contemporary Archives.* London, February–March 1956; December 1956.

Kerr, Malcolm H. *The Arab Cold War: Gamel Abd al-Nasir and His Rivals, 1958–1970,* 3d. ed. Oxford University Press, New York, 1971.

Khuri, Fuad I. "Modernizing Societies in the Middle East." In Morris Janowitz (ed.), *Civil-Military Relations: Regional Perspectives,* Sage, Beverly Hills, 1981, pp. 160–82.

Kolcum, Edward H. "SAM Changes Force New Strategy." *Aviation Week and Space Technology,* 16 November 1970, pp. 16–21.

———. "Soviets Shifting Middle East Balance." *Aviation Week and Space Technology,* 16 May 1970, pp. 18–21.
Korn, David A. *Stalemate: The War of Attrition and Great Power Diplomacy in the Middle East, 1967–1970,* Westview Press, Boulder, Colo., 1992.
*Krasnaya Zvezda.* "Soviet Participation in Arab-Israeli War of Attrition Recalled," 25 March 1989 (in Russian).
Kreis, John F. *Air Warfare and Air Base Defense.* Office
of Air Force History Special Studies, Washington, D.C., 1988.
Labīb, ᶜAlī Muḥammad, Air Brig. *Al-Qūwah al-Thālithah (The Third Power),* Al-Ḥayʾah al-Miṣrīyah al-ᶜĀmah lil-Kitāb, Cairo, 1977.
Lake, Julian S., and Richard V. Hartman. "Air Electronic Warfare." *U.S. Naval Institute Proceedings,* October 1976, p. 48.
Lambert, Mark. "Middle East Market Report—Egypt's Air Force." *Flight International,* 13 March 1975, pp. 419–421.
Lee, David, Air Marshal RAF. *Wings in the Sun: A History of the Royal Air Force in the Mediterranean 1945–1986.* HMSO, London, 1989.
Lloyd, Selwin. "Suez 1956: The Fateful Rendezvous." *Sunday Times,* London, 18 June 1978.
Louchet, F., and F. Ezet. "The Egyptian Air Force After Fifty Years: Muscling Up to Meet the Missions of the Future Challenges." *African Defense,* May 1992, pp. 52–78.
Love, K. *Suez: The Twice-Fought War: A History.* Longman, Harlow, 1970; McGraw-Hill, New York, 1969.
Luttwak, Edward, and Horowitz, David. *The Israeli Army.* Allen Lane, Harmondsworth, 1975; Harper, New York, 1975.
Marshal, Thomas J., Maj. "Israeli Helicopter Forces: Organization and Tactics." *Military Review,* May 1972, pp. 94–98.
Mason, R. A., Air Vice Marshal RAF, and John W. R. Taylor. *Aircraft, Strategy and Operations of the Soviet Air Force.* Jane's, London, 1986.
Matthews, R. G. "Egyptian Defense Industrialization." *Defense Analysis,* 8, no. 2, 1992, pp. 115–31.
Munᶜim, Muḥammad ᶜAbd al-, *Dhiʾb fī Qurṣ al-Shams (Wolf in the Sun's Disc),* Cairo, 1988.
*New York Times.* "Cairo Claims Tank Victory, 'Rather Quiet', Israel Says." 14 October 1973.
———. "Israel Hits Egyptian Targets near Canal," 29 August 1969.
———. "Israelis Report Losing Jet in Dogfight at Canal." 10 February 1070.
———. "Israelis Say They Shot Down 3 Syrian Migs." 9 January 1970.
———. "UAR and Israel Wage Artillery Duel across Suez." 9 March 1969, p. 1.
Nicolle, David. "1967—An Egyptian Testament." *Air Pictorial International,* London, August 1992, pp. 420–25.
———. "The Arab-Israeli East Air Wars: Part One, 1948." *The Illustrated Encyclopedia of World Aviation (vol. 9, issue 103).* Orbis, London, 1983, pp. 2041–44.
———. "The Arab-Israeli East Air Wars: Part Two, Suez." *The Illustrated Encyclopedia of World Aviation (vol. 9, issue 104).* Orbis, London, 1983, pp. 2061–64.
———. "The Arab-Israeli East Air Wars: Part Three, The Six-Day War." *The Illustrated Encyclopedia of World Aviation (vol. 9, issue 105).* Orbis, London, 1983, pp. 2081–84.
———. "The Arab-Israeli East Air Wars: Part Four, Yom Kippur." *The Illustrated Encyclopedia of World Aviation (vol. 9, issue 106).* Orbis, London, 1983, pp. 2101–4.
———. "The Arab-Israeli East Air Wars: Part Five, Lebanon." *The Illustrated Encyclopedia of World Aviation (vol. 9, issue 107).* Orbis, London, 1983, pp. 2121–25.
———. "The Assault on Mount Hermon: An Episode of the October War." *RUSI, Journal of the Royal United Services Institute for Defence Studies,* London, June 1975, pp. 43–46.
———. "The Egyptian Air Force." *Wings over Africa,* Bryanston, South Africa, August 1973, pp. 27–28.

———. "The Egyptian Air Force 1930–1947." *Air Pictorial,* London, October 1988, pp. 402–9.
———. "Egyptian Phoenix." *Air Britain Digest,* London, January–February 1974, pp. 8–13.
———. "Egypt's Aircraft Industry." *Middle East International,* London, February 1976, pp. 30–32.
———. "Egypt's Fury Hero." *Air Pictorial International,* London, June 1993, pp. 282–83.
———. "The Faluja Pocket (an Aspect of the 1947–8 Arab-Israeli War)" (part 1). *The Army Quarterly and Defence Journal,* London, October 1975, pp. 440–56.
———. "The Faluja Pocket (an Aspect of the 1947–8 Arab-Israeli War)" (part 2). *The Army Quarterly and Defence Journal,* London, July 1976, pp. 333–50.
———. "La Forza Aerea Egiziana" (part 1). *JP4,* Florence, July 1974, pp. 17–24.
———. "La Forza Aerea Egiziana" (part 2). *JP4,* Florence, September 1974, pp. 42–50.
———. "Forze Aerea Irakena" (part 1). *JP4,* Florence, January 1976, pp. 45–50.
———. "Forze Aerea Irakena" (part 2). *JP4,* Florence, February 1976, pp. 40–45.
———. "Fury That Fought for Egypt." *Air Pictorial International,* London, June 1993, pp. 278–80.
———. "Heil Ha'Avir: una forza area combat ready." *JP4,* Florence, May 1975, pp. 39–46.
———. "The Holy Day Air War (1973)." *Air Enthusiast International,* London, May 1974, pp. 240–52.
———. "The Imbalance of Power." *Middle East International,* London, June 1973, pp. 20–22.
———. *Lawrence and the Arab Revolts.* Osprey, London, 1989.
———. "Neutral Allies: The Royal Egyptian Air Force in World War Two." *Air Enthusiast Fifty-Two,* London, 1993, pp. 1–16.
———. "Nile Valley Fledglings: Egypt's Aircraft Industry from 1933." *Air Pictorial,* London, January 1976, pp. 26–29.
———. "Syrian Air Force" (part 1). *JP4,* Florence, October 1975, pp. 32–38.
———. "Syrian Air Force" (part 2). *JP4,* Florence, November 1975, pp. 18–24, 32–38.
———. "Uneasy Allies: Egypt's Armed Forces in World War Two." *The Army Quarterly and Defence Journal,* London, October 1979, pp. 429–32.
———. "La Vallata della Morte (Egyptian Vampires in 1956)." *JP4,* Florence, January 1978, pp. 50–51.
Nordeen, Lon O. *Air Warfare in the Missile Age.* Smithsonian Institution Press, Washington, D.C., 1985.
———. *Fighters over Israel.* Orion Books, New York, 1990.
Nutting, Anthony. *No End of a Lesson: The Story of Suez.* Constable, London, 1967; Potter, New York, 1967.
O'Balance, Edgar, *The Arab-Israeli War 1948.* Faber & Faber, London, 1956.
———. *The Electronic War in the Middle East, 1968–1970.* Anchor Books, Hamden, Conn., 1972.
———. *The Sinai Campaign 1956.* Faber & Faber, London, Praeger, New York, 1960.
———. *The Third Arab-Israeli War.* Faber & Faber, London, 1972; Ancor Books, Hamden, Conn., 1972.
Palit, D. K. *Return to the Sinai.* Palit Publishers, New Delhi, 1974.
Perlmutter, Amos. *Egypt: The Praetorian State.* Transaction Books, New Brunswick, N.J., 1974.
———. *Political Roles and Military Rulers.* Frank Cass, London, 1981.
Pimlott, John. *The Middle East Conflicts.* Crescent Books, New York, 1983.
Ranger, Robert J. "The October War." In Daphne Daume and J. E. David (eds.), *Britannica Book of the Year 1974,* Encyclopedia Britannica, Chicago, 1974, pp. 232–33.
Rikhye, Indar Jit. *The Sinai Blunder: Withdrawal of the United Nations Emergency Force Leading to the Six-Day War of June 1967.* Frank Cass, London, 1980.
Robineau, Lucien, Gen. "Les Portes-á-faux de l'Affaire de Suez." *Revue Historiques des Armes,* No. 4, Paris, 1986, pp. 41–50.

Robinson, Clarence A. "MiG-21 Playing Defense Role." *Aviation Week and Space Technology,* 21 December 1981.
Roman, Guy. "Pride of Place." *Israel Air Force Magazine,* 1989, p. 59.
Ropelewski, Robert R. "Tech Training Effort Growing." In *Both Sides of the Suez: Airpower in the Middle East,* special reprint of *Aviation Week and Space Technology* articles, 1975, p. 67.
Rosen, Stephen J., and Martin Indyk. "The Temptation to Pre-empt in a Fifth Arab-Israeli War." *Orbis,* 20, no. 2, Summer 1976, pp. 265–85.
Rothenberg, Gunter E. *The Anatomy of the Israeli Army.* New York, 1979.
Rubinstein, Alvin Z. *Red Star over the Nile.* Princeton University Press, Princeton, 1977.
Russell, T. W. *Egyptian Service 1902–1946.* John Murray, London, 1949.
Sabit, A. *A King Betrayed.* Quartet Books, London, 1990.
Sadat, Anwar. *Revolt on the Nile.* Wingate, London, 1957.
———. *In Search of Identity.* Collins, London, 1978; Harper and Row, New York, 1978.
Sadiq al Misr (Nicolle, David). "Egypt's Migs." *Scale Models,* London, February 1974, pp. 90–94.
Safran, Nadav. "Arab Politics, Peace and War." *Orbis,* 18, no. 2, Summer 1974, pp. 377–401.
———. *From War to War: The Arab-Israeli Confrontation, 1948–1967.* Pegasus, New York, 1969.
Salmawy, Mohammed. "Mubarak: A Man for All Seasons." *South: The Third World Magazine,* London, 16 February 1982, pp. 13–14.
Sansom, A. W. *I Spied Spies.* Harrap, London, 1965.
Schiff, Zeʿev. *A History of the Israeli Army.* London, 1985.
Seale, Patrick. *Asad of Syria: The Struggle for the Middle East.* Tauris, London, 1988.
Sherman, A. *When God Judged and Men Died. A Battle Report of the Yom Kippur War.* Bantam, New York, 1973.
Shores, Christopher. *Fighters over the Desert.* Neville Spearman, London, 1969.
Springborg, Robert. "U.S. Policy Toward Egypt: Problems and Prospects." *Orbis,* 24, No. 4, Winter 1981, pp. 805–18.
Stockholm International Peace Research Institute. *World Armaments and Disarmament: SIPRI Yearbook 1981.* Taylor and Francis, London, 1981.
Streely, Martin. "Middle East Airborne EW." *Jane's Defense Weekly,* 3 February 1990.
"Sunday Times" Insight Team. *Insight on the Middle East War.* Deutsch, London, 1974.
———. *The Yom Kippur War.* Deutsch, London, 1975.
Tachau, Frank (ed.). *Political Elites and Political Development in the Middle East: Seven Cases.* Schenkman, Cambridge, Mass., 1975.
Tanner, Henry. "Cairo Claims Gain." *New York Times,* 15 October 1973.
———. "Cairo Reports a 10 Mile Egyptian Gain." *New York Times,* 10 October 1973.
———. "Egypt Said to Consolidate Position in Sinai." *New York Times,* 14 October 1973.
Theobald, A. B. *Ali Dinar, Last Sultan of Darfur, 1898–1916.* Longmans, London, 1965.
*Time.* "Revenge in the Desert." 1 August 1977.
U.S. Navy. *Office of Naval Intelligence Review,* 14, no. 7, December 1958; no. 9, September 1959.
Vatikiotis, P. J. *The Egyptian Army in Politics: Pattern for New Nations?* Indiana University Press, Bloomington, 1961.
———. *Nasser and His Generation.* Croom Helm, London, 1978; St. Martin's Press, New York, 1978.
*Warplane.* "Mig-25 Foxbat," no. 18, 1985, pp. 346–55.
———. "The Yom Kippur Arab-Israeli War," no. 93, 1983, pp. 1841–45.
Waterbury, John. *The Crossing, American Universities Field Staff, Fieldstaff Reports, North Africa Series.* 18, no. 6, AUFs, Hanover, N.H., December 1973.
Weizeman, Ezer. *The Battle for Peace.* Bantam, London-New York, 1981; Edanim, Jerusalem, 1981.

———. *On Eagles' Wings.* Weidenfeld and Nicolson, London, 1976; Maariv Book Guild, Tel Aviv, 1975.

Westwood, J. N. *The History of Middle East Wars.* Bison Books, Greenwich, Conn., 1984.

Whetten, Lawrence L. "June 1967 to June 1971: Four Years of Canal War Reconsidered." *New Middle East,* June 1971, pp. 15–17.

Whitney, Craig. "Israel Puts Enemy Tank Losses at 220." *New York Times,* 15 October 1973.

Williams, Louis, Lt. Col. *Israeli Defence Forces: A People's Army.* Ministry of Defence Publishing House, IDF Office of the Spokesman, 1989.

Wizārah al-Ḥarbīyah wa-al-Baḥrīyah (Ministry of Military and Marine). *Qirā'at al-kharā'it wa-al-ṣuwar al-jawwīyah,* training text for army and air force cadets. Egyptian General Staff Directorate, Military Operations Division, Cairo, 1949.

Yonay, Ehud. *No Margin for Error: The Making of the Israeli Air Force.* Pantheon, New York, 1993.

Zumbach, Jean. *On Wings of War—My Life As a Pilot Adventurer.* London, 1977.

## Unpublished Sources: Archives, Papers, Broadcasts, and Others

BBC Overseas Broadcasting Arabic Service, D. Nicolle in confidential conversations with members of staff, 1967–71.

Bermudez, Joseph S. *Ballistic Missile Development in Egypt,* unpublished technical paper for the Missile Proliferation Project, Monterey Institute of International Studies, 28 August 1992.

Defense Intelligence Agency. *Free World Air Intelligence Brief, AP-240–3-1–67-INT., United Arab Republic, 1 January 1967,* acquired through a Freedom of Information Act request.

———. *Free World Air Intelligence Brief, 1 February 1969, United Arab Republic,* acquired through a Freedom of Information Act request.

———. *Air Forces Intelligence Study (AFIS), 1 January 1971, United Arab Republic, DIA-240–3-*1–71-Int, *acquired through a Freedom of Information Act request.*

———. *Air Forces Intelligence Study (AFIS), 1 October 1974, Egypt, DI-240-EG-74,* acquired through a Freedom of Information Act request.

Defense Nuclear Agency. *Air Superiority and Airfield Attack: Lessons from History.* BDM Corporation Report prepared for the Defense Nuclear Agency, 1982.

EAF Historical Division, information via Sharīf Sharmī in correspondence with David Nicolle, 1992.

———. *A Biographical Note on the Late Squadron Commander Mohamed Abdul Hamid Abu Zeid.* Manuscript. Cairo, 1974).

———. *Biography of Wing Comdr. ʿAbd al-Ḥamīd Abū Zayd.* Manuscript. Cairo, 1972.

Egyptian Army Historical Division. *The Faluja Battle.* Manuscript. Cairo, 1975.

*Flight International,* aviation news archive, London.

Goldberg, Alfred. *Air Operations in the Sinai Campaign 1956,* formerly secret study, USAF Historical Division.

IDF/AF History Branch. *Operation Kadesh, IDF/AF 1950*–1956, Buildup and Operations, *1986.*

*International Symposium on October 1973 War,* Cairo University 21–27 October 1975, unpublished papers released by the Egyptian Government.

Israeli radio news broadcast, taped by the *BBC Monitoring Service,* Cavesham, 1955.

*Memorial plaque to the dead of the Suez War 1956,* in Egyptian Military Museum, The Citadel, Cairo.
Pajak, Roger F. "The End of Soviet Arms Aid to Egypt?" Unpublished paper delivered at the *8th Annual International Affairs Symposium,* 1973–74.
Plummer, Stephen B., Lt. Col. *The Egyptian Air Force: Insurance for U.S. National Interests in the Middle East.* Technical Paper, U.S. Air Force Air University, 1988, ID no. 89–109–275.
Public Records Office, London, England:
Doc. 1922, AIR 2, no. 1066, Letter from HQ, RAF Middle East, concerning proposed creation of an Egyptian air force.
Doc. 1925, AIR 2, no. 1066, Report from HQ, RAF Middle East, concerning proposed creation of an Egyptian air force.
Doc. 1937, AIR 2, no. 2768, First Half-Yearly Report on the REAF.
Doc. 1937, FO 371, no. 20912, Report on the EAAF and Egyptian Army by the British Ambassador in Cairo for the Foreign Office, supplementary note by Major Gen. Marshall-Cornwall, 1937.
Doc. 1938, FO 371, no. 21939, Report on the REAF by the British Advisory Mission for the Foreign Office in London, 24 May 1938.
Doc. 1938, FO 371, no. 21944, Report on the REAF by the British Advisory Mission for the Foreign Office in London, 24 May 1938.
Doc. 1939, FO 371, no. 23326, "Combined Plan for the Defence of Egypt," report for the Foreign Office, London, August 1939.
Doc. 1939, FO 371, no. 23331, Report on the REAF by the British Advisory Mission for the Foreign Office in London, 14 May 1939.
Doc. 1939, FO 371, no. 23334, Report No. 10 on the Egyptian Army by the British Advisory Mission for the Foreign Office in London, August 1939.
Doc. 1939, FO 371, no. 23337, Report on REAF by the British Advisory Mission for the Foreign Office in London, 30 November 1939.
Doc. 1940, FO 371, no. 24610, British Foreign Office comments added to Report No. 11 on the Egyptian Army by the British Advisory Mission for the Foreign Office, London, December 1939.
Doc. 1940, FO 371, no. 24610, Report No. 13 on the Egyptian Army by the British Advisory Mission for the Foreign Office, London, April 1940.
Doc. 1940, FO 371, no. 24610, Report No. 14 on the Egyptian Army by the British Advisory Mission for the Foreign Office, London, July 1940.
Doc. 1940, FO 371, no. 24612, Report on the REAF by the British Advisory Mission for the Foreign Office, London, 21 April 1940.
Doc. 1940, FO 371, no. 24612, Report on the REAF by the British Ambassador in Cairo for the Foreign Office, London, 12 September 1940.
Doc. 1940, FO 371, no. 24612, Confidential Memo on REAF operations over the Gulf of Suez by the British Advisory Mission for the Foreign Office, London, 6 December 1940.
Doc. 1940, FO 371, no. 24612, Report on the REAF by the British Advisory Mission for the Foreign Office, London, 27 December 1940.
Doc. 1940, FO 371, no. 24613, Confidential Memo from the British Advisory Mission, to the Foreign Office, London, 6 December 1940.
Doc. 1940, FO 371, no. 24626, Report on the Egyptian Army by the British Advisory Mission for the Foreign Office, London, September 1940.
Doc. 1941, FO 371, no. 27383, Reports on sale or transfer of RAF Hurricanes to REAF, to the Foreign Office, London, 31 October 1941–27 July 1942.
Doc. 1941, FO 371, no. 31561, Reports on sale or transfer of RAF Hurricanes to REAF, to the Foreign Office, London, 31 October 1941–27 July 1942.

Doc. 1942, FO 371, no. 31561, Reports on subversion within Egyptian Armed Forces by the British Advisory Mission for the Foreign Office, London, May and June 1942.

Doc. 1942, FO 371, no. 31561, Report on the REAF by the British Advisory Mission to the Foreign Office, London, 21 December 1942.

Doc. 1941, FO 371, no. 31562, Reports on sale or transfer of RAF Hurricanes to REAF, to the Foreign Office, London, 31 October 1941–27 July 1942.

Doc. 1942, FO 371, no. 31562, Report by the British Advisory Mission for the Foreign Office, London, May and June 1942.

Doc. 1942, FO 371, no. 31573, Confidential Memo from the British Advisory Mission for the Foreign Office, London, December 1942.

Doc. 1943, FO 371, no. 35546, Report on defection of REAF pilots by the British Advisory Mission for the Foreign Office, London, 25 July 1943.

Doc. 1943, FO 371, no. 35546, Report of the British Advisory Mission for the Foreign Office, London, 23 July 1943.

Doc. 1943, FO 371, no. 35549, Report on the REAF by the British Advisory Mission for the Foreign Office, London, 28 February 1944.

Doc. 1943, FO 371, no. 35549, Report on the REAF by the British Advisory Mission for the Foreign Office, London, 25 July 1943.

Doc. 1943, FO 371, no. 35550, Report by Wing Commander Burgess of the British Air Ministry on Egyptian offers to take over RAF Hurricanes, for the Foreign Office, London, 18 August 1943.

Doc. 1944, FO 371, no. 41313, Report on the REAF by the British Advisory Mission for the Foreign Office, London, 28 February 1944.

Doc. 1944, FO 371, no, 41313, Report on the REAF by the British Advisory Mission for the Foreign Office, London, 9 August 1944.

Doc. 1944, FO 371, no, 41314, Report on the REAF by the British Advisory Mission for the Foreign Office, London, 9 August 1944.

Doc. 1944, FO 371, no. 40072, Memo on King Faruq's Avro Anson by the British Advisory Mission for the Foreign Office, London, 12 December 1944.

Doc. 1944, FO 371, no. 41313, Memo on King Faruq's Avro Anson by the British Advisory Mission for the Foreign Office, London, 12 December 1944.

Doc. 1944, FO 371, no. 41313, Report on the REAF by the British Advisory Mission for the Foreign Office, London, 18 August 1944.

Doc. 1944, FO 371, no. 41314, Memo by Anthony Eden attached to Report on the REAF by the British Advisory Mission for the Foreign Office, London, 18 August 1944.

Doc. 1944, FO 371, no. 41314, Memo on King Faruq's Avro Anson by the British Advisory Mission for the Foreign Office, London, 12 December 1944.

Doc. 1944, FO 371, nos 41314, Report on the REAF by the British Advisory Mission for the Foreign Office, London, 18 August 1944.

Doc. 1945, FO 371, no. 45477, Report on the REAF by the British Advisory Mission for the Foreign Office, London, 15 January 1945.

Doc. 1945, FO 371, no. 45477, Report on the REAF by the British Advisory Mission for the Foreign Office, London, 21 June 1945.

Doc. 1945, FO 371, no. 45930, Memo on King Faruq's C-47 Dakota and Royal Flight pilot Hassan Aqif, by the British Embassy in Cairo for the Foreign Office, London, 27 March 1945.

Doc. 1945, FO 371, no. 45930, Memo on King Faruq's Avro Anson, by the British Advisory Mission for the Foreign Office, London, 24 January 1945.

Doc. 1945, FO 371, no. 45945, Memo on the sale of military aircraft to Egypt, by the British Advisory Mission for the Foreign Office, London, 4 April 1945.

Doc. 1945, FO 371, no. 45945, Report on the REAF by the British Advisory Mission for the Foreign Office, London, 15 January 1945.

Doc. 1945, FO 371, no. 45945, Report on the REAF by the British Advisory Mission for the Foreign Office, London, 21 June 1945.

Doc. 1945, FO 371, no. 45946, Memo on the sale of military aircraft to Egypt, by the British Advisory Mission for the Foreign Office, London, 4 April 1945.

Doc. 1945, FO 371, no. 45948, Memo on King Faruq's C-47 Dakota and Royal Flight pilot Hassan Aqif, by the British Embassy in Cairo for the Foreign Office, London, 21 June 1945.

Doc. 1945, FO 371, no. 45949, Memo on King Faruq's C-47 Dakota and Royal Flight pilot Hassan Aqif, by the British Embassy in Cairo for the Foreign Office, London, August 1945.

Doc. 1945, FO 371, no. 45988, Memo on King Faruq's C-47 Dakota and Royal Flight pilot Hassan Aqif, by the British Embassy in Cairo for the Foreign Office, London, August 1945.

Doc. 1945, FO 371, no. 45988, Memo on King Faruq's Avro Anson by the British Embassy in Cairo for the Foreign Office, London, 24 January 1945.

Doc. 1946, FO 371, no. 52542, Report on Zionist Terrorism drawn up by the British Foreign and Colonial Office, June 1946.

Doc. 1946, FO 371, no. 53268 Foreign Office Memo on sale of military aircraft to Egypt, London, 7 June 1946.

Doc. 1946, FO 371, no. 53268, Report on sale of military aircraft to Egypt, by the British Advisory Mission for the Foreign Office, London, 2 September 1946.

Doc. 1946, FO 371, no. 53268, Confidential assessment of the REAF by the British War Office, 26 November 1946.

Doc. 1946, FO 371, no. 53268, Letter from the British Ambassador in Egypt to the Foreign Office, London, 21 March 1946.

Doc. 1946, FO 371, no. 53268, Report on the sale of military aircraft to Egypt by the British Advisory Mission for the Foreign Office, London, 27 September 1946.

Doc. 1946, FO 371, no. 53268, Report on the sale of military aircraft to Egypt by the British Advisory Mission for the Foreign Office, London, 18 January 1946.

Doc. 1946, FO 371, no. 53268, Report on the REAF by the British Advisory Mission to the Foreign Office, London, 18 January 1946.

Doc. 1946, FO 371, no. 53268, Report on the REAF by the British Advisory Mission for the Foreign Office, London, 27 June 1946.

Doc. 1946, FO 371, no. 53269, Letter from the British Ambassador on sale of military aircraft to Egypt, to the Foreign Office, London, 21 March 1946.

Doc. 1946, FO 371, no. 53269, Foreign Office Memo on possible sale of military aircraft to Egypt, London, 7 June 1946.

Doc. 1946, FO 371, no. 53269, Report on sale of Spitfires to REAF by the British Advisory Mission for the Foreign Office, London, 27 September 1946.

Doc. 1947, AIR 20, no. 6906, British Air Ministry Report on the strength of the REAF.

Doc. 1947, FO 371, no. 63074, Report on the sale of Spitfires to Egypt, by the British Advisory Mission for the Foreign Office, London, 19 February 1947.

Doc. 1947, FO 371, no. 63074, Foreign Office notes added to Report on the REAF by the British Advisory Mission for the Foreign Office, London, 17 June 1947.

Doc. 1947, FO 371, no. 63074, Foreign Office Memorandum on expected decline in REAF operational standards, London, 18 July 1947.

Doc. 1947, FO 371, no. 63076, Report on the REAF by the British Advisory Mission for the Foreign Office, London, 19 February 1947.

Doc. 1947, FO 371, no. 63076, Report on the REAF by the British Advisory Mission for the Foreign Office, London, 17 June 1947.
Doc. 1947, FO 371, no. 63076, Foreign Office notes added to the Report by the British Advisory Mission to the Foreign Office, London, 17 June 1947.
Doc. 1947, FO 371, no. 63076, Foreign Office Memorandum on expected decline of REAF's operational standards, London, 18 July 1947.
Doc. 1948, AIR 20, no. 6906, Foreign Office Memoranda on Egyptian request for radar sets.
Doc. 1948, FO 371, no. 69188, Foreign Office Memorandum REAF's impounding of British and other civil registered aircraft, London January 1948.
Doc. 1948, FO 371, no. 69188, Memoranda on REAF's impounding of British and other civil registered aircraft, from the British Embassy in Cairo to the Foreign Office, London, April and May 1948.
Doc. 1948, FO 371, no. 69188, Foreign Office Memorandum on REAF staffing of British troops in Palestine, London, 23 May 1948.
Doc. 1948, FO 371, no. 69188, Letter from Col. Ryan on operational capabilities of REAF, from the British Embassy in Cairo to the Foreign Office, London, 27 May 1948.
Doc. 1948, FO 371, no. 69188, Various Foreign Office Memoranda on the Ramat David Incident, London, 1984.
Doc. 1948, FO 371, no. 69188, Confidential information on current strength of REAF, passed from British Air Attaché in Cairo to the Foreign Office, London, 19 October 1948.
Doc. 1948, FO 371, no. 69223, Foreign Office Memoranda on REAF firing on RAF aircraft, London, January 1948.
Doc. 1948, FO 371, no. 69223, Various Foreign Office Memoranda on the Ramat David Incident, London, 1948.
Doc. 1948, FO 371, no. 69223, Foreign Office Memorandum on Ramat David Incident, London, 23 May 1948.
Doc. 1948, FO 371, no. 69223, Report of the British Ambassador's meeting with the Egyptian Prime Minister, in a letter to the Foreign Office, 14 June 1948.
Doc. 1948, FO 371, no. 69223, Memo on the Ramat David and other incidents involving British military personnel, from the British Embassy in Cairo to the Foreign Office, 14 June 1948.
Doc. 1948, FO 371, no. 69287, Foreign Office Memorandum on Egyptian request for long-range fuel tanks.
Doc. 1949, FO 371, no. 24610, Report No. 15 on the Egyptian Army by the British Advisory Mission for the Foreign Office, London, October 1940.
Doc. 1949, FO 371, no. 73573, Foreign Office Memoranda on sale of Sea Otter amphibians to REAF, London, 5 January–17 February 1948 and 4 May 1949.
Doc. 1949, FO 371, no. 73574, Foreign Office Memorandum on REAF expansion plans, London, July 1949.
Doc. 1949, FO 371, no. 73574, Foreign Office correspondence on REAF expansion plans, London, September 1949.
Doc. 1949, FO 371, no. 74984, Foreign Office Memorandum on overhaul of REAF C-47 Dakotas in Italy, London, February 1949.
Doc. 1950, FO 371, no. 73574, Memorandum on senior RAF officer offered advisory job by REAF, from British Embassy in Cairo for the Foreign Office, London, January 1950.
Doc. 1950, FO 371, no. 80466, Foreign Office Memoranda on proposed sale of radar sets to Egypt, London, 1950.
Doc. 1950, FO 371, no. 80473, Foreign Office correspondence on sale of Halifax bomber turrets to Prince Abbas Halim, London, January 1950.

Doc. 1950, FO 371, no. 80474, Memorandum on senior RAF officer offered advisory job by REAF, by the British Embassy in Cairo for the Foreign Office, London, January 1950.
Doc. 1950, FO 371, no. 80474, Memorandum on reported explosion of jet aircraft over Israel, by the British Embassy in Cairo for the Foreign Office, London, 19 April 1950.
Doc. 1950, FO 371, no. 80474, Annual Report No. 1 by the British Air Attaché for the Foreign Office London, Cairo, 23 March 1950.
Doc. 1950, FO 371, no. 80474, Report by Squadron Leader J. R. Baldwin on the Training given to the Royal Egyptian Air Force by the Royal Air Force, April/May, 1949, Appendix B to Air Attaché Report No. 1 on Royal Egyptian Air Force.
Doc. 1951, FO 371, no. 90183, Foreign Office Memoranda on presence of German ex-officers in Cairo, London, 1951.
Doc. 1951, FO 371, no. 90184, Foreign Office Memoranda on presence of German ex-officers in Cairo, London, 1951.
Doc. 1951, FO 371, no. 96993, Annual Report No. 2 on the REAF by the British Air Attaché in Cairo, 24 January 1951.
Doc. 1951, FO 371, no. 90175, Foreign Office Memoranda on proposed Egyptian purchase of anti-aircraft guns in Switzerland, London, 15 August 1951.
Doc. 1952, FO 371, no. 96968, List of military aircraft ordered by Egypt from UK, Foreign Office correspondence, 13 May 1952.
Doc. 1952, FO 371, no. 96993, Annual Report No. 2 by the British Air Attaché in Cairo for the Foreign Office, London, 21 January 1952.
Doc. 1952, FO 371, no. 96993, Foreign Office Memoranda on proposed sale of radar sets to Egypt, London, March 1952 and June 1952.
Doc. 1952, FO 371, no. 96993, Memo from the Air Attaché, British Embassy in Cairo, on the need for Britain to help the REAF, for the Foreign Office, London, 7 August 1952.
Doc. 1953, FO 371, no. 102878, Secret cypher on delaying sale of Meteor jets to Egypt, from the Prime Minister to the Secretary of State at the Foreign Office, London, 27 January 1953.
Doc. 1955, FO 371, no. 113707, Foreign Office Memo on EAF, London, 18 November 1955.
Doc. 1955, FO 371, no. 113707, Foreign Office Memorandum on Egyptian aircraft factory at Hilwan, London, 19 January 1955.
Doc. 1955, FO 371, no. 113707, Foreign Office Memos on British supply of aircraft spares to Egypt, London, 4 April, 30 and 31 August 1955.
Doc. 1955, FO 371, no. 113707, Memo on British supply of aircraft spares to Egypt, by the British Air Attaché in Cairo for the Foreign Office, London, 27 January 1955.
Doc. 1955, FO 371, no. 113708, Foreign Office Memo on British supply of aircraft spares to Egypt, London, 18 November 1955.
Doc. 1955, FO 371, no. 113708, Foreign Office Memos on British supply of aircraft spares to Egypt, London, 3 and 7 September 1955.
Doc. 1955, FO 371, no. 113709, Foreign Office Memo on British supply of aircraft spares to Egypt, London, 1 November 1955.
Doc. 1955, FO 371, no. 113709, Foreign Office Memo on British supply of aircraft spares to Egypt, London, 18 November 1955.
Doc. 1956, FO 371, no. 119007, Memo on arrival of Soviet-supplied military aircraft in Egypt, by the British Embassy in Cairo for the Foreign Office, London, 18 January 1956.
Doc. 1956, FO 371, no. 119008, Memo on supply of military aircraft to Egypt, by the British Embassy in Cairo for the Foreign Office, London, 18 January 1956.
Doc. 1956, FO 371, no. 119009, Memo on the EAF, by the British Embassy in Cairo for the Foreign Office, London, 18 January 1956.

Doc. 1956, FO 371, no. 119009, Annual Report on the EAF No. 6 for the year 1955, by the British Air Attaché in Cairo for the Foreign Office, London, 24 January 1956.

Doc. 1956, FO 371, no. 119011, Memo on training EAF personnel in UK, by the British Embassy in Cairo for the Foreign Office, London, 18 January 1956.

Doc. 1957, FO 371, no. 125489, Assessment of Egyptian Armed Force for Foreign Office, London, 1957.

Doc. 1957, FO 371, no. 125495, Memo on training of EAF personnel by India for Foreign Office, London, 1957.

Doc. 1958, FO 371, no. 131368, Reports on Egyptian rocket bases for Foreign Office, London, 1958.

Doc. 1958, FO 371, no. 131373, Report on EAF anniversary celebrations for Foreign Office, London, 1958.

Quandt, William B., "Soviet Policy in the October 1973 War." *RAND Paper, No. R1864.* Prepared for the Office of the Assistant Secretary of Defense for International Security Affairs, 1976.

Radio Baghdad, 9 October 1973, in *Reports from Arabic,* taped by the BBC Monitoring Service, Cavesham, October 1973.

Radio Cairo, 7 October 1973, in *Reports from Arabic,* taped by the BBC Monitoring Service, Cavesham, October 1973.

Radio Cairo, February 1974, in *Reports from Arabic,* taped by the BBC Monitoring Service, Cavesham, February 1974.

Taha, Sayyid. *Diaries,* supplied by Brigadier-Dr. ʿAbd al-Raḥmān Zakī, 1974.

U.S. Senate, Committee on Armed Services, Subcommittee on Tactical Airpower. *FY 1975 Authorization Hearings,* 11–12 March 1989.

Zaki, Tahsin, Wing Ldr. EAF. *Autobiographical Note,* via Sharīf Sharmī in correspondence with D. Nicolle, May 1991.

## Unpublished Sources: Correspondence

Adkin, F., correspondence with D. Nicolle, July 1990.

Blair, I., Sq. Ldr. RAF, correspondence with D. Nicolle, 1990.

Boyce, L. May, correspondence with D. Nicolle, 1990.

Bradley, Wing Comdr. RAF, correspondence with D. Nicolle, 15 February 1990.

Bushe, Flt. Lt. RAF, correspondence with D. Nicolle, 21 March 1990.

Dicken, A. B., Wing Comdr. RAF, correspondence with D. Nicolle, February 1993.

Humble, William, correspondence with D. Nicolle, 1971.

Indian Air Attaché, Indian Embassy in London, correspondence with D. Nicolle, London, 1975.

Isby, David C., correspondence with L. Nordeen while reviewingan early draft of this book, September 1992.

Israeli Air Attaché, Israeli Embassy in London, correspondence with D. Nicolle, 1968.

Rogers, P. F., Wing Comdr. RAF, correspondence with D. Nicolle, 14 July 1974.

Sharmī, Sharīf, in correspondence with D. Nicolle, 1991 and 1992.

Sharīf, Saʿd al-Dīn , Air Marshal EAF, Acting Chief ADC to President Sadat, correspondence with D. Nicolle, 1973.

Thomason, E., correspondence with D. Nicolle, June 1991.

Zakī, ʿAbd al-Raḥmān, Brig.-Dr. Eg. Army, correspondence with D. Nicolle, September 1971.

Zakī, Taḥsīn, Col. EAF, correspondence with D. Nicolle via Sharīf Sharmī, May 1991.
Zakī, Taḥsīn, Col. EAF, correspondence with D. Nicolle via Sharīt Shamī, May 1991.

## Unpublished Sources: Interviews and Briefings

ʿAbd Allāh, Tamīm Fahmī, Brig. Gen. EAF, interviewed by L. Nordeen, Cairo, 27 March 1987.

ʿĀtif, Aḥmad, Col., interviewed by L. Nordeen, Cairo, 28 March 1989.

Badr, A., Col., Research Department, Egyptian Air Defence Command, briefing to L. Nordeen, Cairo, 10 November 1987.

Fatḥī, Rifʿat, Col., interviewed by L. Nordeen, Ṭanṭá air base Egypt, 30 March 1989.

Ghazzāwī, Fārūq al-, Brig. Gen. EAF, interviewed by L. Nordeen, Cairo, 26, 28, and 31 March 1989.

Ḥāfiẓ, Muṣṭafá, Brig. Gen. EAF, interviewed by L. Nordeen, Cairo, 27 March 1989.

Ḥamdī, ʿAwaḍ, Maj. Gen. EAF, interviewed by L. Nordeen, EAF Headquarters, Cairo, 9 Nov 1987.

Ḥamīd, Qādrī al-, Brig. Gen. EAF, interviewed by L. Nordeen, Cairo, 26 and 29 March 1989.

Ḥarfūsh, Farīd F., Brig. Gen. EAF, interviewed by L. Nordeen, Cairo, 27 and 30 March 1989.

Ḥulwān Government Aircraft Factory, briefing and interview with technical representatives by L. Nordeen, Ḥulwān, 2 April 1989.

Ḥinnāwī , Muṣṭafá al-, Air Force Maj. Gen. EAF, interviewed by L. Nordeen, Cairo, 26 March 1989.

Humble, William, telephone conversation with D. Nicolle, London, 1971.

Ibrāhīm, Muḥammad al-, Col. EAF, interviewed by L. Nordeen, al-Manṣūrah air base Egypt, 29 March 1989.

Jazzārīn, Ibrāhīm Ḥasan, Air Commod. EAF, interviewed by D. Nicolle, Cairo, May 1973.

Masīrī, Muḥammad Nabīl al-, Maj. Gen. EAF, interviewed by L. Nordeen, Cairo, 26 and 28 March 1989.

Mīkhāʾīl, Samīr ʿAzīz, Brig. Gen. EAF, interviewed by L. Nordeen, Cairo, 29 March 1989.

Mīqātī, ʿAbd al-Munʿim, Ạir Vice Marshal EAF, interviewed by D. Nicolle, Cairo, May 1973.

Naṣr, ʿĀdil, Maj. Gen. EAF, interviewed by L. Nordeen, Cairo, 9 Nov. 1987.

Ms. Qūwatlī, granddaughter of Syrian ex-President Qūwatlī, interviewed by D. Nicolle, London, 1975.

Sharārah, ʿĀdil, Col. EAF, interviewed by L. Nordeen, Cairo West air base, 28 March 1989.

Shavit, Y., Brig. Gen. IDF/AF, interviewed by L. Nordeen, Tel Aviv, 17 November 1987.

Sharīf, Saʿd al-Dīn , Air Marshal EAF, Acting Chief ADC to President Sadat, interviewed by D. Nicolle, London, 1973.

Shuwakrī, Nabīl, Maj. Gen. EAF, interviewed by L. Nordeen, Cairo, 9 and 11 November 1987 and 28 March 1989.

Tait, Victor Hubert, Air Vice Marshal RAF, interviewed by D. Nicolle, London, 1974.

Wafāʾī, Aḥmad, Col. EAF, interviewed by L. Nordeen, Egyptian Air Force Headquarters, Heliopolis, 9 November 1987.

Zakī, ʿAbd al-Raḥmān, Brig.-Dr. Eg. Army, interviewed by D. Nicolle, July 1971.

# Index

ʿAbbās II, khedive of Egypt, 10, 11
ʿAbbāsīyah, 39
Abbeville, 26
ʿAbd Allāh, king of Transjordan, 78, 123
ʿAbd Allāh, Ottoman Cavalry 1st Lt., 13
ʿAbd Allāh, Tamīm Fahmī, 183, 185, 189, 191, 236
ʿAbd al-Salīm, 66
ʿAbd al-Wahhāb, 36, 41, 44
Abū Awayjilah, 116, 118, 214
Abū Bakr, 113
Abū Qīr, 36, 42, 58
Abū Rabīʿa, Muḥammad Ibrāhīm, 28, 62, 66, 79, 92, 352
Abū Sidā, Zakaryā, 215
Abū Ṣuwayr, 16–17, 28, 160, 163, 165–66, 200, 207, 213, 324, 346
Abū Zayd, ʿAbd al-Ḥamīd, 40, 54, 56, 69, 81–82, 84, 89, 95, 97, 101–2, 105–6, 114, 352, 356, 358
Abyaḍ, 11
aces, fighter, Egyptian, 373
ʿAdam, al-, 265–66, 318
Adan, Avraham, 284
Aden, 170, 188
Aden Protectorate, 188
advisors: British, 188, 234, (*see also* British Advisory Mission); Chinese, 326; Czechoslovakian, 171; Jordanian, 188; Pakistani, 270; Polish, 171; Russian, 156, 182; Saudi Arabian, 188; Soviet, 171, 173–74, 225, 249–50, 262, 265, 274, 346–47; Soviet bloc/Warsaw Pact, 156, 346; U.S., 319
Aegean, 45
aerial bombs, Soviet types in UARAF, 194
"Aerial Harbour Cairo," 62
aerial photography, 24, 28, 67
aerial reconnaissance, 28
aerial refueling capability, Egyptian, 332
aerial spraying, 24, 128
Aero L-29 Delfin, 7, 226, 230, 273, 296, 305, 312, 320, 331, 333, 335
Aero L-39, 331, 333
ʿAfifī, Dr. Ḥāfiẓ, pasha, 17
ʿAfifī, Maḥmūd Wāʾil, 160
Afrīqī, Ṭāriq al-, 77
AGM-45 Shrike, anti-radar missile, 261
AGM-65 Maverick air-to-ground missile, 324
AH-1Q Cobra, 308
Aḥmad, ʿĀdil, 184
Aḥmad, Imam of Yemen, 188
Aḥmad, Muḥammad ʿAbd al-Munʿim, 352
Ahmad Atef. *See* ʿĀṭif, Aḥmad
AIM-7 Sparrow air-to-air missile, 238, 304, 319, 324, 329
AIM-9B Sidewinder air-to-air missile, 186, 233, 237, 304, 320; AIM-9L Sidewinder, 324–26
air base numbering system, Egyptian, 200–201
aircraft. *See under individual types*
aircraft carriers, 157
aircraft maintainance, role of Egyptian civilian industry, 325
aircraft manufacture in Egypt, 69, 131, 177–79, 227, 316, 326; modification of Soviet aircraft, 305; aircraft spares manufacture in Egypt, 227
aircraft markings: Egyptian high visibility identification, 315; Egyptian "invasion stripes," 88; Egyptian national insignia, 18–19; Ottoman Turkish national insignia, 18–19; Soviet aircraft in Egyptian markings, 228, 262; variation

aircraft markings (*cont.*)
in Israeli camouflage schemes, 292; UAR national insignia, 173
aircraft shelters, hardened, 227, 259, 282, 295, 302
Air Defence Force, Egyptian: 36, 46, 231–32, 242, 245–46, 249, 254–55, 257, 262, 264–65, 267–68, 271–72, 275–76, 279, 281, 285, 287, 298, 300, 310–12, 315, 323, 329; training center, 265
air defense force, Soviet PVO Strany, 157, 183, 195, 231, 253–54
air defense structure, Egypt, 195, 228, 231, 243
air defense tactics, Egyptian, 272, 302–3
airfields, air-bases. *See under individual names*
airfields, construction of new, 227
air force, Abu Dhabi, 270
air force, Algerian, 187, 272, 275, 288–89, 294, 296, 301, 304, 312
air force, Australian, 12
air force, Biafran, 229
air force, British RAF, 7, 16, 20, 32, 36, 41, 43, 54–55, 57, 61, 70, 77, 80, 85–87, 95, 113, 115, 117, 149, 156–58, 162–64, 166, 176–77, 190, 312, 342, 346, 357, 361; Bulstrode Staff Course, 67
air force, British RAF, units: Air Defence East Mediterranean, 53, 55–56; Bomber Command, 162; Central Flying School Cranwell, 22, 312, 361; Middle East Command, 17, 21, 51, 66; Middle East Force, 128; Middle Eastern Officer Training Units, 58; No. 4 Flying Training School, 16, 28; No. 6 Sq., 119, 359; No. 13 Sq., 16, 113, 346, 359; No. 32 Sq., 80, 85–86; No. 48 Sq., 98; No. 80 Sq., 43; No. 113 Sq., 35; No. 208 Sq., 35, 56, 60, 80, 85–86, 118–19, 140, 359; No. 213 Sq., 118–19, 359; No. 971 Balloon Sq., 353; No. 974 Balloon Sq., 353; No. 980 Balloon Sq., 353; No. 1411 Meteorological Flight (joint British-Egyptian unit), 54, 73; No. 70 Officer Training Unit, 59; No. 71 Officer Training Unit, 59; No. 73 Officer Training Unit, 59; No. 74 Officer Training Unit, 59; No. 107 Maintainance Unit, 77–78; No. 205 Group, 117; No. 219 Group, 56; No. 216 Transport and Ferry Group, 57–58, 61; No. 206 Wing, 353; No. 255 Wing, 43; Storage Unit, 36. *See also* Middle East Staff College, British
air force, British RFC, 7, 11, 349
air force, British RFC, units: No. 17 Sq., 11; No. 21 Balloon Company, 12; No. 43 Balloon Section, 12
air force, British Royal Navy Fleet Air Arm, 157, 168
air force, Canadian, 21
air force, Danish, 361
air force, Egyptian: discipline, 52; leadership, 21; morale, 51–52, (*see also* civilian volunteers)
air force, Egyptian, units: No. 1411 Meteorological Flight (joint British-Egyptian unit), 54, 73; No. 1 Sq., 33, 35–36, 39, 41, 44, 60–61, 64, 72, 79, 108, 118, 122, 124, 129, 136, 147, 151–52; No. 2 Sq., 33, 36, 40–41, 43, 53–54, 56–57, 60, 63–64, 67, 69, 72, 79, 108, 118, 122, 136, 151–52, 159, 183, 354; No. 3 Sq., 33–34, 40–41, 44, 61–64, 66, 72, 79, 83–84, 109, 112–13 122, 137, 151–52, 360; No. 4 Sq., 34, 36, 40, 44–45, 61, 64, 73, 83, 96, 99, 109, 122, 137, 152, No. 5 Sq., 40–41, 43, 53–54, 64, 73, 109, 118, 122, 124, 137, 363; No. 6 Sq., 46, 53, 57, 60, 64, 67, 73, 79, 93, 109, 122, 137, 363; No. 7 Sq., 109, 122, 137, 151–52, 200, 360; No. 8 Sq., 79, 109, 122, 137, 200, 360; No. 9 Sq., 98, 109, 137, 142, 147, 151–52, 200; No. 10 Sq., 109, 137, 151, 360, 362; No. 11 Sq., 152, 200; No. 12 Sq., 152, 200; No. 16 Sq. 199, 200; No. 17 Sq., 53, 55–56, 60, 63–64; No. 18 Sq., 199–200; No. 20 Sq., 129–30, 137, 151–52, 200–201; No. 21 Sq., 200; No. 22 Sq., 199; No. 24 Sq., 200; No. 25 Sq., 200; No. 26 Sq., 199; No. 30 Sq., 137, 147, 151–52, 200, 363; No. 31 Sq., 151–52, 200, 363; No. 40 Sq. 199, 201; No. 41 Sq., 199; No. 42 Sq., 199; No. 43 Sq., 200; No. 45 Sq., 198–201; No. 46 Sq., 199; No. 49 Sq., 199; No. 52 Sq., 289; No. 55 Sq., 200; No. 95 Sq., 200; No. 1 Balloon Sq., 353; No. 2 Balloon Sq., 353; No. 3 Balloon Sq., 353; No. 1 Air Regiment, 200; No. 2 Air Regiment, 193, 199–200; No. 5 Regiment, 199; No. 7 Regiment, 199;

No. 9 Regiment, 199, 201; No. 12 Regiment, 200; No. 15 Regiment, 200–201; No. 61 Regiment, 200–201; No. 65 Regiment, 200; 119th Air Division, 311; 129th Air Division, 311; 139th Air Division, 311; 149th Air Division, 311; aerobatic team, 161; Air Academy/Air College/Flying Training College, 126, 137, 141, 146, 149, 152, 180, 187, 208, 210, 226, 296, 312, 326, 361; Air Defence HQ, 205; Air Force HQ, 36, 39, 52, 66, 211, 311, 341; Air Force Intelligence, 361; Air Mechanics School, 36; balloon squadrons, 54–55; Canal Zone HQ, 146; Canal Zone station HQ, 39; Central Region, 195; Eastern Region, 195, 199; Engineering Division, 71; Fighter Training Squadron/Unit, 147, 161; Flying Training School, 34, 36, 40, 41, 46, 52, 59–60, 73, 152; Health and Agricultural Sq./Faruq Health Sq., 128, 137, 152; helicopter squadrons, 239–40, 271, 280, 286, 296; Navigation Training Flight, 73; Northern Region, 195; radio operators, 55; Royal Flight, 40, 61–62, 73, 128, 137, 362; Sinai Air Defence Region, 195; Southern Air District/Division, 311; Target-Towing Flight, 60, 67, 73; Technical Training Institute, 180, 317; trainer squadrons, 296; transport squadrons, 239–40, 286
air force, French, 26,154, 157, 162, 164, 166, 269, 323, 346
air force, French, units: Groupement Mixte No. 1, 168
air force, French Naval, 157, 168
air force, German, 12, 33, 47–48, 54, 114, 116, 350
air force, German, units: 1/JG27, 48; 300th to 304th "Pasha" Sqs., 12
air force, Indian, 179, 312
air force, Indonesian, 187
air force, Iranian, 13, 70
air force, Iraqi, 18, 70, 80, 114, 187, 198, 211, 217, 270, 272, 276, 279, 288, 294, 296, 301, 304, 312
air force, Israeli, 70, 74, 82, 84, 89–90, 93–94, 97, 99–103, 106, 113, 115–16, 143, 145, 149, 154, 158–59, 168, 175–76, 195–8, 203–4, 211, 213–14, 221, 226, 231–32, 234, 238–39, 245–46, 250, 252–54, 257, 262, 268, 271–72, 279–83, 285, 288, 291, 294–95, 298, 300, 302–4, 307–8, 345, 367; attitude to Egyptian pilots, 261; helicopter commando raids, 233; on threat from SAMs, 260–61; morale, 308
air force, Israeli, units: "Negev Squadron," 82; No. 101 Sq., 85, 88, 103, 117
air force, Italian, 9, 31, 35, 41, 43, 45, 92, 99, 360
air force, Italian, units: 41st Gruppo, 45; 172nd Squadriglia, 45
air force, Jordanian, 211, 217, 260
air force, Kuwaiti, 270
air force, Libyan, 187, 270, 272, 290, 292, 318
air force, Nigerian, 187 229–30
air force, North Korean, 272, 275, 294
air force, Norwegian, 143
air force, Ottoman, 12, 350
air force, Ottoman, units: 3rd Bölük, 12; 4th Bölük, 12; 14th Bölük, 12, 350; 4th Army Aircraft Park, 12; Flying School, 13
air force, Pakistani, 270
air force, Saudi Arabia, 139, 269–70
air force, South African, 289
air force, Soviet, 156, 203, 222, 228, 248, 264, 288, 292, 313; in Egypt, 248–49, 254, 258, 262
air force, Spanish, 178
air force, Syrian, 80–81, 147, 162, 173, 175, 181, 193, 196, 198, 211, 217, 244–45, 247, 260, 271–72, 275, 278, 302, 345–47, 365, 373
air force, Syrian, units: Aeronautical Technical Institute, 347
air force, Turkish, 60, 70
air force, UAE, 312
air force, UN Truce Monitor Corps, 93
air force, USAAF, 7, 61–63, 156
air force, USAF, 93, 149, 190, 195, 203, 236, 252, 290–92, 316, 319, 342, 357
air force, U.S. Marine Corps, 291
air force, U.S. Navy, 149, 252, 291
air force, U.S., units: Air Command and Staff Course, 312
air force, Yemeni, 173, 181, 188–90, 365
air force, Zionist Sherut Avir, 70, 76, 80
Air Force Intelligence, U.S., 302
air force leadership, Egyptian, 306

air force organization, Egyptian, 310–11; changes to, 311
air force personnel, Israeli, "Mahal" foreign volunteers, 80, 84, 116
air force training, Egyptian, 155
air losses, Egyptian, 373
air mail, 14
Air Ministry, British, 67
air observation posts, 231,
air observer system, Egyptian, 39, 231, 301, (*see also* coastal observer system, Egyptian)
Airspeed Horsa, 361
Airspeed Oxford 1, 335, 354
airstrips, dispersal, 98, 311
air tactics: Egyptian, 120, 305–6; ground control intercept (GCI) , 305; helicopter-borne commandos, 308; Israeli, 243, 292, 297, 303
Airtech Ltd., 98
air-to-ground rocket, 80mm unguided, 234
air victory claims: 304, 306, 308; Egyptian, 373; gun-camera confirmation, 365
Air War Institute, Egyptian, 312
Airtech Ltd., 98
Airworks Company, 17
ʿAjjūr, 107
ʿAjlūn, 204–5
ʿĀkif, Ḥasan, 22, 62, 66, 134, 352
ʿAlamayn, al-, battle of, 49, 51
Albania, Albanians, 5, 10, 27, 190
*Albion, HMS,* British Royal Navy aircraft carrier, 162
Alcock, Sir John, 16
Aldis lamp, 39
Aleppo, 347
Alexandria, 14, 17–18, 32, 37, 39, 41–43, 45–46, 56–57, 134, 145–46, 148, 151, 166, 195, 228, 231, 249, 267, 307, 313, 325, 346, 353–54, 362
Algeria, 59, 149, 224; Algerian revolution, 153
Algiers, 58
ʿAlī Dīnār, Sultan of Darfur, 11
ʿĀlim, Jalāl ʿAbd al-, 213
Allenby, Gen. Sir Edmund, 12, 14
Allon, Yigal, 107, 111
Almāẓah, 17–18, 21, 39–40, 42–44, 47, 49, 51–54, 60, 66, 70–71, 96, 98, 112, 114, 122, 124, 131–32, 134, 137, 145–47, 151–52, 156, 163, 166, 172, 196, 311, 327, 346, 360
Alon, Modi, 88, 97, 103
ALQ-101, electronic jamming pods, 261
American air attache, 132–33
American Defense Intelligence Agency, 227
American embassy, 133
*Amīrah Fawzīyah, HMES,* Egyptian navy, 45
*Amīr Fārūq, HMES,* Egyptian navy, 105–6
Amman, 88, 90
ʿAmr, ʿAbd al-Ḥakīm, 132, 135, 158, 177, 187, 199, 209, 212, 220
AN/ALE-40 chaff/flare dispenser, 330
AN/ALQ-119 ECM jamming pod, 324, 330
AN/ALQ-131, ECM jamming pod, 324, 330
Anglo-Egyptian Air Defence Agreement, 39
Anglo-Egyptian Treaty, 6, 30–31, 110, 119, 129, 140
anti-aircraft artillery: Austrian mountain gun, 9; Egyptian, 157, 163, 231, (*see also* army units, Egyptian); Israeli, 289
anti-aircraft cooperation, 54, 59–60
anti-aircraft missile defenses, 195–96
anti-drug smuggling patrols, 61
anti-imperialism, 48
anti-malarial campaign, 24
anti-westernism, 7
Antonov An-12, 194, 200, 210, 221, 230, 262, 273, 292, 311, 315, 320, 326, 335
Antonov An-22, 292
Anya Nya, 263
APS-125 radar, 329
ʿAqīr, 83, 87, 115, 160
ʿArabī, 113
Arab League, 72, 319
Arab nationalism, 10, 81, 150, 172
Arab Organization for Industrialization (AOI), 315–16, 326
Arab Revolt, 12, 47
ʿArīsh, al-, 12, 24, 44, 70–71, 78,-79, 81–83, 85–87, 92–97, 99–107, 112–15, 117, 121–22, 124, 126, 131–32, 136, 175, 183, 192–93, 195, 197, 200, 207, 215–16, 259, 261, 276, 294, 342–43
armed forces, America, use of Egyptian bases, 322
armed forces, Egyptian: Armed forces Technical Training Institute, 226; Army Council, 14; strategy in Palestine 1948,

79; subversion in, 47–49, 54; Supreme Command Council, 196
armored cars, Italian, 31
arms embargo, 98–99; British, 346
Armstrong Siddeley aircraft manufacturing company, 34
Armstrong Whitworth Atlas, 16, 28
Armstrong Whitworth Siskin, 16
army, Biafran, 230
army, British, 41
army, British, units: 651st Air Observation Post Squadron, 82; Long Range Desert Group, 40, 46; Suez Canal Brigade, 35
army, Egyptian, 9–14, 17, 20–22, 28, 31, 36, 38–42, 46–47, 64, 74–75, 78, 83–84, 89, 93, 95, 97, 100, 104–5, 107, 111, 121, 130–31, 143, 145, 149, 157–58, 163, 166–67, 169, 189, 199, 203, 208, 211–15, 221, 233, 240, 242, 257, 262–63, 268, 274, 278, 280, 289–90, 293, 295, 309–10, 312, 315, 327, 342–43, 351, 356, 361–62; recreated in 1883, 5; tactics, 32
army, Egyptian, units: 1st Infantry Battalion, 12, 358; 2nd Infantry Battalion, 358; 2nd Infantry Brigade, 104, 358; 6th Infantry Battalion, 358; 7th Infantry Battalion, 358; 9th Infantry Battalion, 95; Air Support Control Organization, 60; air defense units/forces, 197, 227; anti-aircraft regiments, 36, 39, 42–43, 45, 55, 57, 97–98; anti-aircraft units, 195, 234; anti-aircraft Unit 1, 363; anti-aircraft Unit 2, 364; anti-aircraft Unit 3, 364; armored forces, 290; Army of National Liberation, 149; Army Service Corps, 351; artillery, 233, 235, 246, 364; bomb disposal squads, 45; Camel Transport Corps, 10; commandos, 221, 233, 235, 240, 271, 280, 293, 306, 333; Corps of Engineers, 107; Darfur Field Force, 11; Eastern Command, 156; Engineer Corps, 278, 281, 283; Expeditionary Force in Palestine, 93, 78, 104, 110, 114, 121, 358; Expeditionary Force in Yemen, 189–91; Frontier Forces and Frontier Police, 9, 10, 14, 23, 25, 43, 61, 336; HQ, 211; HQ Gaza, 101–2; Labour Battalions, 10; Military Academy/Army Cadet School, 28, 32–33, 141, 225, 257; Mobile Brigade, 39, 42, 44, 351; National Guard militia, 149; parachute battalion/paratroopers, 126, 187; Parachute Training School, 361; radar units, 204; searchlight regiments, 39, 45, 60; Second Army, 299; special forces, 271; Third Army, 297, 299, (*see also* Air Defence Force, Egyptian)
army, German, units: Afrika Korps, 47–48, 353
army, Indian, 41
army, Iraqi, 95, 356
army, Israeli, 80–81, 114, 216, 282, 288, 292–93, 308, (*see also* army, Zionist: Haganah)
army, Israeli, units: artillery, 278; commandos, 143, 233, 237, 239, 245; Negev Brigade, 89; Southern Command, 214; tank corps, 289
army, Italian, 44
army, Jordanian, 214, (*see also* army, Transjordanian Arab Legion)
army, Jordanian, units: radar, 204–5
army, Lebanese, 95, 107–8, 356
army, Libyan, 318
army, Nigerian, 230
army, Palestinian: Arab Liberation Army, 80–81, 108; 143, 145, 149, 153 170, 356, 358; militia, 76–77, 100, 108
army, Saudi Arabian, 356
army, Soviet, 219; troops in Egypt, 249, 255
army, Sudanese, 263; units in Egypt, 163
army, Syrian, 84, 95, 97, 107–8, 272, 274, 285, 287–88, 322, 356; commandos, 278; field bakeries, 274
army, Transjordanian Arab Legion, 84, 88, 104, 107–8, 112, 356, 358, (*see also* army, Jordanian)
army, Yemeni, 188–90; royalist, 189–220, 220
army, Zionist: Haganah, 75, 356, (*see also* army, Israeli)
*Arromanches*, French aircraft carrier, 162
ʿArūsah, Khalaf al-, 116
AS-1 Kennel anti-shipping missile, 186, 194, 200
AS-5 Kelt stand-off missile, 258, 276, 279–80, 288
AS-30 air-to-surface missile, 325
Assad, Hafez al-, 269, 346–47

Association of British Aircraft Manufacturers, 92
Aswan, 39, 226, 311
Aswan High Dam, 148, 170, 175, 195, 249, 266, 311
Asyūṭ, 27
ʿAṭā Allāh, ʿAbd al-Munʿim, 83
ʿĀṭif, Ahmad, 243–44, 297, 319–20
ʿAṭīyah, ʿAlī, 148
Atlantic, 16
Atoll air-to-air missile, 185–86, 206–7, 210, 214, , 233, 237–38, 240, 244, 268, 275–76, 294, 298, 304, 313
Augerten, Rudolf, 112
Austria, 274
Austro-Hungarian Empire, 10
Avia C210, 85, 88, 94–95, 97, 99, 102, 115, 117, 345
Avia S92, 361
*Aviation Week and Space Technology,* magazine, 308
Avro 504N, 16, 28
Avro 618 "Ten," 17, 22–25, 27, 335
Avro 621 Tutor, 25
Avro 626, 22, 24, 26–27, 32, 35, 52, 59, 335, 351, 360
Avro 652, 32, 64, 335, 351
Avro 674, 335
Avro aircraft manufacturing company, 32, 350
Avro Anson, 32, 40, 44, 47, 61, 64, 70, 72–73, 79–80, 83, 90, 98, 109, 122, 127, 137, 142; Mk. 1, 335, 354, 361; Mk. 19, 62, 361; Mk. 22, 354
Avro Commodore, 32, 61, 335, 351
Avro Lancaster, 142, 147, 362; B1, 125, 133, 137, 335, 360
Avro Lincoln, 130, 146
Avro York, 98, 189
AWACS aircraft, 324
Awjah, al-, 114, 140, 160
Axis, German-Italian, 6, 38

BAC Lightning, 269–70
Badr, battle of, 274
Badr, Imām Muḥammad al-, of Yemen, 188
Badrān, Shams al-Dīn, 205
Badrī, Ḥasan al-, 220
Baghdad, 18
Baghdādī, ʿAbd al-Laṭīf, 32, 81, 83, 113, 132, 134–35, 139, 155, 360
Baghdādī, ʿAlī Muṣṭafá, 236, 268
Baghdad Pact, 144
Baḥrīyah Oasis, 27, 36, 39–40, 44, 61
Bakīr, 58
Bakr, 353
Baldwin, J. R., 120–21
Ballāḥ, al-, 118, 122, 359
Balliol trainer, 362
Balūza, 241, 276
Banghāzī, 18
Banī Suwayf, 207, 243, 248, 275–76, 281, 324
Bannatyne, 11
Barakah, Alādaʾ, 185, 222, 328
Barakah, Maḥmūd, 82, 356
Bar-Lev, Haim, 216, 240
Bar-Lev Line, 271, 278, 280
barrage balloons, 54, 259, 282, 302
Barsis, 56
Bat Yam, 88
Battle of Britain, 7, 40
Bayt Affa, 96
Bayt Hānum, 107
Bayt Jibrīn, 94, 103, 107, 111, 358
BBC Arabic Service, 365
bedouin, 23
Beech 1900C, 330, 333, 335
Beech AT-7, 361
Beech Bonanza B35, 82, 122, 151–52, 360
Beech C-45, 346
Beech D-18S, 127, 137, 335
Beech Traveller, 355
Beʾerot Yiẓḥaq, 97, 105
Beersheba, 12, 103, 115, 193, 213
Beʾer Tuviyya, 357
Belgian Congo, 187
Belgian Contracting and Trading Company, 355
Belgium, 10, 341
Ben-Gurion, David, 81, 142
Beriev Be-6, 262, 267
Beriev Be-12, 228
Berque, Jacques, 6
Bhargara, Kapil, 179
Biafra, 229
Bilbays, 126, 137, 141, 149, 152, 163, 166, 180, 187, 208, 210, 222, 226, 259, 289, 296, 312, 326, 361
Binhah, 42, 298
Biʾr al-Thamādah, 197, 200, 209, 276–77, (*See also* al-Mulayz)

Biʾr Arādah, 235, 281
Biʾr Ḥammah/Biʾr Hama, 116, 276
Biʾr Ḥasanah, 278
Birinjīyah, 11
Biʾr al- Jifjāfah, 197, 200, 207–8, 235, 252, 261, 276, 278
Biʾr Lahfān, 114, 175
Biʾr Malit, 11
Blackbushe, 358
Blangy sur Bresle, 26
Board, A. G., 17–18
Boeing 707/320, 332–33, 336
Boeing 737, 332–33, 336
Boeing B-17 Flying Fortress, 96, 102, 104, 361
Boeing C-97 Stratocruiser, 264, 273
Boeing Vertol CH-47C, 332–33, 336
bombing, Israeli, of industrial targets, 246–47
bombs: Egyptian made "dibber," 276–77; electro-optical, 292; German, 93; laser-guided, 292, 307
Bosnia, Bosnians, 5, 190, 333
Boulton Paul Defiant, 67, 73
bowstring, symbolism of, 48
BQM-34 Firebee drone, 264
Bradley, M. G., 140
Brandner, Ferdinand, 178
Brayr, al-, 143
Brazil, 10, 328, 331
Bristol aircraft company, 177
Bristol Beaufighter, 57, 99, 101–2, 106; Mk. X, 355
Bristol Blenheim, 34, 40
Bristol Brigand, 68
Bristol F2B Fighter, 16
Bristol Sycamore, 146
Britain, British, 10, 38, 128–29, 153, 190, 229–300, 316–17, 319, 328, 332
British Advisory Mission, 31–34, 36, 52, 59, 63, 66–68, 79, 341
British air attache, 79, 93, 120–21, 124, 131, 138, 146–47, 361
British ambassador/embassy, 63, 71
British Aviation Insurance Company, 25–26
British Foreign Office, 41, 99, 129
British Military Mission. *See British Advisory Mission*
British occupation of Egypt, 5
British withdrawal from Canal Zone, proposed Plans, 139
Britton, E, 41
Brown, Sir A. Whitten, 16
Bücker Bü-181D Bestmann, 131, 177
Bulgaria, Bulgarians, 10, 145, 172
*Bulwark, HMS,* British Royal Navy aircraft carrier, 162
Burayj, al-, 82
Burgundy, 26
Bushe, C., 140

C-5, 307
C-141, 307
Cairo, *passim*
Cairo Aviation Company, 355
Cairo International Airport, 163, 200, 209, 311
Cairo Museum, 25
Cairo Police Force, 356, (see also police, Egyptian)
Cairo-Suez road dispersal airstrip, 166
Cairo West, 98, 122, 137, 146, 151–52, 163, 165–66, 185, 200, 205–7, 209, 211, 213, 223, 236, 313, 319, 329
caliph, caliphate, 15, 350
Calshot, 16
Camel Transport Corps. *See under* army, Egyptian, units
Camp David Peace Accords, 3, 310, 316, 322, 326–27
Canada, 143, 328
Cant 1007bis, 45
capitulations, legal system, 23
Carter, President Jimmy, 316
Casetiato, Guido, 92
Catania, 18, 58
Caucasus mountains, 12
Central Intelligence Agency (CIA), 345
Central Narcotics Intelligence Bureau, Egyptian, 23
Centurion tank, 215
CH-53A, 291
Chad, 331
Chaparral anti-aircraft missile system, 308, 329
Chick, "Johnny," 52
China, Chinese, 10, 317–19, 322–23, 325, 327–28, 332
"Chinese Farm," battle of, 293
Churchill, Winston, 4, 46, 138–39, 362
Circassians, 5, 9

civil airlines: Aeroflot, 230; BOAC, 71; British Airways, 362; Imperial Airways, 18; Misrair, 39, 44, 113, 350; United Arab Airlines, 22, 353
civilian volunteers/mechanics, in Egyptian air force, 22, 342, 357
civil service, Egyptian, 42
coastal observer system, Egyptian, 281, (*see also* air observer system, Egyptian)
coastal railway, Egyptian, 39
Coastguard and Fisheries Administration, Egyptian, 10, 45, 105, 350, (*see also* navy, Egyptian)
combat experience, of EAF, 4
Combined Plan for the Defence of Egypt, 39
Commander of the Order of the British Empire (CBE), 49
communications, Israeli ability to monitor Egyptian, 258
communism, 20
Congo, Belgian, 189, (*see also* Zaire)
Constantinople (Istanbul), 15
Continental 220 hp engine, 331
Convair PBT-5 Catalina, 336, 361
Coote, P. B., 41
Coptic Christians, in Egyptian air force, 353
Corder, John, 302
Corsica, 59
Council of Ten, 133, (*see also* Free Officers movement)
Cranwell, 361
Crotale surface-to-air missile, 315, 323, 329
CT 20 drone, 330
culture, Arab-Islamic, 2
Curtiss aircraft company, 63, 66
Curtiss C-46 Commando, 63, 68, 79, 90, 96–97, 100, 102, 109, 113, 122, 137, 142, 151–52, 342, 357, 360–62
Curtiss P-40 Tomahawk IIA, 53, 55, 336, 354
Curtiss P-40E Kittyhawk, 58
Cyprus, 80, 85, 87, 154, 157, 162, 164, 168, 359
Czechoslovakia, 85, 127, 145, 172, 230, 331

Ḍabʿah, 42–43, 48
Daimler-Benz DB 605, 115
Dākhilah (Alexandria), 21, 34, 36, 39, 122, 124, 137, 146, 151–52, 166, 354
Dākhīlah Oasis, 27
Damascus, 12, 80–81, 89–90, 158, 200, 247, 269, 274, 288, 313, 345, 347
Darāw, 226
Dardanelles, 12
Darfur, 11
Dassault-Dornier Alpha Jet, 316, 326, 331, 333, 336
Dassault Mirage: 192, 205, 207, 210–11, 213–17, 221, 223, 235–41, 242–43, 245–47, 252, 254–55, 264–65, 269–70, 272–73, 275, 279–81, 283, 285, 290–91, 295, 297–98, 303–5, 316–19, 328, 330–31; III, 184, 198, 269; IIIC, 195, 269; IIIE, 270; III SDE, 336; V SDD, 314, 333, 336; V, 269–70, 275, 296–97, 314–15, 325, 330, 333; V SDE, 314; VR, 333
Dassault Mirage 2000: 324, 326, 328, 333, 336; 2000BM, 325; 2000EN, 325
Dassault Mystère Falcon, 331, 333, 336
Dassault Mystère: 175–76, 195, 207–8, 241; IVA, 154, 158–61, 168
Dassault Ouragan, 145–46, 154, 158–61, 168, 175–76, 205–6
Dassault Super Mystère: 183–84, 198, 206, 213, 216, 241, 273, 367; B2, 176
Dayan, Moshe, 221, 296
Dayr al-Ballāḥ, 117
Dayr Haim, 357
Dayr Sunayd, 84
Dayr Yāsīn, 76
Defective Arms Scandal, 62, 136
Defense Intelligence Agency, American, 232, 238
defense manufacturing, in Egypt, 329
De Havilland aircraft company, 17–18, 22, 129, 178
De Havilland Canada DHC-1 Chipmunk, 137, 141, 152, 336, 345–46, 361–62
De Havilland Canada DHC-5D Buffalo, 331, 333, 336
De Havilland Comet, 22
De Havilland Devon, 69
De Havilland DH 9, 16
De Havilland DH 60T Moth Trainer, 17, 22–25, 52, 61, 336, 351, 360
De Havilland DH 82 Tiger Moth, 17
De Havilland DH 89A Dragon Rapide, 82, 99, 114, 336
De Havilland DH 98 Mosquito, 61, 113, 158, 355, 359

De Havilland DH 104 Dove, 69, 90, 109, 122, 127, 137, 142, 336, 355, 361–62
De Havilland Vampire: 69, 142–43, 159, 161, 178, 314, 360–61; FB5, 361; FB52, 125–26, 130–31, 133, 137, 139–40, 142, 145–47, 151–52, 156, 160, 168–69, 336, 346, 361–64; NF10, 126, 362; T55, 140–41, 152, 336, 361
De Havilland Venom, 167
Deversoir, 140, 151–52
Deversoir Gap, 293, 296, 300
Dibus, 215
Dickson, Sir William, 118, 128–29
Dighaydī, ʿAbd al-Ḥamīd, 22, 24, 26
Dijon, 26, 269
Dimona, 193
Dingwall, 26
Director of Egyptian Military Aviation, 17
discipline: Egyptian military, 2
dispersal airstrips, 166–67
Dixon, 33
Dorot, 88, 107, 357
Douglas C-47 Dakota, 45, 62–63, 68, 71–73, 79, 81–83, 86, 88–90, 97, 99–102, 109, 112–15, 122, 124, 134, 137, 141–42, 151–52, 187–89, 194, 200, 273, 336, 346, 355, 358, 360–62
Douglas DC-3, 236, 355
Doyle, J., 116
drill, Egyptian soldier's love of, 5
drugs: addiction, 22–23; hashish, 22; heroin, 22; smuggling, 63; trafficking, 23
Ḍumayr, 200
Dumyat, 241
Duss, Shuhdī, 26
Duwaynī, Saʿd Ṣādiq al-, 86

E-3A (AWACS), 323
E-300 jet engine, 179
*Eagle, HMS,* British Royal Navy aircraft carrier, 162
Eastchurch, 16
école de Guerre, French, 312
Eden, Anthony, 139, 145, 159, 353
Egypt: ability to take losses, 8; religious character of society, 8
Egyptian air attache in London, 135, 148
Egyptian Aircraft Construction Factory, 131, 177
Egyptian ambassador in London, 17, 25
Egyptian Antiquities Department, 24
Egyptian Chamber of Deputies, 41
*Egyptian Gazette,* newspaper, 80
Egyptian Military Museum, 48
Egyptian National Assembly, 219, 257
Eisenhower, President Dwight, 170
Ekron, 364
*Elat,* Israeli navy destroyer, 224
Elazar, David, 275
electronic counter measure (ECM) systems, 204, 236, 261, 277, 280, 291, 301, 307, 325, 330
electronic listening intelligence systems (ELINT), 258
EMB-312 Tucano, 326, 331, 333, 336
emergency code word "Grapes," 205
English Channel, 25–26
English Electric Canberra, 150, 158–59, 163, 166–67, 346
Enugu, 230
Eritrea, 44
Ethiopia, 11, 31, 327; Italian invasion, 32–34
"ethnic cleansing," 76
Europe, 322
Evans, David L, 133
Eẓyon, 261

F-6 (Chinese-built MiG-19): 318, 325–26, 328, 331, 333; FT-6, 318, 325, 333
F-7 (Chinese-built MiG-21), 325–26, 328, 330–31, 333
Factory 57, 178
Factory 333, 179–80
Fahmī, Muḥammad ʿAlī, 231, 285
Faʾid, 118, 140–41, 159, 161, 166–67, 197, 200, 206–7, 209, 213, 296, 359
Fairchild F-24 Argus, 89, 346
Fairey Gordon, 34, 36, 52, 336, 351
Falcon business jet, 326
fallāḥūn, Egyptian peasantry, 4, 23
Faluja, 84, 100, 103–4, 112, 358
Faluja Pocket, 22, 104, 107, 110–11, 115–16, 120–21, 343
Fan Song B radar, 249
Farāfirah, 27
Faraj, Muḥammad, 28
Farīd, Ṣalāḥ, 53
Farid F. Harfoush, *See* Ḥarfūsh, Farīd)
Farouke El Gazawy. *See* Ghazzāwī, Fārūq al-

Fārūq, King of Egypt, 6, 18, 22, 31, 41, 48, 62, 64, 66, 93–94, 96, 120, 123, 125, 128, 130–32, 136, 173, 342, 354, 362
fārūqīyah, air force cap, 36
Fascism, 20, 50
Fashar, al-, 349
Fatḥī, Rifʿat, 269, 290, 294
Fawzī, Muḥammad, 219, 231, 246
Federation of Arab Republics, 269
Ferranti, electronics industry, 317
Fiat aviation company, 99, 114
Fiat G55: A/1, 99, 109, 113, 118, 122, 124, 127, 137, 141, 337, 341, 361; B, 99, 114
Fiat G59–2A, 99, 345
Fiesler Fi-156 Storch, 339
fighter control system, Egyptian, 148
Firdān, al-, 195, 222
Firebee drone, 330
Fire Service, Egyptian, 22
Fitchett, Leonard, 106
Five Year Defence Plans, Egyptian, 322–23, 328
Fokker E types, 12
Fokker Trimotor, 17
Ford "Tin Goose", 17
Foreign Office, British, 27, 66, 68, 70
formation flying, Egyptian, 28, 68, 120
Foschini, Ettore, 360
Fouga Magister, 212
"Fourteen Points". *See* Wilson, President Woodrow
France, 10, 26, 38, 58, 128–29, 145, 148–49, 153–54, 179, 269, 300, 312, 314, 316–17, 325, 328, 332, 345; Vichy regime, 47; missile assistance to Israel, 179
Free Officers Movement, 81, 123, 131–35, 156, (*see also* Revolutionary Council; Council of Ten)
FROG missile, 288, 327; 7A, 180
frontier fence, Egypt-Libya, 31
Frontier Forces and Frontier Police, Egyptian. *See* army, Egyptian, units
Frunze General Staff Academy, 172
Fuʾād, King of Egypt, 13, 14, 18, 25, 27, 31

Galatina, 99
Galilee, 108
Gal-On, 84
Gan Yavne, 88
gas, bombing, 190–91
Gat, 84, 111
Gavish, Yehu, 214
Gaza, 12, 77, 82, 94, 99–105, 143, 145, 170, 342
Gaza Strip, 108, 110, 119, 143, 146, 153
Gazerine, I. H. *See* Jazzārīn, Ibrāhīm Ḥasan
Gedera, 88
General Dynamics F-16: 185, 316, 323–24, 326, 328, 330–33; F-16A, 320, 324, 333, 337; F-16B, 320, 324, 333; F-16C, 320, 324, 337; F-16D, 320, 324, 333
genocide, 30
Germany, Germans, 9, 16, 38, 63
Ghazzāwī, Fārūq al-, 160, 165–66, 185, 214
Gibraltar, 18
Gloster aircraft company, 69
Gloster Gladiator, 36, 39–40, 43, 45, 48–49, 53–54, 64, 351, 360; Mk. I, 337; Mk. III, 337
Gloster Meteor, 69, 142–43, 145, 147, 156, 158–59, 161, 167–69, 314; F4, 111, 125–27, 129–30, 133, 137, 151–52, 337, 360–61; F8, 138–40, 151–52, 337, 346, 362; FR9, 140; NF13, 126, 140–42, 152, 158, 163, 337, 346–47; T7, 126, 137, 141, 152, 337, 346, 361–62, 360
Golan Heights, 174, 196, 216–17, 245, 271–72, 278, 287–88, 290, 345
government documents, British, 50
Great Bitter Lake, 254, 272, 287, 292
Great Sand Sea, 40
Greece, 10, 13, 27
ground-controlled interception (GCI) system, Egyptian, 159
Grumman Avenger, 361–62
Grumman E-2C Hawkeye, 295, 327, 329, 332, 337
Grumman Mallard, 137, 152, 337
Guardrail, 330
guided missile development, Egyptian, 227
Gulf of Aqaba: 142–43, 158, 198; blockade of, 153
Gulf of Suez, 44, 157, 236, 239, 243, 245, 277, 299, 323
Gulfstream III, 332–33
Gulf War, 329

H-3, 217
Ha-200 Al-Qāhirah, 178, 210, 337
Ha-300, 179, 209–10, 227, 337
Ḥāda, ʿAlī, 296
Hadda, 191

Ḥāḍir, ʿAbd al-Qadir, 279
Ḥafiẓ, Muṣṭafá, 208, 277, 294–95
Haganah. *See* army, Zionist
*Haganah,* Israeli escort vessel, 105–6
Haganah Plan D, 76
Haifa, 77, 90, 92–93, 107
Haifa Enclave, 77, 80, 85, 87, 90
Ḥajjāj, Fuʾād ʿAbd al-Ḥamīd, 16, 18, 24, 26
Ha Jumhūrīyah, 131, 141, 152, 177–78, 187, 226, 337, 312, 320, 331, 333
Ḥalīm, Prince ʿAbbās, 136
Haluza, 213
Hama, 347
Ḥamādah, Kamāl al-Dīn, 352
Ḥamdī, ʿAwaḍ, 185, 207
Ḥamīd, Muṣṭafá Ṣabrī ʿAbd al-, 114
Ḥamīd, Qādrī al-, 198 205, 208, 221, 232–33, 282, 291
Hammarskjöld, Dag, 167
Hamshū, Bassām, 244
Hancock, N. P., 56
Handley Page Halifax, 109, 133, 137, 142, 155, 337, 360; A9, 125, 360–61; C8, 116, 122, 125, 361
Ḥaqqī, Ibrāhīm ʿAbd al-Raḥman, 22, 26, 92
Ḥarfūsh, Farīd, 222
Hārūn, Ḥaqqī, 352
Hārūn, Saʿd Allāh, 66
Ḥashshād, 58, 353
Ḥasīb, Shafīq, 45, 98, 106, 113
Hawker aircraft company, 71, 129, 350
Hawker Audax, 32–33, 35–36, 59, 61, 64; "Egyptian" Audax, 335, 337, 350, 360; "Egyptian" Audax Mk. VI, 351; "Egyptian" Audax Mk. X, 351
Hawker Fury (Sea Fury), 71, 80, 114, 147, 337, 362; Fury prototype, 70–72, 84, 89, 106, 337, 355; FB11, 125, 129–30, 136, 156, 161
Hawker Hart, 41, 61, 64, 337, 360
Hawker Hind, 35
Hawker Hunter, 146, 190, 217, 270, 272–73, 276, 279, 288, 305
Hawker Hurricane: 43, 53–55, 63–64, 68, 73, 360; Mk. I, 46, 55, 60; Mk. II, 354; Mk. IIC, 55–57, 60–61, 337
Hawker Tempest, 118–19, 359
Ḥaydar Pasha, 117, 178, 342
Hawk surface-to-air missile, 194, 236, 240, 243, 252, 255, 259, 264–65, 277–80, 289
Hayes, G. W., 93, 117, 119–20
Hebrew, speakers in Egyptian Air Force, 258
Hebron, 104
Hejaz, 12
Heliopolis, 9, 26, 92, 122, 131, 137, 151, 177; Air Display, 70
Heller UH-12E, 331, 337
Henlow, 28, 55
Herzliyya, 82, 88, 94, 103
Hetz, Shmuel, 253
HF-24 Marut, 179
High Commissioner, British, 14
highway dispersal strips, 227
Hill 113, 93
Ḥillah, 11
Hiller helicopter, 152
Ḥilmī, Baghāt Ḥasan, 160
Ḥilmī, Muḥammad ʿAbd al-Ḥamīd, 327
Ḥinnāwī, Muṣṭafá Shalabī al-, 115, 164–65, 225, 236, 359
Hispano Ha-200, 178
historical aircraft collection, 125
Hod, Mordechai, 232
Hodges, L. M., 163
Horton Airways, 355
HOT anti-tank missile, 332
Ḥulwān, 43, 60, 125, 136–37, 178–80, 208–10, 226, 237, 317, 326, 331
Humble, Bill, 70–71
Humphrey, L. F., 41
Hungary, 41
Hurghadah, 44, 198, 201, 205, 211, 232–33
Ḥusayn, Aḥmad Saʿūdī, 48
Ḥusayn Kāmil, Sultan of Egypt, 10–11, 13, 349
Hussein, King of Jordan, 136

Ibrāhīm, Dr. ʿAlī Pasha, 113
Ibrāhīm, Dr. Ḥasan, 113
Ibrāhīm, Ḥasan, 81, 132, 134–35
Ibrāhīm, Muḥammad, 275, 295
Ibrāhīm, Tutmus Kāmīl, 86
ʿId al-Aḍḥá, Muslim religious festival, 100
identification friend or foe (IFF), 271, 276, 279; failures of, 301
Idkū, 56–57
Ifṭār, Muslim religious festival, 96
Ilyushin Il-10, 189
Ilyushin Il-14: 145, 158, 162, 187–90, 194, 199–200, 206, 209, 226, 264, 273, 280, 311–12, 320, 326, 337

Ilyushin Il-28, 145–47, 151–52, 155–58, 162–66, 169, 171–72, 175, 184, 186, 189–90, 194–95, 200, 207, 211, 215–16, 220, 230, 241, 258–59, 263, 320, 338, 346
Ilyushin Il-38, 262
Imbābah, 226
Improved Hawk anti-aircraft missile, 323, 329
ʿInān, ʿAbd al-Raḥmān, 86–87
India, 35, 179, 312, 317
Inshāṣ, 61, 163, 166, 201, 207, 209–10, 222–23, 236, 243, 277, 311, 320, 324
Inshāwi, 210
Inspector General of the Egyptian Armed Forces, 46. *See also* sirdār
Institute of War Studies, Egyptian, 196
instructors: Czech, 147, 162, 180; East German, 180: Indian, 180; Pakistani, 270; Polish, 180; Soviet, 147, 162, 180. *See also* advisors
intelligence service, British, 47
International Institute for Strategic Studies, 203
Iran: 327; hostage crisis, 322
Iran-Iraq War, 321
Iraq, 6, 12–13, 15, 38, 62, 124, 170, 203, 208, 307, 321, 326–27, 333
ʿIrāq al-Manshīyah, 12, 102, 104, 107, 358
ʿIrāq Suwaydan, 88–89, 96, 101, 103–4, 107, 358
Irgun Gang, 76, 356. *see also* terrorism, Zionist
irrigation schemes, Nile, 24
Irving Type X parachute, 126
Isdūd, 88–89, 102–3, 358
Islam, Shia Muslims, 350
Islam, Sunni Muslims, 15, 350
Islām Bey, ʿAlī, 33, 36
Islamic studies, 332
Islamism, political movement, 20, 48
Ismāʿīl, General, 271
Ismāʿīl Daʾūd, Prince, 42–43
Ismāʿīlīyah, 45, 156, 195, 213, 216, 223–24, 295, 353
Israel, Israelis, 45, 153, 327
Israeli Defence Ministry, 143
Istanbul, 15
Italian East Africa, 46
Italy, Italians, 10–11, 14, 34–35, 37, 38–39, 41, 58, 92, 126, 139, 341, 362; invasion of Libya, 9; surrender, 55
Iwāsī, ʿAbd al-Munʿim Ḥāfiẓ Muḥammad, 167
ʿIzz, Madhkūr Abū al-, 219, 222
ʿIzzat, 353
ʿIzzat, Aḥmad, 296
ʿIzzat, Ḥusayn, 233

Jabal al-Bāsūr, 226
Jabāl Libnī, 200, 207
Jabal Ruzzah, 47
Jabara, Muḥammad, 222
Jaghbūb, 31, 35, 42
Jamāl, ʿAmr al-, 83
Jamasī, Muḥammad al-, 274
Jamīl (Port Said), 167
Janzūrī, Muḥammad al-, 69, 71, 88, 97
Japan, Japanese, 10
Jawāʿī, 166
Jazīrat al-Akhḍar, 239
Jazzārīn, Ibrāhīm Ḥasan, 28, 35, 40, 48, 59, 131, 352
Jericho surface-to-surface missile, 327
Jerusalem, 117, 160, 310
Jews, Egyptian, 46
Jibrāʾīl, 78
Jiddī Pass, 271, 290
jihād, 10–11
Jindī, Muḥammad Fakrī al-, 247, 278
Jiyānklīs, 226, 281, 314
Jordan, 12, 110, 123, 136, 153, 170, 175, 197, 202–4, 208, 210, 217, 224, (*see also* Transjordan)
Jordan River, Israeli diversion of, 175
Judean hills, 100, 104, 112
Jūlis, 358
Junkers F13a, 14
Junkers Ju-52, 346
Junkers Ju-86, 33
Junkers Ju-88, 54
Jurah, 102

Kadry Abd El Hamid. *See* Qādrī, ʿAbd al-Ḥamīd
Kafāfī, Muḥammad ʿAdlī, 45, 83, 98, 113–14
Kafāfī, Munīrah, 78, 83, 96
Kafr al-Zayyāt, 42
Kāmil, Dr. Ḥasan, 180
Kano, 230
Kasfarīt, 78, 166
Kāstīnā, 160

Kavalla, 27
Kawm Awshim, 243, 248, 296
Kefar Am, 88, 357
Kefar Sirkin, 83
Kemal Atatürk, 15
Kennedy, President John F., 190
Kennel air-to-surface missile. *See* AS-1
Kenya, 189
Kfir, 315
Khalīfah, ʿAbd al-Ḥalīm, 22, 26, 352
Khalīl, ʿIṣām al-Dīn Maḥmūd, 172–73, 178
Khamis, Muḥammad ʿAlī, 213
Khānkah, al-, 52, 54, 246
Khān Yūnus, 102
Khārijah Oasis, 27, 40
Khartoum, 11, 63, 224, 229, 263
Khaṭāṭibah, al-, 311
Khayrī, Ḥusayn, 173
khedives, rulers of Egypt, 5
Khirbat al-Maghaz, 100
Khutāmīyah (Ṭanṭá/Birma), 281, 289
Khutmīyah Pass, 290
kibbutz settlements, fortified, 83
Kibrīt, 145, 151–52, 159, 163, 192, 199–200, 207–8, 297
Kissinger, Henry, 242
Kitaf, 191
Kleinwachter, Dr. Emil, 180
Knesset, 293, 310
Korea, 218; Korean War, 168, 183
Korotyuk, Konstantin Andreyevich, 248
Kosygin, Alexei, 247
Kufrah Oasis, 30, 39–40
Kunming, 358
Kutakhov, Pavel S., 253
Kuwait, 322, 333

Labīb, ʿAlī Muḥammad, 42, 56, 66, 163, 220
Lampson, Sir Miles, 46–47
Lance tactical missile, 327
Land Rover, 315
Lavon Affair, 142
Lawrence of Arabia, 47
Le Bourget, 26
League of Nations, 34, 75
Lebanon, Lebanese, 95, 107, 170, 208, 322, 327, 329–30, 346, 356
Lend Lease, 53
Libya, Libyans, 9, 11, 14, 18, 30, 39, 42, 45–46, 78, 229, 265, 269–70, 290, 292, 317–18, 322–23
Link Trainer, 130
Lloyd, Lord, 14
Lockheed C-130H, 291, 315, 317, 331, 333, 338
Lockheed EC-130, 330
Loyalty, of Egyptian officer corps, 6
Luxembourg, 10
Luxor, 166, 201, 211, 233
Lydda, 89, 102
Lympne, 18, 25

M-1 tank, 323
M-60A3 tank, 323
M113 armored vehicle, 232
Maʾadim Road, 284
Maʿāṭin Baqqūsh, 39
Macchi aircraft company, 126, 139
Macchi MB308, 92, 338, 362
Macchi MC202, 92
Macchi MC205V, 92, 109, 112–13, 115–18, 122, 124, 127, 137, 338, 341, 345, 360–62
Māhir, ʿAlī, 133
Māhir, Muṣṭafá, 353
Maḥmūd, Ḥasan, 24, 135
Maḥmūd, Maḥmūd Ṣidqī, 66–67, 136, 172, 196, 199, 209, 219–20
*Mahrūsa,* Egyptian royal and presidential yacht, 134
maintainance personel, training, 266
Maitland-Wilson, 39
Majdal, al-, 88, 97, 101–7, 358
Makkāwī, 353
Malayjī, Maḥmūd Ṣidqī al-, 78–79, 97, 105, 115
Malayjī, Midḥat al-, 213
Malta, 18, 154, 157, 162
Manchester, 25
Mann, Maurice, 102
Manqabād, 131
Mansūr, Ṣalāḥ, 193
Mansūrah, al-, 208, 240, 268, 281–82, 285, 291, 301, 311
Marāghī, Murtaḍá al-, 173, 223
Marconi electronics company: 317–18, 353; AMES 13 radar, 361; AMES 14 radar, 361; AMES 21 radar, 361
Marīyah, al-, 43
Markham command and control system, 258
Marsá Maṭrūḥ, 18, 35, 37, 39, 42, 57, 226, 265, 313

Marseilles, 18
Marshall-Cornwall, 32, 34
martial law, 42
Martin, Maurice, 93,
Martin B-26 Marauder, 229–30
Masīrī, Muḥammad Nabīl al-, 165–66, 192, 205, 211, 217, 226, 232, 234, 246, 252, 265, 276, 282
Masri, Aziz Ali al-. *See* Miṣrī, ʿAzīz ʿAlī al-
Matra 530 air-to-air missile, 193, 237, 315, 325
Matra 550 Magic air-to-air missile, 315, 325
Maverick TV-guided air-to-ground missile, 291–92, 307
Mazzah, al-, 193, 247, 269, 278, 347
McCarthy, Ramsey. *See* Mīqātī, Ramzī)
McDonnell Douglas A-4 Skyhawk, 237, 239, 241–42, 245–47, 253, 257, 259, 261, 268, 273, 281, 291, 298, 307,
McDonnell Douglas F-4 Phantom: 237–38, 242–47, 250, 253, 254, 257, 259, 261, 264, 266, 268, 273, 279–82, 285, 288, 291, 293, 297–99, 303–4, 307, 317, 319, 325, 330–331, 333; F-4E, 316, 338
McDonnell Douglas F-15, 316, 320
McDonnell Douglas Helicopter AH-64 Apache, 332–33, 338
MD500, 308
Mecca, 274
Medal of Excellence, Egyptian award, 233
medical tests, 16
Medina, 13
Mediterranean, 18, 31, 40, 228, 295, 329
Megiddo, 85, 116
Meir, Golda, 247–48, 272, 275
Messerschmitt, Willi, 178
Messerschmitt Bf-109, 48, 85
Messerschmitt Me-262 A-1, 361
MFI-9B (Swedish light aircraft), 230
Middle East Defence Organization, 129
Middle East Staff College, British, 59, 61
Mikaati, A. M. *See* Mīqātī, Muḥammad ʿAbd al-Munʿim
Mīkhāʾīl, Samīr ʿAzīz, 206, 209, 235, 240–41, 283
Mikoyan-Gurevich MiG-9, 157
Mikoyan-Gurevich MiG-15: 145–47, 151–52, 156–57, 159–63, 165–69, 172, 187, 189–90, 192, 195, 199, 226, 230, 234, 241, 311, 320, 338, 346; MiG-15bis, 200; MiG-15UTI, 145, 162 206, 208, 230, 312, 320, 338, 346
Mikoyan-Gurevich MiG-17: 157, 161, 168–69, 171–72, 175–76, 183, 186–87, 189–90, 193–95, 199–200, 209–11, 214, 216, 220, 223, 226–27, 230, 232, 234, 237–38, 240–41, 243, 246, 258–59, 263, 265, 269, 272–73, 275–78, 288–89, 293, 304–5, 318, 320, 325, 328, 347, 373; MiG-17F, 200, 312, 338; MiG-17PF, 208, 347
Mikoyan-Gurevich MiG-19: 183–84, 193, 195, 198–200, 205, 207, 210–11, 214–16, 318, 331; MiG-19SF, 338
Mikoyan-Gurevich MiG-21: 179, 183–84, 195–96, 198, 206, 208–9, 214–16, 223, 226–27, 231–34, 236–41, 243, 247–48, 250, 253, 258, 261, 265, 268–69, 271–72, 275–76, 278, 281–85, 289–90, 293–96, 299, 301–6, 311, 313–14, 317–18, 320, 324–26, 328, 330–33, 373; MiG-21BFM, 221; MiG-21F-13, 184–85, 198–99, 210, 237, 244, 258, 338; MiG-21FL, 185, 199, 207, 210, 213, 221, 235–36, 258; MiG-21H, 280; MiG-21M, 236, 258; MiG-21MF, 248–49, 254, 267–68, 275, 277, 283, 295, 314, 338; MiG-21PF, 338; MiG-21R, 333; MiG-21UM, 338; MiG-21UTI, 320
Mikoyan-Gurevich MiG-23: 248, 263, 266, 313, 328; MiG-23B, 338; MiG-23BN, 313; MiG-23MS, 313; MiG-23U, 313
Mikoyan-Gurevich MiG-25, 264, 313
Miks, al-, 43
Milāzī, al-, 288
Miles Aerovan, 355
Miles aircraft manufacturing company, 34
Miles M14A Hawk (Magister), 34, 41, 52, 59, 64, 72–73, 90, 137, 152, 339, 351, 355, 361
Miles M19 Master II, 60, 339, 354
Miles Martinet, 58, 60
military career: status of in Egypt, 5–6
military dictatorship, 3
military doctrine, Soviet, 157
military operations and exercises: "Bright Star," 323; "Exercise Contentment," 129; "Exercise Gestic," 361; "Operation 750," 155; "Operation Archer," 162; "Operation Badr," 274; "Operation Dust," 100; "Operation Focus," 202–3; "Operation Hard Surface," 190;

"Operation Horev," 114; "Operation Peace Pharaoh," 316; "Operation Qahir," 197; "Operation Tussle," 60
militia, Palestinian. *See* army, Palestinian
Mil Mi-1, 189, 226
Mil Mi-2, 338
Mil Mi-4, 189–90, 200, 226, 273, 312, 320, 338
Mil Mi-6, 194, 200, 273, 320, 338
Mil Mi-8, 200, 263–64, 273, 278–80, 296, 306, 312, 315, 320, 332–33, 339
Ministry of Defence/War, Egyptian, 20, 33, 36, 39, 52, 56, 111, 205, 341
Minyā, al-, 208, 226, 294
Mīqātī, Muḥammad ʿAbd al-Munʿim, 16, 18, 24–26, 36, 63, 86, 92, 94, 129, 131, 133–34
Mīqātī, Ramzī, 362
Mishmar HaʿEmeq, 116
Mishmar HaYarden, 97
Miṣrī, ʿAzīz ʿAlī al-, 9, 46–48, 132, 352
missile boats, Egyptian, 224
Mitlā Pass, 158–60, 164, 214–15, 235, 271, 277, 290
Mitwallī, Muḥammad, 62, 66–67
Mohammed Nabil El Messiry. *See* Masīrī, Muḥammad Nabīl al-
Moizo, Riccardo, 9
Montenegro, Montenegrans, 10
Morane 502, 137, 152, 339
Morocco, 59, 78
Moscow, 246, 248, 262
Mount Hermon, 278
Mraz M-1D Sokol, 127, 137, 152, 339, 361
Muʿāfī, ʿAlī, 43, 49
Mubarak, Husni, 186, 222, 268, 285, 300–301, 305–6, 311–13, 320, 321–23, 326, 328, 333
Muegyetemi Sportreulo Egyuselet (MS and EM) M-24, 41, 339
Muḥammad, Ḥāfiẓ Muḥammad, 56, 101
Muḥammad, Prophet, 15, 274
Muḥammad ʿAlī Barrage, 42
Muḥammad ʿAlī, ruler of Egypt, 4, 27, 361
Mukhtār, ʿIzz al-Dīn, 294
Mulayz, al-, 197–98, 205, 207, (*see also* Biʾr al-Thamādah)
Munich Crisis, 35
Munʿim, 26
Munʿim, Muḥammad Shākir, ʿAbd al-, 313
Mursī, ʿAbd al-Bakr, 113
Muscat, 170
Muslim Brotherhood, political movement, 77, 135, 321
Mussolini, Benito, 30–31, 46
Mustafa al Hinnawi. *See* Ḥinnāwī, Muṣṭafá Shalabī al-
Mustafa Hafaz. *See* Ḥāfiẓ, Muṣṭafá
Muwawī, Muḥammad Kamil al-, 184

Najāt, 113
Najʿ Ḥammādī, 233
Nājī, Aḥmad, 24, 26, 352
Najīb, Muḥammad, 133–35, 216, 362
Najīlah, al-, 105
Nakhl, 215
napalm, 93
Naples, 18
Napoleon, Emperor, 1
Nāṣir, Jamal ʿIrfān Ṣāfī al-, 84
Naṣr al-Dīn, 86
Naṣr, ʿĀdil, 181, 213
Nasser, President Gamal Abd El-, 6, 22, 28, 32, 47, 81, 102, 123, 129, 131–32, 134–36, 138, 144, 148–50, 153, 157–58, 163–64, 170–73, 175, 177–78, 180, 182, 187, 191, 196–99, 202, 204, 211–12, 218–20, 222, 224, 227–29, 231, 234, 236, 247–48, 251, 254, 256, 261, 266, 309, 352, 354, 362
Nastenko, Yuri, 251
Natke, 284
NATO, 260
naval base, Alexandria, 37, 39
navigator training, Egyptian, 226, 312
navy, British Royal Navy, 57, 188
navy, Egyptian, 105, 130–31, 134, 145, 166, 224, 240–41, 262, 277, 312, 315: Faruqiyah Naval School, 14, (*see also* Coastguard and Fisheries Administration, Egyptian)
navy, Israeli, 105
navy, Italian, 44; submarines, 352
navy, NATO fleets, 228
navy, Soviet, 228, 262
navy, U.S.: 166, 228, 262, 264, 294; Sixth Fleet, 262, 264
Nayrab, 346
nazism, 20, 50
Negba, 84, 88, 93–95, 357
Negev, 83, 93–94, 97, 100, 104–5, 175
Nesher, 292
Nevatim, 213

Nieuport 6, 9
Nigeria, 229, 239
Nile, 39, 233, 248, 263; Delta, 43, 79, 199, 231, 247, 249, 290; flood, 28; valley, 40
Nile Delta dispersal airstrips, 167
Nirʿam, 82, 88, 107, 357
Nixon, President Richard, 242, 247, 315
Nobil Shoakry. *See* Shuwakrī, Nabīl
*Nogaw,* Israeli naval vessel, 105
noncommissioned officers, British in Egyptian forces, 33
Nord Noratlas, 162, 214, 273
Norduyn Norseman, 99–100, 339, 361
North Africa, 58, 356
North American AT-6 Harvard, 59–60, 66, 73, 80, 90, 117, 137, 143–44, 152, 189, 339, 345, 354–55, 361
North American B-25 Mitchel, 68, 229–30
North American F-86 Sabre, 143, 156
North American P-51 Mustang, 58–59, 99, 104, 115–17, 158–59, 161, 168, 364
North American Super Sabre F-100, 190
Northrop F-5E/F, 316
North Vietnam. *See* Vietnam
Norway, 143
nuclear weapons, Israeli, 327
Nuhūd, al-, 11
numbers: "bean counts," irrelevance of, 3
Nuqrāshī Pasha, 67

officers: Arab-Egyptian, 5; British: 4; British in Egyptian forces, 33; Egyptian Air Defence Force attached to Air Force, 260; German, 125, 179, 360; Iranian, 13
officers, Egyptian: 5–6, 10, 31–32, 49, 51, 99, 123, 131, 174, 177, 323; oath of allegiance, 6, 31; "royalist secret organization," 47–48
oil embargo, 299
oil refinery, Port Said, 37
Oldham, 26
"One Thousand and One Nights," 42
Onitsha, 230
opium, 22–23
organized sports and physical exercise, in Egyptian armed forces, 21–22
Ottoman Empire, 5, 9–11, 47

P-12 Barlock radar, 245
P-15 Flat Face radar, 258
Palestine, Palestinians, 10, 12, 14, 22–23, 40, 44–45, 57, 59, 61–62, 74–76, 310, (*see also* refugees, Palestinian)
Palmach, 356
pan-Arabism, 150, (*see also* Arab Nationalism)
parachute mines, 43
paranoia, of Arab states, 7
Paris, 18, 26
Park, Sir Keith, 62
Patch, Sam, 16
patronage, in Arab-Islamic culture, 5
Payne Field, 62–63, (*see also Cairo West*)
Pearl Harbor, 7, 202
Peled, Benjamin, 283
Percival Proctor, 68, 86
Percival Q6, 40, 61, 72, 339
Peres, Shimon, 143
Pfalz A II, 12
photo interpretation, 60–61
photo reconnaissance, 60
Pike, Sir Thomas, 28
pilot experience, Egyptian, 239, 305
pilot losses: Egyptian, 169, 307; Israeli, 308
pilot morale, Egyptian, 302–3, 308, 342–43
pilot rescue system, Egyptian lack of, 289–90
pilot training: Egyptian, 140–42, 145, 149–50, 198, 203, 225–26, 231, 238–39, 252, 260, 265, 296, 302–3, 307, 312, 363; Egyptian helicopter, 312; Egyptian test pilots, 312; foreign students in Egypt, 187, 331; Israeli, 232, 303; Soviet concepts, 191–92, 215; Syrian, 346–47,
pilots: American, 292; Czech, 346; Egyptian, 130–31, 304–5; Egyptian helicopter, 240; Egyptian on secondment to Nigerian air force, 230; Egyptian on secondment to Yemeni Air Force, 189; Egyptian seconded to Syrian air force, 347; Egyptian shortage of, 164; European volunteers in Nigerian air force, 230; European volunteers in Syrian air force, 345; Iraqi defection, 198; Israeli, 292, 301; Nigerian, 229; Soviet, 346; Soviet in Egypt, 248–49, 253, 264; Swedish volunteers in Biafran air force, 230; Syrian, 147, 162, 345
Piper L-8 Cub, 115–16, 159, 161, 168, 234, 345
Piper Super Cub, 152, 339

Pisa, 18
PMP pontoon bridge, 278
Podgorny, Nikolai, 220
Poland, 156, 172, 322, 347
police, Egyptian, 22–23
Pope, V. A., 41
Port Fuʾad, 221, 236, 277
Port Said, 37, 148, 166–67, 195, 213, 224, 231, 236, 241, 283, 285, 288, 293, 303, 353
Port Sudan, 45
Portugal, 10
Pratt and Whitney engine company, 177
prejudice: Israeli, 250; western: 2, 6–7
"Project 776," air defense command and control system, 323, 328–29
Pyramids of Giza, 9

Qaʾābis, 27
Qaddafi, Muammar, 265, 318
Qādrī, ʿAbd al-Ḥamīd, 215
Qāhir surface-to-surface missile, 180
Qantarah, al-, 221, 224, 234, 240
Qaṣabah, 42
Qaṣdī, Midḥat Muḥammad, 57, 352
Qatar, 315
Qaṭṭarah Depression, 40
Qāwuqjī, Fawzī al-, 81
Quwaysina, 211, 277

radar, 126, 131, 148, 163, 195, 204, 231, 249, 280, 301, 361
Radio Cairo, 166
Rafaḥ, 12, 14, 96, 102, 213
Raḥabah, 189
Raḥmān, Aḥmad ʿAbd al-, 332
Raḥmān, Muḥammad ʿAbd al-, 222
Rāʾid, al-, surface-to-surface missile, 180
Ramadan, Muslim month of fasting, 271, 274
Ramat Dawid, 77, 80, 85–87, 89, 101, 160
Ramat Raziʾel, 160
Raʾs al-ʿUshsh, battle of, 221
Raʾs Banās, 201, 211
Raʾs Naṣrānī, 276–79, 298
Raʾs Sudr, 246, 279
Raʾūf, Munʿim ʿAbd al-, 47, 132, 135
Rāziq, Aḥmad ʿAbd al-, 16, 18, 24, 28, 55
red aircraft dope/paint, 52
Red Cross, 70; International, 191
Redeye surface-to-air missile, 295, 308
Red Sea, 32, 45, 148, 174, 198, 211, 232, 272, 279–80, 323, 333
Refidim, 279, (*see also* Biʾr Jifjāfah)
refugees, Palestinian, 76–77, 99, 107, 123, 346
religious tensions in Egypt, 309
remotely piloted vehicles (RPV), 327, 330
Republic F-84: F-84F, 154, 158, 161, 166; RF-84, 162, 346
Republic P-47 Thunderbolt, 339, 342, 357, 361
Republic RC-3 Sea Bee, 82
Revolutionary Council, 148, (*see also* Free Officers Movement)
Rifāʿī, 58, 353
Rifāʿī, ʿAbd al-Majīd, 96, 193
Rishon LeZion, 84, 88
Riyāḍ, ʿAbd al-Munʿim, 234
Riyāḍ, Maḥmūd, 319
road-strips. *See* dispersal airstrips
Roberts, 26
Rogers, William, 253
Rolls Royce engines: Merlin, 342, 345; Spey, 317
Romana, 241
Rome, 18
Rommel, Field Marshall, 44, 47–49, 353
Roosevelt, President Franklin D., 44
Rosen, Count Carl Gustav von, 230
Royal Aircraft Factory BE2C, 11
Royal Egyptian Aero Club, 27, 70
royal palace, Cairo, 96
Ruḥāmah, 89, 100, 105, 357
Rumania, 10, 171, 322, 332
Rumpler C I, 12
Russell, Pasha, 23
Russia, Russians, 10, 38, 145
RWD-13, 82
Ryan, Frank, 91–92, 357

S-5K 57 mm unguided air-to-ground rocket, 213, 215–16, 295–96,
S-5M unguided air-to-ground missile, 296
SA-2 surface-to-air missile: 192, 196, 208, 227, 231, 237, 241, 245, 249, 252, 257, 259, 261, 264, 272, 275, 302; SA-2D, 249; SA-2F, 249
SA-3 Low Blow radar, 249
SA-3 surface-to-air missile, 249–50, 252, 254, 258–59, 261–62, 267, 272, 275, 302

SA-6 surface-to-air missile, 267, 271–72, 275–81
SA-7 surface-to-air missile, 272, 275, 281, 295, 301
Saab B18B, 143
Saad El Din Sherif. *See* Sharīf, Saʿd al-Dīn
Saʿadist political party, 41
Sābit, Saʿīd, 58
Sabrī, ʿAlī, 132–33, 262, 267, 354
Sabrī, Ḥusayn Dhū al-Fikr, 47
Sādāt, ʿĀdil, 280
Sadat, President Anwar, 47–49, 132–33, 246, 256–57, 261–63, 266–70, 279–80, 292, 306, 309–10, 313–14, 316, 318–20, 321, 354, 362
Ṣādiq, Aḥmad, 270
Saʿīd, 113,
Saʿīd, ʿAbd al-Fatāḥ, 116
Saʿīd, Mukhtār Maḥmūd, 106
Saʿīd, Nabīl, 239
sail-makers, 21
SAKR EYE surface-to-air missile, 329
Ṣalāḥ al-Dīn, 353
Ṣāliḥ, Ṣāliḥ Maḥmūd, 43–44, 79–79, 86, 101, 103, 353
Ṣāliḥīyah, 291
Salīm, Fatḥī, 214
Salīm, Jamal al-Dīn Muṣṭafá, 32, 132, 134
Salīm, Muḥammad Riḍwān, 48–49
Salīm, Ṣalīḥ, 132
Sallūm, 14, 57
Samir Aziz Michael. *See* Mīkhāʾīl, Samīr ʿAzīz
Sanʿāʾ, 188–89
Sanhūr (Qārūn), 54
San Stefano. *See* Yesilköy
Sanūsī, Muslim sect, 11, 30–31
Ṣaqr 80, surface-to-surface missile, 327
Ṣaqr factory. *See* Factory 333
Sarafand, 88
Sardinia, 58
satellite airstrips, Sinai, 98
Saudi Arabia, 12, 47, 78, 162, 166, 173, 188, 190–91, 194, 220, 269, 314–16, 327, 333, 363
Savoia Marchetti trimotors, 35; SM 79, 43; SM 81, 43; SM 82, 45
Scarab drone (RPV), 330
Schneider Trophy, 21
Schonau, 274
Schwechat airport, 172
SCUD surface-to-surface missile, 263, 267, 327–28
secret service, Israeli, 180
secularism, political movement, 20
security, Egyptian concern about, 2
Śedeh Dov, 82–83
Selenia IHS-6 ELINT and jamming system, 330
Selenia SL/ALQ-234 jamming pods, 330
Senior, Boris, 112
Serbia, 10
Shabānah, Muḥammad Luṭfī, 316, 323
Shabāwa, Yaḥyá al-, 83
Shāfiʿī, Ḥusayn, 199
Shafrir air-to-air missile, 233, 237, 304
Shakīb, ʿAmr , 79, 83–84, 112–13
Shalabī, Aḥmad Sa īd al-, 352
Shalabī, Zuhayr, 179, 209–10
Shanghai, 358
Shannāwī, Yaḥyá al-, 83
Sharābah, Muḥammad Ṣabrī, 352
Shararah, ʿĀdil, 277, 288, 298
Shaʿrāwī Bey, 67, 94, 124
Sharīf, Saʿd al-Dīn, 58, 185, 353–54
Sharm al-Shaykh, 167, 170, 246
Sharmi, ʿAlī, 161
Shawqī, Aḥmad, 352
Shāzlī, Saʿd al-, 189
Sheherazade, 42
Shell Oil Company, 18
Shihāta, Muḥammad, 213
Shinnāwī, ʿAbd al-Munʿim al-, 172
Short Stirling: 341, 343, 358, 360; Mk. IV, 339, 361; Mk. V, 98, 109, 114, 116, 122, 136
Shrike anti-radar missile, 252
Shuwakrī, Nabīl, 210, 216, 230, 328, 330
Sicily, 58
Sīdī Barrānī, 43
Ṣidqī, 289
Ṣidqī, Muḥammad, 16, 350
Sidqi Tanman, 350
Sikorsky CH-53, 273
Sikorsky S-58, 214
Sikorsky S-61, 315, 339
Sikorsky UH-60 Black Hawk, 332–33, 339
Simarī, Aḥmad al-, 213, 222
Sinai, 12, 23–24, 39, 110, 114, 117–19, 121, 131–32, 146, 149, 153–54, 156–59, 162–64, 167, 170–71, 202–4, 207–9, 212, 214, 217, 220–22, 224, 228, 233–37,

239–41, 242–43, 250, 253, 255, 260, 264, 266, 269–71, 275–76, 278–80, 282, 284–85, 287–93, 299, 304–5, 316, 327, 342
Sinai 23, mobile air-defence system, 329
sirdār, 14, 21, 31, 33
Siwa Oasis, 35, 40, 42–44, 46
Skyeye drone (RPV), 330
Skyguard surface-to-air missile, 323, 329
Slessor, J. C., 11
Smirnov, Alexei, 248
Smith, 26
Smiths, electronics industry, 317
soldiers: Egyptian, 4–5; Sudanese, 4
Somalia, 11, 331
Somaliland, British, 45
*Sonja, SS,* merchant ship, 92
South Africa, 230
Soviet ambassador, 144, 321
Soviet Bloc, 130, 150
Soviet influence in Egypt, 153, 174
Spain: 143; Civil War, 92
Spinks, Pasha, 17, 33
Stack, Sir Lee, 14
Stahlschmidt, Hans-Arnold 48
*Stalingrad,* Soviet freighter, 145
Sternberg, Freiherr Speck von, 33
Stern Gang, 92, 356, (*see also* terrorism, Zionist)
Stinson AT-19 Reliant, 361
*St. Louis Post Dispatch,* American Newspaper, 190–91
Stocks, S. J., 18, 24, 27
Strait of Tiran, 197
submarines, Italian, 44
Sudan, 11, 31, 40, 77–78, 175, 224, 229, 263–64, 317, 322–23, 327, 331, 356, 362
Sud Aviation Super Frelon, 273
Sud S.E. Alouette, 273
Sud S.O. Vautour, 176, 210–11, 241
Sud SA-342 Gazelle, 326, 331–33, 339
Suez, 39, 42–45, 47–48, 53–54, 60, 195, 223 224, 231, 234, 285, 303, 353; oil refinery, 236
Suez Canal, Canal Zone, 10–11, 13, 16, 24, 28, 35–37,39, 41, 43–45, 53–54, 59, 67, 80, 92, 110, 117–18, 121, 127–28, 130, 132–33, 139–40, 145, 148–49, 153–59, 161, 167–68, 170, 178, 197, 199, 203, 213, 221–22, 228, 232, 235, 239–40, 242, 245, 248–54, 260, 264, 268, 271, 274, 276, 278–83, 285, 287–88, 291–93, 295, 297, 303, 323, 342, 346, 359, 361
Sukhoi Su-7: 161, 184, 186–87, 195, 198, 200, 209, 211, 213–14, 216, 222–23, 226, 233–35, 238–41, 243–44, 246, 250, 258–59, 265, 267–69, 272–73, 275–78, 288–89, 293, 296, 305–6, 311, 320, 325, 328; Su-7BMK, 339; Su-7U, 320
Sukhoi Su-11, 184
Sukhoi Su-13, 184
Sukhoi Su-15, 184, 253–54, 258
Sukhoi Su-20, 267–68, 273, 275–76, 290, 295, 305, 311, 313–15, 318, 320, 325, 328, 339
Supermarine Sea Hawk, 164–65, 167
Supermarine Sea Otter, 99, 127, 137, 339, 361
Supermarine Spitfire: 56, 63, 67–71, 94, 97, 99, 100–2, 104–6, 115, 342–43, 355, 357, 359–60; Mk. V, 53, 57–58, 63, 73, 79, 90, 93, 97, 108, 339, 354–55; Mk. VIII, 58, 355; LF9, 57–58, 66, 72, 78–79, 81, 83–86, 89–90, 93, 108, 112, 115–19, 122, 124, 127, 339, 354–56, 361; T9, 126, 340, 362; Mk. 18, 362; FR18, 85–86; F22, 125–26, 136, 141, 156, 161, 339, 345; Mk. 24, 355
support/technical personel, training: Egyptian, 317, 323, 360; foreign in Egypt, 317
surface-to-air missile (SAM) belt, 243
surface to air missile sites, movement of, 260
Surt, 18
Sūsah, Muḥyī al-Dīn, 70
Swamp command and control system, 258
Sweden, 143
Switzerland, 127
Syria, 10, 12–13, 123–24, 126, 139, 153, 158, 162, 166, 173, 175–76, 180, 193, 200, 202–3, 208, 210, 217, 224, 247, 254, 269–71, 274, 292, 299, 307, 322, 327, 345, 363

T-54 tank, 240
Ṭāhā, Sayyid, 111
Ṭāhir, Ḥusnī, 49, 52
Tait, Victor Hubert, 21, 23, 25–27, 32–33, 40, 67
tall, 358
Tall Nūf, 101
Tāmīya, 243
Tangiers Charter Company, 358

Ṭanṭá, 195, 290–91, 294, 314, 319
ṭarbūsh, "fez" hat, 36
target towing, 34
Tāsā, al-, 292
Tawāfiq, 175
Tawfīq, 66, 124
Ṭawīl, Ṣaḥbī al-, 179
Taylorcraft aircraft company, 69
Taylorcraft Auster: J-1 Autocrat, 89, 93; VII, 69
technical assistance, Soviet, 181
technical officer training, Egyptian, 226
technical personel, Egyptian pool of, 177
technicians, British ex-RAF in REAF, 68
Tel Aviv, 82–83, 88, 100, 121, 176, 197, 250, 357–58
terrorism: Israeli, 142, 180; Libyan, 318; Palestinian, 274; Zionist, 75–76, 81, 92, 356
third world, 144
Thornhill, 47
Tiberius, 84
Timurtāsh, 355
Tour of the Oases Air Rally, 27, 32
TOW anti-tank missile, 291
training systems, development in Egypt, 227, 311–12
Trans-Air of Melsbroeck, 358
Transjordan, 70, (*see also* Jordan)
Treasury, British, 55
trials, for dereliction of duty, 22
Trieste, 14
Tripartite Declaration, 128
Tripoli, 18
Tubruq, 18, 318
Tunis, 58
Tunisia, 27, 78
Tupolev Tu-16, 184, 186, 190, 194–95, 200, 205, 207, 211, 217, 228, 241, 248, 250, 258, 262–63, 267–68, 273, 275–76, 278–80, 288, 305, 311, 318, 320, 340
Tupolev Tu-22, 262–63, 266–67
Ṭūr, al-, 39, 276
Turkey, Turks, 5, 48, 129, 324
Type 63 Early Warning Radar, 118

ᶜUbayd, Muḥammad Ibrāhīm, 71
Uganda, 11
UH-1, 273
UK-12E, 333
Ukraine, 38
ᶜUmar, Ibrāhīm, 205
ummah, 350
ᶜUmrah, Muslim Lesser Pilgrimage, 274
UN, 75, 78, 167, 170, 260–61, 333; Atomic Energy Commission, 22; cease-fire, truce in Palestine, 88–89, 91 93, 99, 105, 107, 111, 113, 118, 124; ceasefire, on Suez Canal, 254; demilitarized zone, 140; Partition Plan for Palestine, 78, 84, 89, 97, 358; peacekeeping force in Sinai, 197, 307, 310, 327; peacekeeping force, in Zaire, 187; peacekeeping forces, Egyptian contribution to, 333; peace-keeping forces, Truce Supervision teams, 118, 170, 174; Resolution 338, ceasefire call, 297; Security Council, 105, 107
uniforms, Egyptian Air Force, 35–36, (*see also* fārūqīyah, tarbūsh)
Union of Arab States. *See* United Arab Republic
United Arab Emirates (UAE), 315
United Arab Republic (UAR), 173, 188; collapse of, 176
United Nations. *See* UN
United States, 10, 62, 67, 128–29, 143–44, 148, 168–69, 170, 202, 238, 242, 246–48, 261, 275, 299–300, 307–8, 310, 312, 319–20, 321–25, 328–30, 332; resupply of Israel, 307; sympathy for Israel, 124
Upavon, 22
Upper Egypt, 199
ᶜUrābī, 6
U.S. Embassy, 133
U.S. Senate, 302
U.S. State Department, 170
USSR, 128, 144–46, 148, 150, 156, 167, 169, 171–72,179–81, 191, 202, 220, 222, 224–25, 229, 242, 247–49, 254, 256, 262–63, 266, 268, 270, 292, 299–300, 307, 310, 313–14, 317, 319, 324, 346–47, 362; leaders demand control of advanced weaponry in Egypt, 263, 266; resupply of Egypt and Syria, 307, (*see also* Soviet bloc)
ᶜUwaynah, al-, 58

Valle Olona, 92
Venegono, 92
Versailles Peace Conference, 13
VHF radio, 53, 56, 60–61

Vickers Valentia, 34
Vickers Valiant, 163, 166
Vickers Vimy, 16
Vickers Warwick Mk. V, 355
Vickers Wellington, 63, 66, 355
Vienna, 172
Vietnam, 186, 195, 236, 245, 251–52, 254, 261, 305
Villacoubley, 26
volunteers, Muslim in Palestine War, 77
Voss, Dr. Wilhelm, 179
Vulcan 20mm anti-aircraft gun, 308
Vultee A-31 Vengeance, 152
Vultee BT-13 Valiant, 128, 137, 361

Wādī Ḥalfā', 39
Wādī Qina (Wādī Abū Shīhāt), 277
Wādī Sayyidnā, 263
Wādī Sukhrīr, 88
Wafā'ī, Aḥmad, 277, 285, 293, 298, 300
Wafd political party, 128
Wahhāb, ʿAbd al-, 49
Wahhāb, Muṣṭafá Kāmil ʿAbd al-, 116
Wajaʿī, ʿAlī, 373
warrant officers, British in Egyptian forces, 33
Warsaw Pact, 260, 267, 346
Wāṣif, Yūsuf, 353
water hoses, to breach sand defences, 278
Webster, S. N., 21, 25–26
*Wedgewood*, Israeli Navy corvette, 105
Weeckman, Robert, 95
West Bank, 108, 110, 216–17
Western Desert, 35, 39–40, 43, 112, 265
Westland aircraft company, 268–69
Westland Commando, 315, 317, 320, 332–33, 330, 340
Westland Lynx, 316
Westland Lysander, 35–36, 39–40, 42–44, 60–61, 64, 68, 72–73, 79, 94, 340, 351, 360–61
Westland Sea King, 315, 333, 340
Westland-Sikorski S-51 Mk. 1B Dragonfly, 137, 152, 340, 361–62
Westland Wapati, 16
Westland Wessex IV, 27, 52, 340, 351
Whitlock, 26
Williamson F28 aerial camera, 28
Wilson, President Woodrow, 13,
Woodford, 25
*Worcester, HMS,* British Royal Navy, 14

Yakovlev Yak-11, 145, 189–90, 226
Yakovlev Yak-11, 340
Yakovlev Yak-15, 157
Yakovlev Yak-18, 320, 340
Yakovlev Yak-28, 184
Yarqon River, 82, 357
Yemen, 13, 173, 176, 180, 188, 203, 220, 226, 239, 247; Revolution, 188
Yesilköy, 13
Yom Kippur, 271, 274–75
Yugoslavia, 322
Yusrī, Rifʿat, 268

Ẓafar surface-to-surface missile, 180
Zaire, 187
Zakharov, Matvei, 219
Zakī, ʿAbd al-Raḥmān, 48, 352
Zakī, Ḥalīm Ṭāhir, 71, 113
Zakī, Taḥsīn, 161, 187, 192–93, 196, 213, 215–16, 221, 223
Zamālik, al-, 195
Zaqīlah, 113
Zaytūn, Saʿīd ʿAlī, 58, 352
Ziebal, Zvi, 115
Ziftá, 42
Zimbabwe, 331
Zionism, 63, 65, 74–75, 77, 80
Zlin 126, 189
Zlin 226T, 340
ZSU-23 anti-aircraft gun, 267, 271–72, 281
Zumbach, Jean, 230